Resource Utilization of Solid Waste and Circular Economy

Resource Utilization of Solid Waste and Circular Economy

Guest Editors

Haoqi Yang
Ganyu Zhu
Jun-Ye Zhang

MDPI

Basel • Beijing • Wuhan • Barcelona • Belgrade • Novi Sad • Cluj • Manchester

Guest Editors
Haoqi Yang
College of Electrical Energy and Power Engineering
Yangzhou University
Yangzhou
China

Ganyu Zhu
National Engineering Laboratory for Hydrometallurgical Cleaner Production Technology
Chinese Academy of Sciences
Beijing
China

Jun-Ye Zhang
Department of Chemistry
Northwestern University
Evanston, IL
USA

Editorial Office
MDPI AG
Grosspeteranlage 5
4052 Basel, Switzerland

This is a reprint of the Special Issue, published open access by the journal *Applied Sciences* (ISSN 2076-3417), freely accessible at: https://www.mdpi.com/journal/applsci/special_issues/SRNFJAUPD4.

For citation purposes, cite each article independently as indicated on the article page online and as indicated below:

Lastname, A.A.; Lastname, B.B. Article Title. *Journal Name* **Year**, *Volume Number*, Page Range.

ISBN 978-3-7258-5105-8 (Hbk)
ISBN 978-3-7258-5106-5 (PDF)
https://doi.org/10.3390/books978-3-7258-5106-5

Contents

applied sciences

MDPI

Editorial

Editorial: Resource Utilization of Solid Waste and Circular Economy

Haoqi Yang

College of Electrical, Energy and Power Engineering, Institute of Technology for Carbon Neutralization, Yangzhou University, Yangzhou 225127, China; yanghq@yzu.edu.cn

1. Introduction

The rapid pace of industrialization and urbanization worldwide has led to an unprecedented increase in the generation of solid waste, posing serious challenges to environmental sustainability and resource security [1]. Traditional disposal practices such as landfilling and incineration not only consume valuable land and energy but also release hazardous pollutants, undermining global efforts toward carbon neutrality and sustainable development [2,3]. Against this backdrop, the concept of a circular economy, which promotes the reduction, reuse, and recycling of waste, has emerged as a transformative paradigm for closing material loops and enhancing resource efficiency [4].

Solid waste, once considered an environmental liability, is now increasingly viewed as a valuable secondary resource [5]. Technological innovations in materials science, chemical engineering, environmental management, and digital intelligence have paved the way for the efficient valorization of various types of solid waste—including industrial by-products, municipal waste, biomass residues, and e-waste—into construction materials, energy carriers, functional materials, and other high-value products [6–8]. However, despite these promising developments, the transition toward a fully circular economy remains constrained by technical, economic, regulatory, and social barriers.

This Special Issue of *Applied Sciences*, titled "Resource Utilization of Solid Waste and Circular Economy", aims to present the latest research progress in solid waste valorization and circular economy practices. It brings together 11 contributions from multidisciplinary teams across the globe, exploring innovative materials, green technologies, system optimization methods, and policy-oriented strategies that support waste minimization and sustainable resource use.

2. An Overview of Published Articles

2.1. Valorization of Solid Waste in Construction Materials

Several studies in this Special Issue explore the conversion of industrial waste into functional construction materials. Yu et al. investigated the effects of water-reducing agents on the mechanical performance of foamed phosphogypsum, a promising material for lightweight and sustainable construction (Contribution 1). Similarly, Haddad et al. assessed the use of concrete washing fines in mortar mixtures, highlighting their impact on microstructure and carbon footprint (Contribution 2). Zhou et al. developed a dual chemical–physical approach for curing soluble phosphorus and fluorine in phosphogypsum, providing insights into stabilization mechanisms and potential field applications (Contribution 6). Meanwhile, Souza Oliveira et al. introduced a clinker-free cement concept that eliminates Portland clinker altogether, showcasing a low-carbon alternative to traditional cementitious materials (Contribution 5).

https://doi.org/10.3390/app15169098

2.2. Waste-to-Energy and Carbonization Processes

A noteworthy contribution by Yang et al. presents a method to repurpose graphite waste from Acheson furnaces as anode materials for lithium-ion batteries (Contribution 3). This study not only enhances the value of carbonaceous waste but also advances sustainable battery design through circular materials engineering. Cavali et al. reported on the co-hydrothermal carbonization of sawdust and sewage sludge, producing hydrochar with effective adsorption properties while assessing the ecotoxicity of process water (Contribution 7). Complementing this, Yu et al. provided a comprehensive review of hydrothermal oxidation technologies for sludge treatment, discussing operational challenges, pollutant removal efficiency, and energy recovery potential (Contribution 4).

2.3. Circular Economy and Digital Waste Management

Beyond material transformation, this Special Issue also examines system-level strategies. AlJaber et al. analyzed circular economy implementation in the building sector, emphasizing stakeholder roles, feasibility factors, and lifecycle considerations (Contribution 8). Pinho et al. addressed wood waste management in the furniture industry, outlining pathways for revalorization and regulatory alignment toward circularity (Contribution 9). Digital technologies are playing a crucial role in modern circular economy systems. Vesga Ferreira et al. proposed a low-cost capacitive sensing system coupled with machine learning algorithms for automatic waste classification (Contribution 11). This study demonstrates how AI-powered sensing can reduce human intervention and improve recycling accuracy. Finally, Moreno-Bermedo et al. explored the reuse of tea waste as a catalyst support in the electro-Fenton process for removing emerging organic contaminants (Contribution 10). This work exemplifies the potential of biomass residues in environmental remediation within circular economy frameworks.

3. Conclusions and Outlook

The contributions gathered in this Special Issue showcase a broad spectrum of innovative approaches to solid waste utilization, spanning from construction materials and energy systems to digital waste management and pollutant removal technologies. These studies reflect a growing global consensus on the urgency of transitioning from a linear to a circular economy, wherein waste is not an end product but a resource to be continuously reintegrated into productive cycles.

While significant technical progress has been achieved, challenges persist in scaling up laboratory findings, ensuring economic feasibility, aligning regulatory frameworks, and fostering cross-sector collaboration. Future research should focus on:

- Integrating life cycle assessment (LCA) and techno-economic analysis in material design;
- Enhancing digitalization for smart waste monitoring and logistics;
- Strengthening policy tools and public engagement to promote behavioral shifts;
- Developing standards and certification systems for waste-derived products.

We hope that this Special Issue will serve as a valuable resource for researchers, engineers, policymakers, and industry practitioners striving to build more sustainable, resource-efficient, and resilient societies through the lens of circular economy and solid waste valorization.

Acknowledgments: The Guest Editor sincerely thanks all the authors for their valuable contributions and the reviewers for their constructive feedback, which significantly enhanced the quality of this Special Issue. Gratitude is also extended to the editorial team at *Applied Sciences* for their professional support throughout the publication process.

Conflicts of Interest: The authors declare no conflicts of interest.

List of Contributions:

1. Yu, J.; Wang, H.; Wu, F.; Yu, H.; Guo, J. Effects of Water-Reducing Agents on the Mechanical Properties of Foamed Phosphogypsum. *Appl. Sci.* **2024**, *14*, 8147. https://doi.org/10.3390/app14188147.
2. Haddad, B.; Alassaad, F.; Affan, H.; Mohamad, A.; Sebaibi, N. Characterization of Mortars Incorporating Concrete Washing Fines: Impact on Mechanical Properties, Microstructure and Carbon Footprint. *Appl. Sci.* **2024**, *14*, 8381. https://doi.org/10.3390/app14188381.
3. Yang, I.; Choi, S.; Kim, S.; Ha, M.; Park, S.; An, J. Utilizing Graphite Waste from the Acheson Furnace as Anode Material in Lithium-Ion Batteries. *Appl. Sci.* **2024**, *14*, 11353. https://doi.org/10.3390/app142311353.
4. Yu, H.; Liu, Y.; Guo, N.; Piao, W.; Pan, Z.; Zhu, B.; Zhu, Y.; Wu, L.; Wan, J.; Wei, H. Recent Advances in Hydrothermal Oxidation Technology for Sludge Treatment. *Appl. Sci.* **2024**, *14*, 11827. https://doi.org/10.3390/app142411827.
5. Souza Oliveira, M.; Carvalho Zago, S.; Vernilli Junior, F. Clinker-Free Cement: A New Reality. *Appl. Sci.* **2024**, *14*, 11898. https://doi.org/10.3390/app142411898.
6. Zhou, J.; Yang, Y.; Li, H.; Zhu, G.; Yang, H. Synergistic Chemical Modification and Physical Adsorption for the Efficient Curing of Soluble Phosphorus/Fluorine in Phosphogypsum. *Appl. Sci.* **2025**, *15*, 780. https://doi.org/10.3390/app15020780.
7. Cavali, M.; Hennig, T.; Libardi Junior, N.; Kim, B.; Garnier, V.; Benbelkacem, H.; Bayard, R.; Woiciechowski, A.; Matias, W.; de Castilhos Junior, A. Co-Hydrothermal Carbonization of Sawdust and Sewage Sludge: Assessing the Potential of the Hydrochar as an Adsorbent and the Ecotoxicity of the Process Water. *Appl. Sci.* **2025**, *15*, 1052. https://doi.org/10.3390/app15031052.
8. AlJaber, A.; Martinez-Vazquez, P.; Baniotopoulos, C. Exploring Circular Economy Strategies in Buildings: Evaluating Feasibility, Stakeholders Influence, and the Role of the Building Lifecycle in Effective Adoption. *Appl. Sci.* **2025**, *15*, 1174. https://doi.org/10.3390/app15031174.
9. Pinho, G.; Calmon, J.; Medeiros, D.; Pinho, C. Furniture Wood Waste Management Towards the Circular Economy. *Appl. Sci.* **2025**, *15*, 1360. https://doi.org/10.3390/app15031360.
10. Moreno-Bermedo, L.; Correa-Puerta, J.; González-Fuentes, C.; Escalona, N.; Onfray, C.; Thiam, A. Tea Waste as a Sustainable Catalyst Support for Enhanced Removal of Contaminants of Emerging Concern via the Electro-Fenton Process: A Circular Economy Approach. *Appl. Sci.* **2025**, *15*, 1418. https://doi.org/10.3390/app15031418.
11. Vesga Ferreira, J.; Perez Waltero, H.; Vesga Barrera, J. Design of a Waste Classification System Using a Low Experimental Cost Capacitive Sensor and Machine Learning Algorithms. *Appl. Sci.* **2025**, *15*, 1565. https://doi.org/10.3390/app15031565.

References

1. Khan, S.; Anjum, R.; Raza, S.T.; Bazai, N.A.; Ihtisham, M. Technologies for municipal solid waste management: Current status, challenges, and future perspectives. *Chemosphere* **2022**, *288*, 132403. [CrossRef] [PubMed]
2. Roy, H.; Alam, S.R.; Masud, R.B.; Prantika, T.R.; Pervez, M.N.; Islam, M.S.; Naddeo, V. A review on characteristics, techniques, and waste-to-energy aspects of municipal solid waste management: Bangladesh perspective. *Sustainability* **2022**, *14*, 10265. [CrossRef]
3. Hsu, H.; Binyet, E.; Nugroho, R.A.A.; Wang, W.; Srinophakun, P.; Chein, R.; Demafelis, R.; Chiarasumran, N.; Saputro, H.; Alhikami, A.F.; et al. Toward sustainability of waste-to-energy: An overview. *Energy Convers. Manag.* **2024**, *321*, 119063. [CrossRef]
4. Corvellec, H.; Stowell, A.F.; Johansson, N. Critiques of the circular economy. *J. Ind. Ecol.* **2022**, *26*, 421–432. [CrossRef]
5. Xu, P.; Liu, C.; Zhang, X.; Zheng, X.; Lv, W.; Rao, F.; Yao, P.; Wang, J.; Sun, Z. Synergic mechanisms on carbon and sulfur during the selective recovery of valuable metals from spent lithium-ion batteries. *ACS Sustain. Chem. Eng.* **2021**, *9*, 2271–2279. [CrossRef]
6. Sun, B.; Zhang, J.; Wang, M.; Yu, S.; Xu, Y.; Tian, S.; Gao, Z.; Xiao, D.; Liu, G.; Zhou, W.; et al. Valorization of waste biodegradable polyester for methyl methacrylate production. *Nat. Sustain.* **2023**, *6*, 712–719. [CrossRef]
7. Mir, A.E.; Hwalla, J.; Hassan, H.; Assaad, J.J.; Dieb, A.; Shehab, E. Valorization of waste perlite powder in geopolymer composites. *Constr. Build. Mater.* **2023**, *368*, 130491. [CrossRef]
8. Xia, G.; Han, W.; Xu, Z.; Zhang, J.; Kong, F.; Zhang, J.; Zhang, X.; Jia, F. Complete recycling and valorization of waste textiles for value-added transparent films via an ionic liquid. *J. Environ. Chem. Eng.* **2021**, *9*, 106182. [CrossRef]

applied sciences

MDPI

Article

Effects of Water-Reducing Agents on the Mechanical Properties of Foamed Phosphogypsum

Jian Yu [1,2], Hongxia Wang [2,*], Fade Wu [2], Haiyan Yu [1] and Junhua Guo [2]

1 School of Materials Science and Engineering, Tianjin Chengjian University, Tianjin 300384, China; yujian20220222@163.com (J.Y.); yuhy858@tcu.edu.cn (H.Y.)
2 China Building Materials Academy Co., Ltd., Beijing 100024, China; 15822068725@163.com (F.W.); guojunhua870108@163.com (J.G.)
* Correspondence: wxx022@126.com

Abstract: In the present experiment, two types of water-reducing agents, naphthalene (FDN) and polycarboxylic acid (PCE), were selected, and their effects on the mechanical properties of foamed phosphogypsum were evaluated. It was shown that when the water-reducing agent contents were increased, the strength of the foamed phosphogypsum first increased and then gradually decreased, and that the dry density of the foamed phosphogypsum first decreased and then gradually increased. The FDN samples had better mechanical properties and a lower dry density than the PCE samples. The effect of the water-reducing agent dose on the apparent viscosity and shear stress of the phosphogypsum slurries was in the order of 0% > 0.4% > 0.3% > 0.5% > 0.2% > 0.1%. The apparent viscosity and shear stress of the gypsum slurry mixed with 0.4% FDN or PCE were the highest. FDN and PCE both enlarged the pore size distribution range, increased the size and proportion of large pores, and decreased the total pore content of foamed phosphogypsum; however, the effect of PCE was more significant. The foamed phosphogypsum slurry mixed with 0.4% FDN had the highest total pore content. Among the samples, the total pore content of foamed phosphogypsum A was able to reach 91% and the total pore content of foamed phosphogypsum B reached 77%; the lowest proportion of large pores for foamed phosphogypsum A and B separately reached 17% and 7%, respectively. The water-reducing agents mainly reduced the water consumption of the phosphogypsum slurries; improved the viscosity and shear stress of the slurries; affected the stability of the foam in the gypsum slurries; influenced the pore size and distribution in the foamed phosphogypsum samples; and caused a difference in the strength and dry density of the foamed phosphogypsum samples. The viscosity of the gypsum slurry doped with 0.4% FDN better matched that of the foam; therefore, it had the highest macro-strength and the lowest dry density.

Keywords: foamed phosphogypsum; water-reducing agent; rheology; pore structure distribution

1. Introduction

Phosphogypsum is a by-product of phosphate fertilization. According to statistics, approximately 4–6 tons of phosphogypsum are produced from 1 ton of phosphoric acid. This large amount of phosphogypsum occupies land resources and causes pollution to the surrounding environment, seriously restricting the green development of the phosphate industry. Therefore, the resource utilization of phosphogypsum has become the focus of the Government and the phosphate industry [1–8].

Paper gypsum boards are one of the main absorption media of phosphogypsum. When phosphogypsum is used as the gypsum base raw material in paper gypsum boards, it can alleviate the pressure of efficient resource utilization for phosphogypsum and solve the shortage problem of gypsum resources. Therefore, under current green, low-carbon, and other strict environmental requirements, it is of great significance to carry out the research and production of phosphogypsum paper boards [9–11].

Phosphogypsum paper boards are mainly composed of a phosphogypsum board core and a protective paper [12–14]. During the preparation of gypsum board cores, to save production energy consumption and reduce cost, water-reducing agents are often added to ensure that gypsum slurries keep a certain fluidity with reduced water consumption [15–20]. Moreover, the decrease in the water consumption of gypsum slurries increases their surface density; thus, foaming agents are used to reduce the surface density of gypsum boards [21–27]. Water-reducing agents and foaming agents are indispensable raw materials in the production of paper gypsum boards. Superplasticizers and foam are highly active organic materials. When they work together in gypsum slurries, strong chemical interactions occur between them. If the degree of interactions between them is small, they manifest little influence on the macro-performance of the gypsum board core. On the contrary, if the interaction degree is strong, they yield an adverse effect on the macro-performance of the gypsum board core. Feng, Julian et al. [28] analyzed the interaction between the foaming agent and the water-reducing agent for desulfurized gypsum and found that the effect of the foaming agent on 0.3% of FDN was less than that on 0.1% of PCE. Although the effects of foaming agents on water-reducing agents are considered during the preparation of gypsum slurries, the influences of water-reducing agents on the formation and distribution of foam are generally neglected, causing the poor stability of paper gypsum boards [29,30]. Zeng, Z. et al. [31] studied the compatibility between different water-reducing agents and phosphogypsum-based cementitious materials, including the dependence of mechanical properties and rheological properties. The test results showed that the FDN has strong compatibility with the gypsum-based cementitious materials of phosphorus buildings. Gao, F. et al. [32] studied the effects of polycarboxylic acid and sulfonated melamine superplasticizer on the fluidity, setting time, and dry compression strength of phosphogypsum; the results showed that the 0.48% dosage of polycarboxylic acid was more suitable for phosphogypsum. However, in these studies, foamed phosphogypsum was not considered. Zhang, Y. et al. [33] studied the effects of hydroxypropyl methyl cellulose ether (HPMC) on the rheological properties, pore structure and mechanical properties of EPS–phosphorus gypsum composite paste. The results showed that, with the increase in HPMC content, the yield stress and plastic viscosity of paste increase gradually and that the pore structure and mechanical properties of gypsum paste are improved. The optimum HPMC content for EPS–phosphorus gypsum composite slurry is 0. 20%. Wang, J. [34] focused on the preparation of foamed phosphogypsum and the effect of silicone on the properties of foamed phosphogypsum. However, the effects of water-reducing agents on the mechanical properties of foamed phosphogypsum were not considered. Hence, to improve the stability of phosphogypsum paper boards, two types of water-reducing agents, naphthalene and polycarboxylic acid, were selected in the present work to study their effects on the mechanical properties of foamed phosphogypsum. The effects of the water-reducing agents on foamed phosphogypsum were discussed based on the correlation between the bubble size distribution in the foam and the slurry viscosity. This research aims to provide theoretical and technical support for the development of light phosphogypsum paper boards and lay a theoretical foundation for the efficient resource utilization of phosphogypsum.

2. Experimental Section

2.1. Materials

Phosphogypsum A and B samples were produced in Hubei Province, China, and their chemical composition and phase composition are presented in Table 1 and Figure 1, respectively. In Figure 1, (a) and (b) both show broadened diffraction peaks at 14.8°, 25.7°, and 29.7°, corresponding to the (101), (110) and (202) crystal face of $CaSO_4{\cdot}0.5H_2O$, respectively. Both had a strong peak strength of 29.7°, indicating that $CaSO_4{\cdot}0.5H_2O$ mainly existed in the (202) crystal face and presented a plate-like crystal structure. The SEM image is shown in Figure 2, which shows that the phosphogypsum exhibits irregular layers of plate-like structures. The particle size distribution characteristics of the phosphogypsum samples are presented in Figure 3 and Table 2, and their physical and mechanical properties are listed in Table 3. Sodium dodecyl sulfate, a K12-type foaming agent, was procured from Kao (Japan). Naphthalene (water-reducing agent; FDN) was produced in Tianjin (China); its water reduction rate was 23%. Polycarboxylic acid (superplasticizer; PCE) was purchased from Shandong (China); its water reduction rate was 27%. Pure tap water was used for sample mixing.

Table 1. Chemical composition of phosphogypsum powder.

Sample	Chemical Composition (wt.%)												
	Na_2O	MgO	Al_2O_3	SiO_2	P_2O_5	SO_3	K_2O	CaO	TiO_2	Fe_2O_3	SrO	F	Others
A	0.12	0.22	0.2	6.77	1.09	49.8	0.34	39.6	0.16	0.51	0.17	0.01	1.01
B	0.37	0.12	0.77	5.65	1.2	48.9	0.21	40.3	0.14	0.42	0.1	0.7	1.12

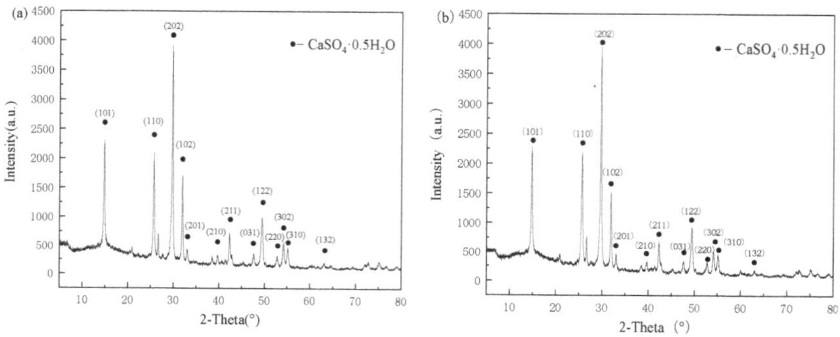

Figure 1. XRD patterns of the phosphogypsum samples. (**a**) Phosphogypsum A. (**b**) Phosphogypsum B.

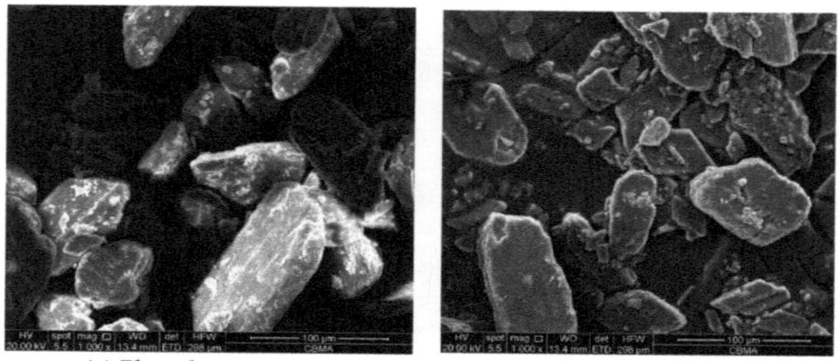

(**a**) Phosphogypsum A (**b**) Phosphogypsum B

Figure 2. SEM image of a phosphogypsum samples.

Table 2. Particle size distribution characteristics of the phosphogypsum samples.

Sample	d (0.1; μm)	d (0.5; μm)	d (0.9; μm)	Specific Surface Area (m^2/kg)
A	2.648	19.284	39.997	963
B	3.776	31.191	128.867	660

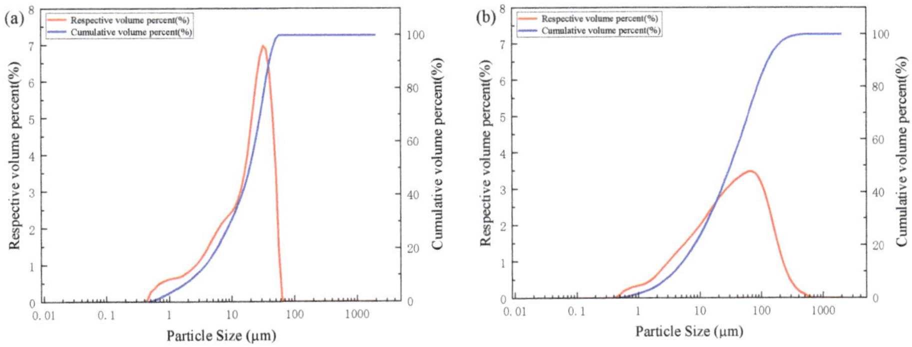

Figure 3. Particle size distribution curves of the phosphogypsum samples. (**a**) Phosphogypsum A. (**b**) Phosphogypsum B.

Table 3. Physical properties of the phosphogypsum samples.

Sample	Standard Thickness (%)	Initial Setting Time (min)	Final Setting Time (min)	Two-Hour Flexural Strength (MPa)	Absolute Flexural Strength (MPa)	Two-Hour Compressive Strength (MPa)	Absolute Compressive Flexural Strength (MPa)
A	0.76	3.20	6.20	2.7	4.2	4.8	11.8
B	0.76	3.15	5.20	2.5	3.3	4.3	9.0

2.2. Sample Preparation

Preparation of foam: The foaming agent and purified water were weighed at a ratio of 1:250 and poured into a mixing pot (pot A).

The rotational speed of the mixer was set to eight gears; pot A was quickly removed after 2 min of stirring.

Preparation of phosphogypsum slurries: The contents of the water-reducing agent were 0.1%, 0.2%, 0.3%, 0.4%, and 0.5% of the total gypsum content.

According to the GB/T17669.3-1999 Standard [35] the fluidity of gypsum slurries should reach 180 ± 5 mm. The water/gypsum ratio and fluidity of the phosphogypsum A and B samples are presented in Table 4. Pure water, the water-reducing agent, and phosphogypsum powder were poured into a mixing pot (pot B) similar to pot A. The rotational speed of the mixer was first set to two gears for a stirring time of 10 s. Then, the corresponding values were adjusted to four gears and 15 s, respectively.

Table 4. Water/gypsum (W/G) ratio and fluidity of the phosphogypsum.

Type	W/C Ratio	Fluidity (mm)
A	0.760	180
A-0.1%FDN	0.690	178
A-0.1%PCE	0.720	178
A-0.2%FDN	0.680	178
A-0.2%PCE	0.700	180
A-0.3%FDN	0.670	180
A-0.3%PCE	0.690	178
A-0.4%FDN	0.660	178
A-0.4%PCE	0.680	178
A-0.5%FDN	0.675	180
A-0.5%PCE	0.695	180

Table 4. *Cont.*

Type	W/C Ratio	Fluidity (mm)
B	0.760	180
B-0.1%FDN	0.700	178
B-0.1%PCE	0.720	178
B-0.2%FDN	0.670	178
B-0.2%PCE	0.700	180
B-0.3%FDN	0.650	180
B-0.3%PCE	0.690	178
B-0.4%FDN	0.640	178
B-0.4%PCE	0.680	178
B-0.5%FDN	0.660	180
B-0.5%PCE	0.695	180

Preparation of foamed phosphogypsum samples: The as-prepared foam with a mass of 5% of the total phosphogypsum content was poured into pot B. The rotational speed of the mixer was first set to four gears for a stirring time of 10 s. Then, the corresponding values were adjusted to six gears and 15 s, respectively. The evenly mixed foamed phosphogypsum was then quickly poured into a mold of 40 mm × 40 mm × 160 mm in size. The wet strength, absolute dry strength, and dry density of the gypsum samples were then determined according to the GB/T17669.3-1999 Standard,.

Furthermore, to characterize the microstructure of phosphogypsum samples, 20 mm × 20 mm × 20 mm samples were made; the hydration of these samples was terminated after 2 h and the samples were placed in a sealed container for future use. The flowchart of the foamed phosphogypsum sample preparation is shown in Figure 4.

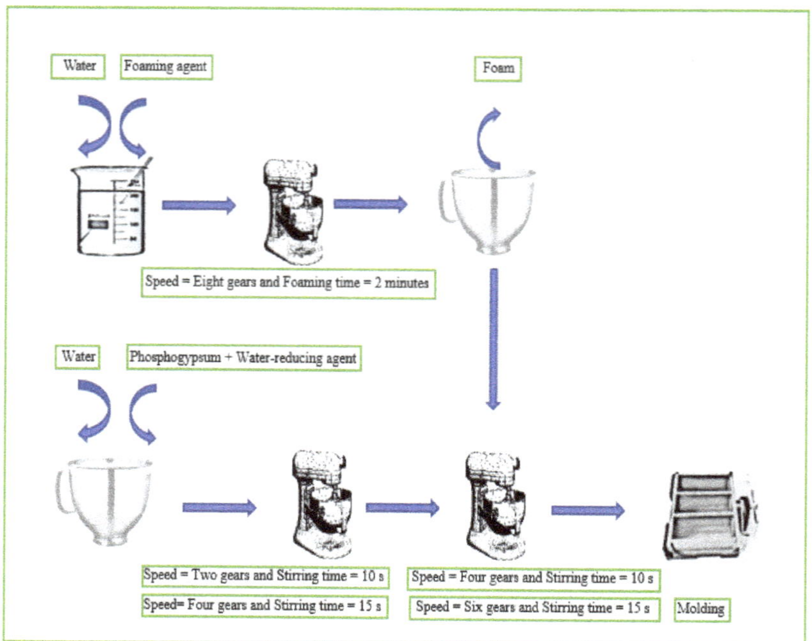

Figure 4. Flowchart of the foamed phosphogypsum sample preparation.

2.3. Test Methods

The rheological properties of the phosphogypsum slurries were measured using a TADHR-3 rheometer (Waters Corporation, Newcastle, DE, USA) under a shear stress range of 0–325 Pa, a shear rate range of 0–200 s^{-1}, and a viscosity range of 1×10^{-3}–3×10^{5} Pa. The pore structure characteristics of the foamed phosphogypsum samples were tested using a Mack AutoPore V 9600 mercury injection instrument (Micromeritics Instruments Corporation, Norcross, GA, USA).

3. Results and Discussion

3.1. Effect of the Water-Reducing Agents on the Mechanical Properties of the Foamed Phosphogypsum Samples

It can be seen in Table 5 that, under the action of both FDN and PCE, the development trends of the 2 h strength and absolute dry strength of the phosphogypsum A samples were almost similar. With the increase in the water-reducing agent content, the strength of the phosphogypsum A samples first increased and then gradually decreased. When the content of FDN or PCE was 0.4%, the 2 h strength and absolute dry strength of the phosphogypsum A samples reached a maximum. However, the 2 h strength and absolute dry strength of the foamed phosphogypsum samples mixed with FDN were higher than those of the samples mixed with PCE under the same water-reducing agent content. It is evident from Table 6 that under the action of both FDN and PCE, the strength development law of the phosphogypsum B samples was similar to that of the phosphogypsum A samples. However, in comparison to PCE, FDN yielded a more significant effect on the strength of the foamed phosphogypsum samples; when the FDN content was 0.4%, the strength of the corresponding sample was much higher than that of the sample with the same amount of PCE. These results indicate that there was an optimal dosage of the water-reducing agents for the foamed phosphogypsum samples. In this experiment, the optimal dosage of the PCE and FDN was selected as 0.4%. The foamed phosphogypsum samples mixed with FDN had a higher strength than the samples mixed with PCE.

Table 5. Mechanical properties of the foamed phosphogypsum A samples with different water-reducing agent dosages.

Sample	Dosage of Water Reducer (%)	Two-Hour Flexural Strength (MPa)	Absolute Flexural Strength (MPa)	Two-Hour Compressive Strength (MPa)	Absolute Compressive Flexural Strength (MPa)
Phosphogypsum A doped with FDN	0.1	0.9	1.88	1.42	2.95
Phosphogypsum A doped with PCE	0.1	0.89	1.73	1.36	2.90
Phosphogypsum A doped with FDN	0.2	0.92	1.89	1.43	3.00
Phosphogypsum A doped with PCE	0.2	0.9	1.82	1.38	2.95
Phosphogypsum A doped with FDN	0.3	0.96	1.94	1.49	3.15
Phosphogypsum A doped with PCE	0.3	0.92	1.88	1.44	3.05
Phosphogypsum A doped with FDN	0.4	0.98	1.96	1.52	3.25
Phosphogypsum A doped with PCE	0.4	0.94	1.92	1.47	3.15
Phosphogypsum A doped with FDN	0.5	0.94	1.92	1.46	3.10
Phosphogypsum A doped with PCE	0.5	0.91	1.85	1.41	3.00

Table 6. Mechanical properties of the foamed phosphogypsum B samples with different water-reducing agent dosages.

Sample	Dosage of Water Reducer (%)	Two-Hour Flexural Strength (MPa)	Absolute Flexural Strength (MPa)	Two-Hour Compressive Strength (MPa)	Absolute Compressive Flexural Strength (MPa)
Phosphogypsum B doped with FDN	0.1	0.73	1.16	1.1	2.13
Phosphogypsum B doped with PCE	0.1	0.69	1.05	1.08	1.84
Phosphogypsum B doped with FDN	0.2	0.74	1.17	1.15	2.16
Phosphogypsum B doped with PCE	0.2	0.70	1.1	1.10	2.06
Phosphogypsum B doped with FDN	0.3	0.8	1.29	1.23	2.33
Phosphogypsum B doped with PCE	0.3	0.72	1.21	1.14	2.14
Phosphogypsum B doped with FDN	0.4	0.82	1.48	1.28	2.60
Phosphogypsum B doped with PCE	0.4	0.73	1.25	1.16	2.20
Phosphogypsum B doped with FDN	0.5	0.77	1.23	1.19	2.22
Phosphogypsum B doped with PCE	0.5	0.71	1.16	1.12	2.10

3.2. Effect of the Water-Reducing Agents on the Dry Density of the Foamed Phosphogypsum Samples

Figure 5 presents the dry densities of the foamed phosphogypsum A and B samples with different water-reducing agent dosages. The dry density of the foamed phosphogypsum samples first decreased and then increased when the water-reducing agent contents were increased. The dry density of the samples was found to be the lowest when the water-reducing agent content was 0.4%. The lowest dry densities of the phosphogypsum A and B samples doped with FDN were 0.65 g/cm^3 and 0.66 g/cm^3, respectively. The samples doped with FDN had lower dry densities than the foamed phosphogypsum samples mixed with PCE with the same water-reducing agent contents.

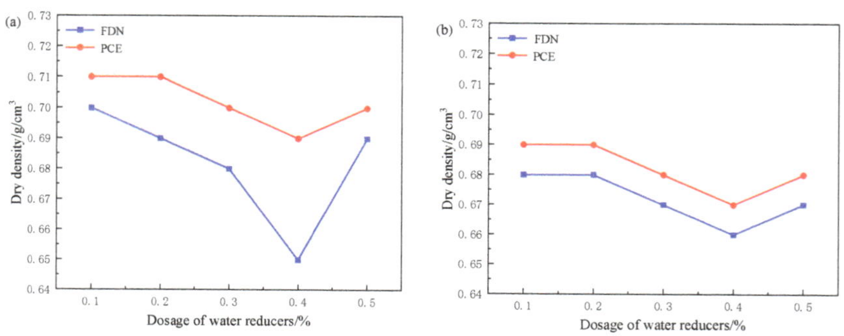

Figure 5. Dry densities of the foamed phosphogypsum samples with different water-reducing agent dosages. (**a**) Phosphogypsum A. (**b**) Phosphogypsum B.

3.3. Discussion on the Effect of the Water-Reducing Agents on the Foamed Phosphogypsum Sample

The addition of water-reducing agents manifests a certain effect on the rheology of gypsum slurries; the bubble size distribution in foam is strongly correlated with the rheology of gypsum slurries. Therefore, in this section, the effects of the water-reducing agents on the rheological properties of the phosphogypsum slurries are discussed.

3.3.1. Effect of the Water-Reducing Agents on the Rheological Properties of the Foamed Phosphogypsum Slurries

Figures 6 and 7 display the apparent viscosity curves and shear stress curves of the foamed phosphogypsum A and B pastes mixed with different water-reducing agents, respectively. It can be seen in Figure 6 that the apparent viscosity of the phosphogypsum slurries gradually and rapidly decline with the increase in the shear rate under the same water-reducing agent content. When the shear rate was in the range of 1–50 s^{-1}, the apparent viscosity of the slurries significantly decreased. When the shear rate was large, the variation in the apparent viscosity tended to be flat. At the same shear rate, the effect of the water-reducing agent content on the apparent viscosity of the phosphogypsum slurries was in the order of 0% > 0.4% > 0.3% > 0.5% > 0.2% > 0.1%. In comparison to the apparent viscosity of the phosphogypsum slurry doped with 0.4% PCE, the apparent viscosity of the slurry mixed with 0.4% FDN was closer to that of the slurry with no water-reducing agent.

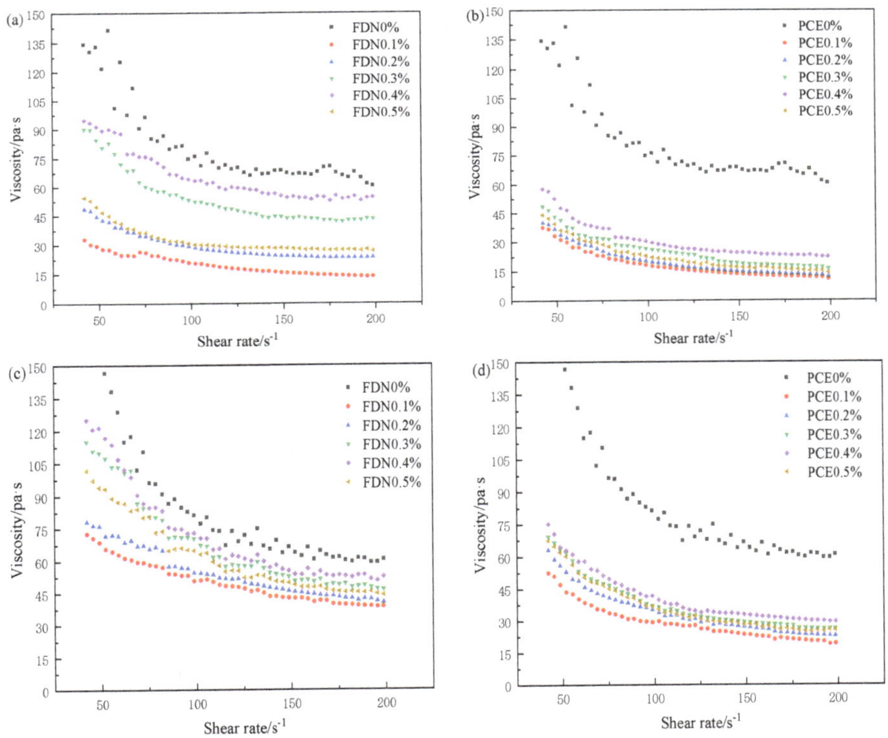

Figure 6. Apparent viscosity curves of the phosphogypsum slurries mixed with different water-reducing agents. (**a**) Phosphogypsum A doped with FDN. (**b**) Phosphogypsum A doped with PCE. (**c**) Phosphogypsum B doped with FDN. (**d**) Phosphogypsum B doped with PCE.

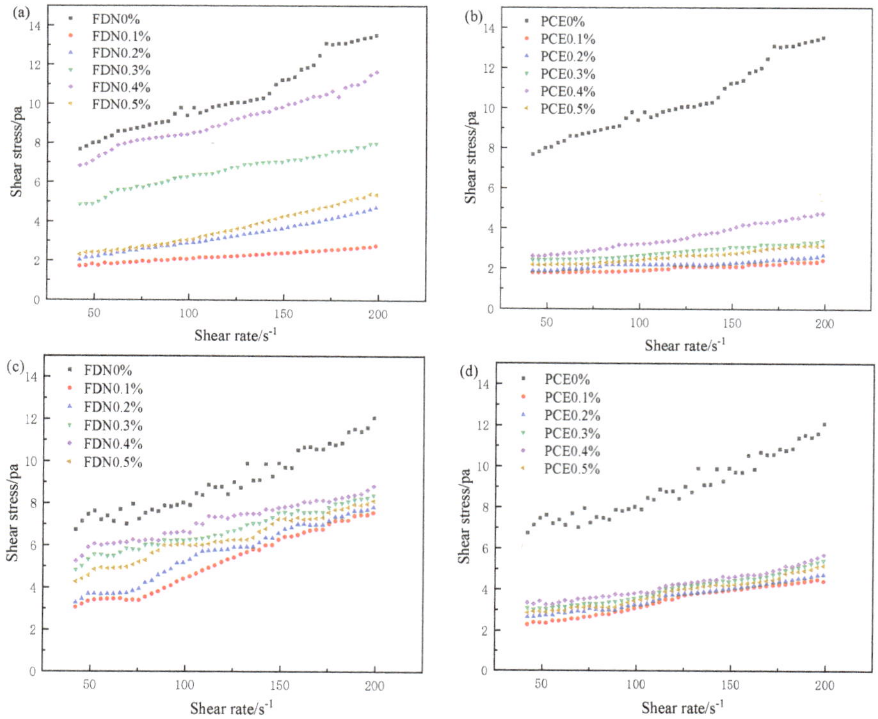

Figure 7. Shear stress curves of the phosphogypsum slurries mixed with different water-reducing agents. (**a**) Phosphogypsum A doped with FDN. (**b**) Phosphogypsum A doped with PCE. (**c**) Phosphogypsum B doped with FDN. (**d**) Phosphogypsum B doped with PCE.

Figure 7 reveals an approximately linear relationship between the shear stress and shear rate of the slurries. The shear stress of the slurries increased with the increase in the shear rate under the same water-reducing agent amount. At the same shear rate, the effect of the water-reducing agent content on the shear stress of the phosphogypsum slurries was in the order of 0% > 0.4% > 0.3% > 0.5% > 0.2% > 0.1%. In comparison to the shear stress of the phosphogypsum slurry doped with 0.4% PCE, the shear stress of the slurry mixed with 0.4% FDN was closer to that of the slurry with no water-reducing agent. With the increase in the water-reducing agent content from 0.1% to 0.4%, the water consumption of the gypsum slurries significantly decreased, causing an increase in the viscosity of the slurries; thus, greater shear force was required to maintain the same shear flow rate. When the water-reducing agent content was 0.5%, the water consumption of the gypsum slurry was between 0.2% and 0.3%. Therefore, the apparent viscosity and shear stress of the phosphogypsum slurries mixed with 0.4% of the water-reducing agents were closer to those of the slurry with no water-reducing agent. The phosphogypsum slurries mixed with PCE had lower apparent viscosity and shear stress than the slurries doped with FDN under the same water-reducing agent dosage due to their slow coagulation by PCE.

3.3.2. Effect of the Water-Reducing Agents on the Pore Structure Distribution of the Foamed Phosphogypsum Samples

Figure 8 displays the pore size distribution curves of the foamed phosphogypsum samples doped with FDN and PCE. The pore size of both foamed phosphogypsum samples followed a bimodal distribution; the pore size distribution was mainly concentrated in the range of 100–3000 nm. FDN and PCE both enlarged the pore size distribution range of foamed phosphogypsum and increased the size and proportion of large pores; however, the influence of PCE was found to be more significant. The large pore size in the foamed phosphogypsum samples doped with PCE was around 100,000 nm, whereas the corresponding value in the foamed phosphogypsum samples doped with FDN was around 10,000 nm. The peak heights in the pore distribution curves shown in Figure 8 represent the proportions of pores with corresponding sizes. It is noticeable from Figure 8d that the peak height of macropores increased in the phosphogypsum B samples doped with PCE, indicating that PCE significantly increased the proportion of macropores. In addition, the water-reducing agent content had a significant effect on pore size distribution. When the dosages of FDN and PCE were both 0.4%, the pore proportions of 100–3000 nm were the highest, whereas the pore proportions of 1000 nm and 10,000 nm in the corresponding samples were the lowest.

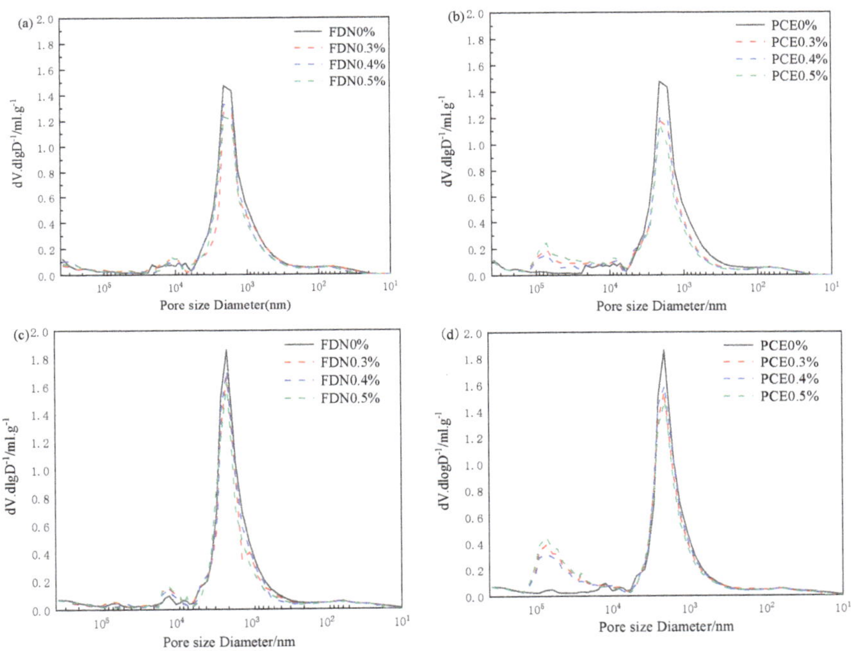

Figure 8. Pore size distribution curves of the foamed phosphogypsum samples mixed with different water-reducing agents. (**a**) Phosphogypsum A doped with FDN. (**b**) Phosphogypsum A doped with PCE. (**c**) Phosphogypsum B doped with FDN. (**d**) Phosphogypsum B doped with PCE.

Figure 9 exhibits the cumulative pore size distribution curves of the foamed phosphogypsum samples doped with FDN and PCE. FDN and PCE both reduced the total pore amount in foamed phosphogypsum; however, the effect of PCE was found to be more significant. The total pore content in the corresponding sample was the highest when the content of FDN or PCE was 0.4%. These results indicate that the type and dosage of the water-reducing agents had significant effects on the total pore content and the distribution of different pore sizes in the foamed phosphogypsum.

Figure 10 is a cross-sectional view of the foamed phosphogypsum. It can be clearly seen that the sample doped with FDN has a smaller macroscopic pore size and more stomatal content than the sample with PCE.

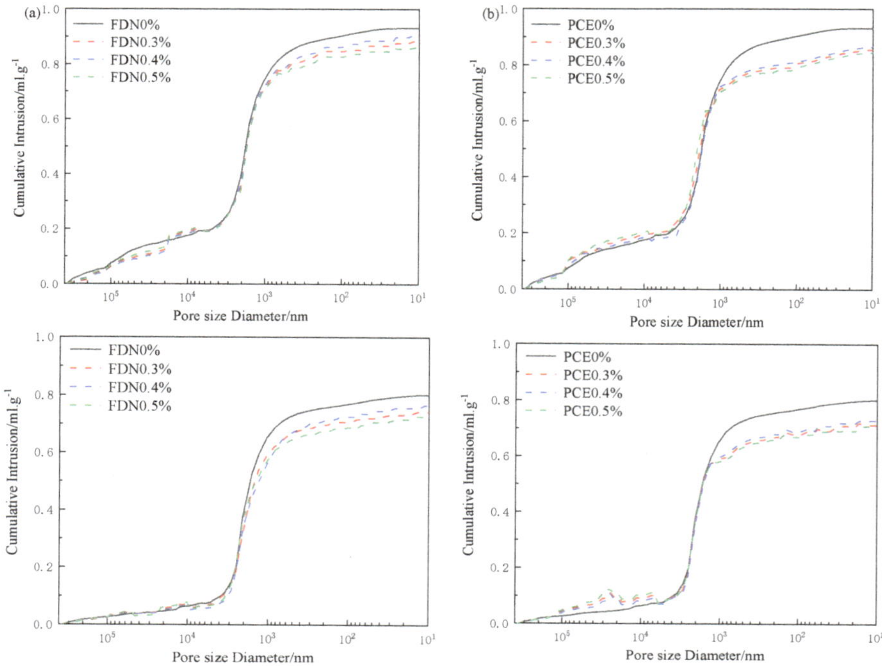

Figure 9. Cumulative pore size distribution curves of the foamed phosphogypsum samples mixed with different water-reducing agents. (**a**) Phosphogypsum A doped with FDN. (**b**) Phosphogypsum A doped with PCE. (**c**) Phosphogypsum B doped with FDN. (**d**) Phosphogypsum B doped with PCE.

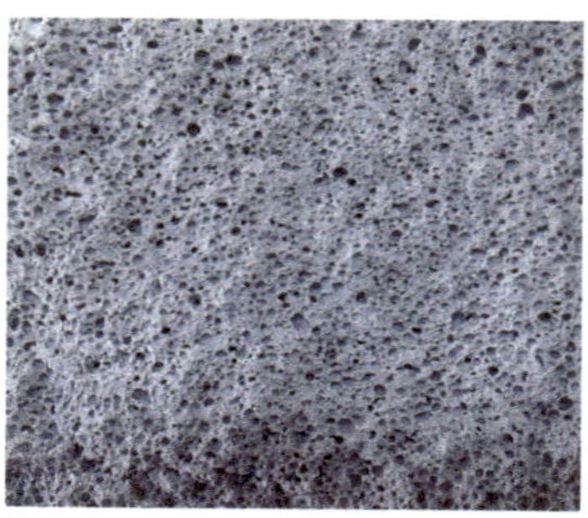

(**a**) Phosphogypsum A doped with 0.4% FDN

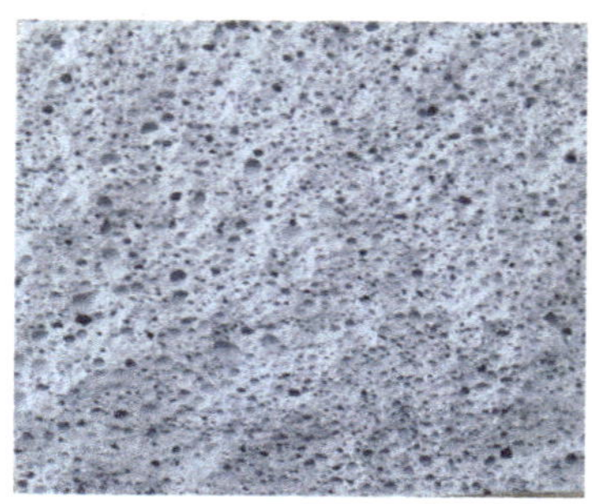

(**b**) Phosphogypsum B doped with 0.4% FDN

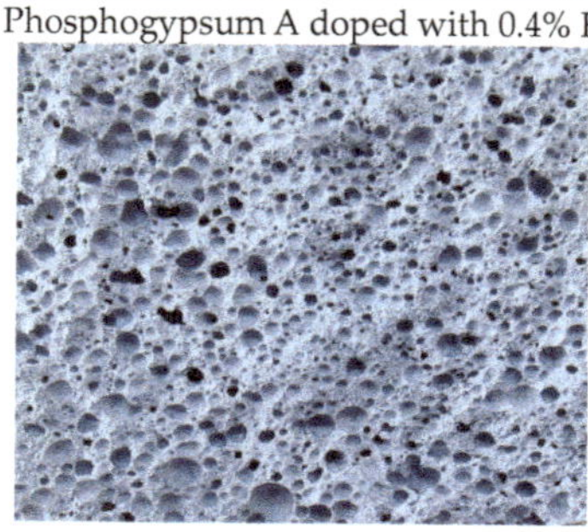

(**c**) Phosphogypsum A doped with 0.4% PCE

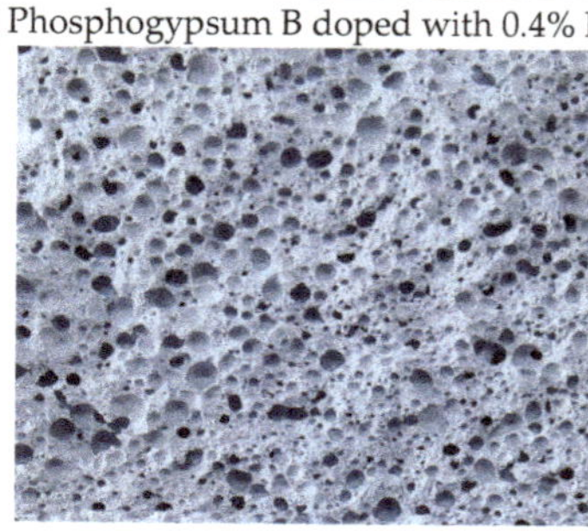

(**d**) Phosphogypsum B doped with 0.4% PCE

Figure 10. Cross-sectional view of foamed phosphogypsum.

3.4. Discussion

It can be concluded that the types and dosages of water-reducing agents have a significant effect on the properties of foamed phosphogypsum; the effects vary greatly between PCE and FDN. Compared with PCE, the match between FDN and foamed phosphogypsum is better; the optimal content of FDN is 0.4%.

Compared with the foamed phosphogypsum doped with PCE, the water consumption of the foamed phosphogypsum doped with FDN was significantly lower, which meant that the viscosity and shear stress of the slurry doped with FDN were larger, the foam was tightly wrapped by the gypsum slurry, and the foam flow range was relatively small. Therefore, the foamed phosphogypsum doped with FDN had a relatively low proportion of large pores and a large number of uniformly distributed medium pores; hence, the foamed phosphogypsum samples doped with FDN had a lower dry density and higher strength. The foamed phosphogypsum slurry mixed with 0.4% FDN consumed the least amount of water; hence, its viscosity and shear stress were relatively large. Moreover, the slurry mixed with 0.4% FDN had the lowest number of large pores and the highest number of medium pores; thus, it yielded the best mechanical properties. The water-reducing agents mainly reduced the water consumption of the phosphogypsum slurries; changed the viscosity and shear stress of the gypsum slurries; affected the stability of foam in the gypsum slurries; changed the pore size and distribution in the gypsum slurries; and influenced the strength, dry density, and other macro-properties of the foamed phosphogypsum samples. The viscosity of the gypsum slurry doped with 0.4% FDN better matched that of the foam; therefore, it had the highest macro-strength and the lowest dry density.

4. Conclusions

(1) When the water-reducing agent content was increased, the strength of the foamed phosphogypsum first increased and then decreased and the dry density of the foamed phosphogypsum first decreased and then gradually increased. There was an optimal dosage range for water-reducing agent in phosphogypsum. The optimal content of FDN and PCE in this experiment was found to be 0.4%. Compared to PCE, FDN is a better match for the foamed phosphogypsum slurry;

(2) The effect of the water-reducing agent content on the apparent viscosity and shear stress of the phosphogypsum slurries was in the order of 0% > 0.4% > 0.3% > 0.5% > 0.2% > 0.1%. The phosphogypsum slurry doped with 0.4% FDN had higher apparent viscosity and shear stress than the slurry mixed with 0.4% PCE;

(3) FDN and PCE both enlarged the pore size distribution, increased the size and proportion of large pores, and decreased the total pore content of foamed phosphogypsum; however, the influences of PCE were found to be more significant. The foamed phosphogypsum slurry mixed with 0.4% FDN had the highest total pore content. Among the samples, the total pore content of foamed phosphogypsum A reached 91%, and the total pore content of foamed phosphogypsum B reached 77%. The lowest proportion of large pores was 17% for foamed phosphogypsum A and 7% for foamed phosphogypsum B;

(4) The water-reducing agents mainly reduced the water consumption of the phosphogypsum slurries; improved the viscosity and shear stress of the slurries; influenced the stability of foam in the gypsum slurries; changed the pore size and distribution in the phosphogypsum slurries; and affected the strength and dry density of the foamed phosphogypsum samples. The viscosity of the gypsum slurry doped with 0.4% FDN better matched that of the foam, having the highest macro-strength and the lowest dry density.

Author Contributions: Conceptualization, J.Y.; methodology, J.Y. and H.W.; validation, J.Y. and H.W.; writing—original draft preparation, J.Y.; writing—editing, J.Y. and H.W.; review, H.Y. and J.G.; formal analysis, F.W. and H.Y.; data curation, F.W.; Chart processing, J.G.; funding acquisition, H.W. All authors have read and agreed to the published version of the manuscript.

Funding: This study was funded by the 14th Five-Year "National Key Research and Development Plan Project" (Grant No. 2022YFC3902704).

Data Availability Statement: Data are contained within the article.

Acknowledgments: We would like to express our heartfelt gratitude to China Building Materials Academy Co., Ltd. for providing the experimental instruments and workstations, as well as to Gao Chunyong and Zhao Ruixia for their contributions to the experimental research.

Conflicts of Interest: Hongxia Wang, Fade Wu and Junhua Guo were employed by the company China Building Materials Academy Co., Ltd. Jian Yu and Haiyan Yudeclare that the research was conducted in the absence of any commercial or financial relationships.

References

1. Awad, S.; Essam, M.; Boukhriss, A.; Kamar, M.; Midani, M. Properties, Purification, and Applications of Phosphogypsum: A Comprehensive Review Towards Circular Economy. *Mater. Circ. Econ.* **2024**, *6*, 9. [CrossRef]
2. Bellefqih, H.; Bourgier, V.; Bilal, E.; Dumitraş, D.-G.; Marincea, Ş.; Mazouz, H.; Haneklaus, N. Effect of $HPO_4{}^{2-}$ and brushite on gypsum reactivity and implications for utilization of phosphogypsum in plaster production. *J. Clean. Prod.* **2024**, *451*, 142013.
3. Li, W.; Ma, L.; Qiu, S.; Yin, X.; Dai, Q.; Du, W. Sustainable Utilization of Phosphogypsum in Multi-Solid Waste Recycled Aggregates: Environmental Impact and Economic Viability. *Sustainability* **2024**, *16*, 1161. [CrossRef]
4. Guan, Q.; Wang, Z.; Zhou, F.; Yu, W.; Yin, Z.; Zhang, Z.; Chi, R.; Zhou, J. The Impurity Removal and Comprehensive Utilization of Phosphogypsum: A Review. *Materials* **2024**, *17*, 2067. [CrossRef] [PubMed]
5. Saadaoui, E.; Ghazel, N.; Ben Romdhane, C.; Massoudi, N. Phosphogypsum: Potential uses and problems–a review. *Int. J. Environ. Stud.* **2017**, *74*, 558–567. [CrossRef]
6. Rashad, A.M. Phosphogypsum as a construction material. *J. Clean. Prod.* **2017**, *166*, 732–743. [CrossRef]
7. Zhou, Y.; He, B. Status quo and suggestions for comprehensive utilization of phosphogypsum in China. *Phosphate Fertil. Compd. Fertil.* **2023**, *38*, 11–16.
8. Chernysh, Y.; Yakhnenko, O.; Chubur, V.; Roubík, H. Phogypsum recycling: A review of environmental issues, current trends and prospects. *Appl. Sci.* **2021**, *11*, 1575. [CrossRef]
9. Yang, Z. Steadily promote the comprehensive utilization of phosphogypsum to help realize the implementation of the "double carbon" strategy. *Phosphate Fertil. Compd. Fertil.* **2023**, *38*, 4.
10. Zhou, J.; Li, X.; Zhao, Y.; Shu, Z.; Wang, Y.; Zhang, Y.; Shen, X. Preparation of paper-free and fiber-free plasterboard with high strength using phosphogypsum. *Constr. Build. Mater.* **2020**, *243*, 118091. [CrossRef]
11. Dong, Z.; Zhai, Y.; Ren, Z.; Yan, B. Research Progress on Resource Utilization of Phosphogypsum Building Materials. *Inorg. Chem. Ind.* **2022**, *54*, 5–9.
12. Jang, M.; Kang, C.-S.; Moon, J.H. Estimation of 222Rn release from the phosphogypsum board used in housing panels. *J. Environ. Radioact.* **2005**, *80*, 153–160. [CrossRef]
13. Murali, G.; Azab, M. Recent research in utilization of phosphogypsum as building materials. *J. Mater. Res. Technol.* **2023**, *25*, 960–987. [CrossRef]
14. Prasad, M.N.V. Resource potential of natural and synthetic gypsum waste. In *Environmental Materials and Waste*; Academic Press: Cambridge, MA, USA, 2016; pp. 307–337.
15. Zhao, X.; Xu, G. Research on the application of polycarboxylate superplasticizer in the production of gypsum materials. *Chem. Eng. Equip.* **2021**, 3–4+24. [CrossRef]
16. Pundir, A.; Garg, M.; Singh, R. Evaluation of properties of gypsum plaster-superplasticizer blends of improved performance. *J. Build. Eng.* **2015**, *4*, 223–230. [CrossRef]
17. Cao, W.; Yi, W.; Yin, S.; Peng, J.; Li, J. A novel low-density thermal insulation gypsum reinforced with superplasticizers. *Constr. Build. Mater.* **2021**, *278*, 122421. [CrossRef]
18. Neuville, M.; Bossis, G.; Persello, J.; Volkova, O.; Boustingorry, P.; Mosquet, M. Rheology of a gypsum suspension in the presence of different superplasticizers. *J. Rheol.* **2012**, *56*, 435–451. [CrossRef]
19. Zhi, Z.; Ma, B.; Tan, H.; Guo, Y.; Jin, Z.; Yu, H.; Jian, S. Effect of competitive adsorption between polycarboxylate superplasticizer and hydroxypropylmethyl cellulose on rheology of gypsum paste. *J. Mater. Civ. Eng.* **2018**, *30*, 04018141. [CrossRef]
20. Zhang, M.; Wang, L.; Zhang, X.; Mi, X.; Sun, F.; Fang, Z.; Pei, M. Effect of Polycarboxylate Superplasticizer Modified by Sodium Hypophosphite on the Properties of β-Calcium Sulfate Hemihydrate. *J. Mater. Civ. Eng.* **2023**, *35*, 04022362. [CrossRef]
21. Capasso, I.; Pappalardo, L.; Romano, R.A.; Iucolano, F. Foamed gypsum for multipurpose applications in building. *Constr. Build. Mater.* **2021**, *307*, 124948. [CrossRef]
22. Zhao, Y. Optimal Design and Performance of Gypsum Foaming Agent and Foamed Gypsum. Master's Thesis, East China University of Science and Technology, Shanghai, China, 2018.
23. Zhong, Z.; Chen, Y.; Zhao, H.; Kang, X. Effect of Foaming Agent on Physical and Mechanical Properties of Foamed Phosphogypsum. *J. Mater. Civ. Eng.* **2024**, *36*, 04023611. [CrossRef]
24. Çolak, A. Density and strength characteristics of foamed gypsum. *Cem. Concr. Compos.* **2000**, *22*, 193–200. [CrossRef]
25. Bazelova, Z.; Pach, L.; Lokaj, J. Effect of surface active substance concentration on the properties of foamed and non-foamed gypsum. *Ceramics-Silikáty* **2010**, *54*, 379–385.

6. Isern, E.R.; Messing, G.L. Messing. Direct foaming and seeding of highly porous, lightweight gypsum. *J. Mater. Res.* **2016**, *31*, 2244–2251. [CrossRef]
7. Mortada, N.; Phelipot-Mardelé, A.; Lanos, C. Impact of surfactant type on performances of gypsum foams. In *CIGOS 2021, Emerging Technologies and Applications for Green Infrastructure: Proceedings of the 6th International Conference on Geotechnics, Civil Engineering and Structures, Ha Long, Vietnam, 28–29 October 2021*; Springer: Singapore, 2022.
8. Feng, J.; Chen, H.; Zhou, M.; Wang, X.; Yang, Q.; Wang, L.; Song, C. Discussion on the compatibility of gypsum board admixtures. In Proceedings of the China Building Materials Federation Gypsum Building Materials Branch (Preparation) Annual Meeting and the Fourth National Gypsum Technology Exchange Conference and Exhibition Proceedings, Wuhan, China, 19 April 2009; Gypsum Building Materials Branch of China Building Materials Federation (in preparation), Editorial Department of Gypsum Building Materials: Beijing, China; Technology Center of Beijing New Group Building Materials Co., Ltd.: Beijing, China, 2009.
9. Xing, T.; Qu, S.; Wu, J. Main factors influencing the batching process of gypsum board. *New Build. Mater.* **2021**, *48*, 109–112.
0. Shen, R. Interaction of admixtures in the production of gypsum board on paper. *Juye* **2014**, *7*, 59–63.
1. Zeng, Z.; Zhao, Z.; Quan, S.; Wu, J.; Liang, Y.; Luan, Y. Study on Compatibility between Superplasticizer and Phosphor Building Gypsum-Based Cementitious Material. *Non-Met. Mines* **2017**, *40*, 32–35.
2. Gao, F.; Zhao, Z.; Quan, S.; Zhang, W.; Yao, Y.; Luan, Y. Different Water Reducing Agents on the Properties of beta Type Building Phosphogypsum. *Bull. Chin. Ceram. Soc.* **2017**, *36*, 960–964+978.
3. Zhang, Y.; Dai, S.; Ma, B.; Sun, Z.; Li, Y. Study on the regulation of slurry rheology in EPS-phosphogypsum composite system. *Bull. Chin. Ceram.* **2020**, *39*, 206–212+218.
4. Wang, J. *Preparation and Basic Properties of Foamed Phosphogypsum Lightweight Materials*; Harbin Institute of Technology: Harbin, China, 2016.
5. Zheng, J.; Wang, Z.; Liu, Y. GB/T 9776—2008 Revision Instructions for Building Gypsum. *New Build. Mater.* **2009**, *36*, 13–16.

applied sciences

MDPI

Article

Characterization of Mortars Incorporating Concrete Washing Fines: Impact on Mechanical Properties, Microstructure and Carbon Footprint

Bechara Haddad [1], Farjallah Alassaad [2], Houssam Affan [1,*], Abdelrahman Mohamad [3] and Nassim Sebaibi [1]

1 Builders Lab, Builders Ecole d'ingénieurs, COMUE Normandie Université, 1 Rue Pierre et Marie Curie, 14610 Epron, France
2 CMEG, Z.A. de Cardonville, Rue Compagnie D, 14740 Thue et Mue, France
3 Vicat, 4 rue Aristide Berges, 38080 L'Isle-d'Abeau, France
* Correspondence: houssam.affan@builders-ingenieurs.fr

Abstract: This study examines the potential use of wash fines, a waste product from concrete plant cleaning, as supplementary cementing materials (SCMs) in mortars. The main objective is to assess the feasibility and benefits of this incorporation in terms of technical performance and environmental impact. Extensive tests were carried out on different mortar formulations, incorporating varying rates of washing fines (0%, 10%, 20%, 30%) as a partial replacement for cement. This choice of replacement is prompted by the fineness of washing fine particles. The properties studied included compressive and flexural strength, porosity, density, water absorption, shrinkage and fire resistance. The results show that the incorporation of washing fines increases porosity and decreases mortar density. There was also a decrease in mechanical strength and fire resistance as the substitution rate increased. However, the use of washing fines enables a significant reduction in the mortar's carbon footprint, reaching up to 29% for the formulation with 30% substitution. This study demonstrates the potential of washing fines as an alternative SCM, as part of a circular economy approach to reducing the environmental impact of the concrete industry. However, it underlines the need to optimize formulations to maintain acceptable technical performance.

Keywords: washing fines; cement additives; waste recycling; carbon footprint

1. Introduction

Concrete is the world's most widely used construction material, with annual production exceeding 10 billion tons [1]. This ubiquity is due to its exceptional durability, remarkable strength and unrivaled versatility. Composed mainly of cementitious binder, aggregates, water and chemical admixtures, concrete offers a unique combination of properties that make it the preferred choice for a multitude of construction applications.

However, despite its countless advantages, the production of its main component, cement, generates around 7% of the world's carbon dioxide (CO_2) emissions, making a significant contribution to global warming, one of the greatest environmental challenges of our time [2]. The cement manufacturing process is not only energy-intensive, requiring substantial quantities of fossil energy, but also releases considerable volumes of CO_2 during the calcination of limestone, the crucial stage in the production of clinker, the key component of cement.

Faced with these growing environmental concerns and the urgent need to act to mitigate the impacts of climate change, intensive efforts have been made within the concrete industry to develop alternative cementitious materials and optimize production processes. The main objective is to significantly reduce the carbon footprint associated with concrete manufacture while maintaining its outstanding technical performance and durability.

Supplementary cementing materials (SCMs) are at the heart of this innovative strategy to reduce the environmental impact of concrete [3–13]. By partially replacing Portland cement clinker, the main but energy-intensive component of cement, with SCMs, it is possible to significantly reduce the CO_2 emissions associated with concrete production [14,15]. This partial substitution not only reduces clinker demand and related emissions but also improves concrete performance.

The use of cement substitute materials (SCMs) plays a crucial role in promoting the circular economy by adding value to industrial by-products that would otherwise be considered waste. By incorporating these materials into concrete formulations, we not only reduce the quantities of waste to be disposed of but also make more efficient use of available resources, thus contributing to a more sustainable future for the construction sector.

Among the most commonly used SCMs are fly ash, silica fume and metakaolin. These materials offer significant environmental benefits by reducing CO_2 emissions and waste generation, while improving the mechanical properties and durability of concrete thanks to their pozzolanic properties and beneficial effects on the microstructure of cement paste. Fly ash, derived from coal combustion, is widely recognized for its ability to improve the workability and long-term strength of concrete. Its incorporation into concrete formulations reduces the need for Portland cement, while increasing the strength and durability of the final material [16]. Silica fume, a by-product of silicon production, stands out for its significant effect on concrete strength and durability. Due to its extreme fineness and high reactivity, it contributes to the formation of an additional hydrated calcium silicate gel (C-S-H), thus improving the density and durability of concrete [17]. Metakaolin, obtained by calcining kaolinite clay, improves not only the mechanical strength of concrete but also its resistance to chemical attack. Its fine granulometry and pozzolanic reactivity make it an excellent reinforcement for concrete mixes, increasing their longevity and performance [18,19].

In addition, other SCMs, such as granulated blast furnace slag (GGBFS), are gaining importance in research and practical applications. Granulated blast furnace slag, derived from the steel-making process, is a by-product rich in silica and alumina, which can replace part of the Portland cement in concrete mixes. It improves the durability of concrete, notably its resistance to sulfates and chloride penetration, and reduces the carbon footprint of concrete [20,21].

SCMs offer promising opportunities for reducing the carbon footprint of cementitious matrices. For example, a 22% reduction in CO_2 emissions has been observed for a 40% replacement of cement by fly ash in a cementitious paste [22]. Similarly, replacing 15% of cement with silica fume in concrete results in a 12% reduction in CO_2 emissions [23]. In the case of metakaolin, a 20% substitution results in around 17% reduction [24]. GGBS can reduce emissions by up to 47% with a substitution rate of 70% in concrete [25]. These results demonstrate the potential of SCMs to significantly reduce the environmental impact of cementitious materials.

At present, research is focusing on optimizing the substitution rates of these SCMs, assessing their impact on the mechanical and durability properties of concrete and analyzing the life cycle of the materials. However, major gaps remain, notably concerning the variability of the chemical composition of these wastes and their long-term effects on concrete structures. Further study is needed to better understand these effects and improve SCM design and application practices in the concrete industry.

By judiciously exploiting SCMs, the concrete industry can reduce its carbon footprint while optimizing the technical performance of its products, thus contributing to a more sustainable future for the construction sector. The use of concrete industry waste as SCM is a rapidly expanding field. The current research focuses on optimizing substitution rates, assessing impacts on mechanical and durability properties and life-cycle analysis. However, there are still significant gaps concerning the variability of the chemical composition of these wastes and their long-term effects on concrete structures. Some of these studies have explored recycling concrete industry waste as SCM. For example, the reuse of fines from washing ready-mix concrete has been studied, showing a reduction in compressive strength but an improvement in durability [26]. Similarly, recycled concrete fines obtained by mechanosynthesis have been valorized for construction material production [27].

Among the promising new sources of SCMs, washing fines represent a particularly interesting and innovative opportunity. These fines are waste products from the cleaning operations of concrete plants and truck-mixers after the production of ready-mix concrete. Composed mainly of fine silica and calcium particles, these wash fines are often considered non-value-added waste and simply disposed of, generating additional costs and undesirable environmental impacts.

However, their incorporation into mortar and concrete formulations as a partial substitute for Portland cement offers a significant double advantage. On the one hand, it would reduce cement consumption and, consequently, the CO_2 emissions associated with its energy-intensive production. On the other hand, it would transform a previously unused waste product into a valuable resource, reclaiming these washing fines and reducing the environmental impact associated with their management and disposal.

The use of washing fines as an alternative SCM fits perfectly with the concrete industry's strategy to reduce its carbon footprint while promoting a more sustainable circular economy. The judicious recovery of this waste not only reduces waste management costs but also reduces demand for Portland cement, the production of which is particularly energy-intensive and emits greenhouse gases.

This innovative approach represents a further step towards a more environmentally friendly concrete industry by fully exploiting the potential of all available by-products and waste as valuable alternative resources. Wash fines, once considered a worthless waste product, could play a key role in the transition to more sustainable construction.

A great deal of research has been carried out into the use of washing fines as a substitute for sand. The results of this research indicate that washing fines can be used as a partial substitute for fine aggregates (sand) in concrete, sometimes even improving certain properties of the material [26–28]. However, studies focusing specifically on their use as a partial cement substitute remain limited [29].

The main objective of this study is to assess the feasibility and potential benefits of using wash fines as SCMs in mortars. A series of extensive tests were carried out to examine the impact of incorporating these fines on various key mortar properties, such as compressive strength, flexural strength, porosity, bulk density, water absorption, shrinkage and fire resistance. These rigorous tests make it possible to determine the extent to which wash fines can improve or modify mortar characteristics, and thus to assess their viability as promising SCMs.

The results of this research provide a detailed overview and in-depth understanding of the valorization of wash fines in cement-based construction materials. They demonstrate how more sustainable waste management can be seamlessly integrated into concrete production, contributing to a greener, more efficient construction industry. This study paves the way for innovative industrial practices that valorize waste while meeting the growing need for sustainability and superior performance in the construction sector. It represents an important step towards the transition to a more environmentally friendly circular economy in the concrete industry.

2. Materials and Methods

This section presents a detailed description of the materials used in this study, based on various methods.

2.1. Cement

The binder used in this study is Portland cement type CEM I 52.5R, known for its rapid hardening and high strength. This cement is commonly used in building construction where high mechanical performance is required. Table 1 shows the cement's chemical composition and its key mechanical properties, including compressive strengths at different hardening ages.

Table 1. Cement characteristics as given in technical sheet.

Data	CEM I 52.5R–SR5 CE
SO_3	2.1%
MgO	0.6%
Na_2O	<0.3%
Cl^-	<0.04%
Loss on ignition at 950 °C over time	1.1%
Compressive strength at 1 day	21–27 MPa
Compressive strength at 2 days	40–48 MPa
Compressive strength at 28 days	66–76 MPa

Cement is a fine powder. Figure 1 illustrates the particle size distribution of the cement used in this study.

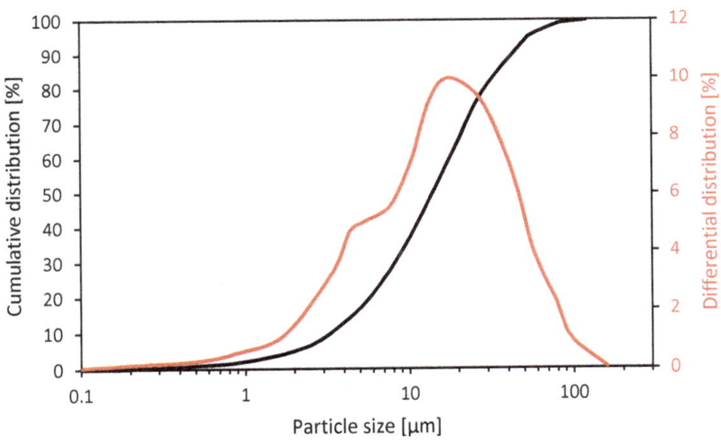

Figure 1. Particle size distribution of the used cement.

2.2. Sand

Natural washed sand with a particle size distribution conforming to fine aggregate standards was used in this study. It is clean, free of impurities and has a density of 2.63. The properties described include density, water absorption and other relevant characteristics.

For this study, washed natural sand complying with aggregate standards was used. The sand is clean and free of impurities. Its particle size distribution complies with the normative requirements applicable to fine aggregates. Table 2 presents the key properties of this sand relevant to its use in mortar and concrete formulations.

Table 2. Sand physical properties as given in technical sheet.

Physical Properties	0/4 Sand
Density [kg/m^3]	2630
Water absorption [%]	0.1
Sand equivalent	81.3
Fineness modulus	2.3

2.3. Fresh Concrete Washing Fine

In this study, fresh concrete washing fine comes from a precast concrete plant washing process. Preparation of washing fines for use as supplementary cementing materials

(SCMs) begins with their collection from production sites and storage in hermetically sealed containers at room temperature. The fines are then dried at 65 ± 5 °C to a constant mass, ensuring complete dewatering. The dried sludge is then ground by hand to obtain a fine powder, which is sieved to 100 µm to remove the coarsest particles; see Figure 2. Wash fines can be used without heating, but heating is chosen to make mixing easier.

Figure 2. Fine washing powder.

To characterize the thermal stability and composition of these washing fines, thermogravimetric analyses (TGA) were carried out. Measurements were carried out in an inert nitrogen atmosphere, varying the temperature from ambient to 1000 °C, in a Netzsch STA449 F3 Jupiter® furnace (Netzsch Gerätebau GmbH, Selb, Germany). Several critical phases are revealed at different temperature ranges (see Figure 3). Up to 120 °C, a decrease in mass is observed, corresponding to the evaporation of residual free water. Between 120 and 600 °C, another significant mass loss is identified, probably linked to the release of chemically bound water during the decomposition of calcium hydroxide ($Ca(OH)_2$), with a peak between 400 and 500 °C. Finally, a further decrease in mass occurs between 600 and 1000 °C, suggesting the release of carbon dioxide (CO_2) resulting from the decomposition of calcium carbonate ($CaCO_3$), with a notable peak between 800 and 900 °C.

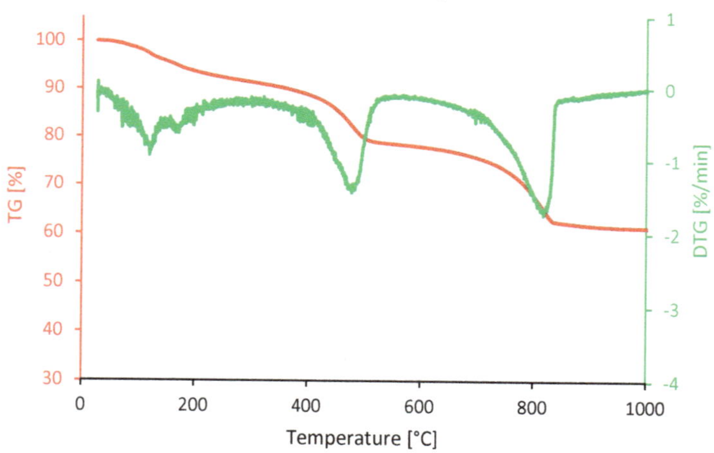

Figure 3. Thermogravimetric analyses of washing fine. TG (Thermogravimetric Analysis, shown in red) and DTG (Derivative Thermogravimetry, shown in green).

In summary, washing fine TGA reveals mass loss processes associated with three main phenomena: residual free water evaporation up to 120 °C, chemically bound water release between 120 and 600 °C (probably related to calcium hydroxide decomposition) and calcium carbonate decomposition between 600 and 1000 °C with carbon dioxide release. It provides crucial information for understanding the chemical composition and thermal stability of washing fines, guiding the appropriate processing and recovery strategies for these materials.

In addition, washing fines were examined using laser granulometry. The results in Figure 4 show a distribution similar to that of cement but with certain differences. Wash fines contain finer particles, probably resulting from additions to the concrete mix, such as filler. At the same time, coarser particles are present, originating from the fines in the sand used in the concrete. This particle distribution depends on the nature of the concrete produced. Indeed, a formulation containing more sand, cement or supplementary cementing materials, as well as admixtures, may affect the results. Studies on several sources could be of interest for future work.

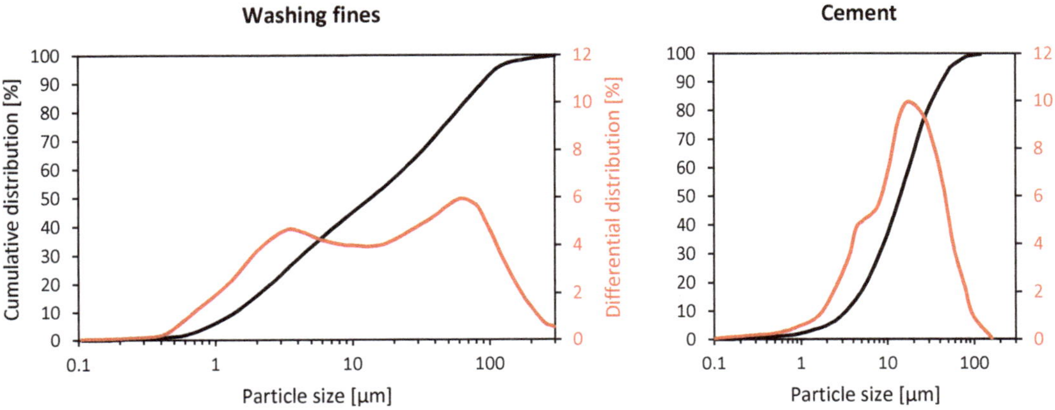

Figure 4. Washing fine particle size distribution compared to cement. Cumulative distribution (black line) and differential distribution (red line).

2.4. Porosity and Density

Mortar samples were tested for water-accessible porosity and density in accordance with standard NF P18-459 [30]. This test, commonly known as the "vacuum water test", measures both sample open porosity and bulk density. Samples are first oven-dried to a constant mass, and then placed in a desiccator connected to a vacuum pump. A high vacuum is applied for 24 h to extract the air contained in the open pores of the samples. The desiccator is then filled with demineralized water, allowing for complete impregnation of the pores under vacuum for a further 24 h. After resaturation, the immersed mass of the samples is measured by hydrostatic weighing. The samples are then superficially wiped with a damp cloth to determine their saturated mass. The combination of these different mass measurements gives the water-accessible porosity and the apparent density according to the formulas prescribed by standard NF P18-459.

2.5. Water Absorption by Immersion

The water absorption of mortars was measured using a simple direct immersion protocol. Samples of hardened mortar were first oven-dried to constant mass to remove any residual water. Their dry mass was then carefully recorded. The samples were then fully immersed in demineralized water for a period of 24 h to allow for total saturation by capillary absorption. After this immersion period, the samples were removed from the water, superficially wiped with a damp cloth, and their saturated mass was measured. The water absorption rate, expressed as a percentage, was calculated as the increase in sample mass due to water absorption, relative to their initial dry mass.

2.6. Flexural Strength

Mortar flexural strength was assessed using a three-point bending test in accordance with the French standard NF EN 196-1 [31]. Prismatic specimens measuring 40 mm × 40 mm × 160 mm were cast for each mortar formulation. After a standardized curing period, these specimens were subjected to a three-point bending test, in which the specimen is simply placed on two lower supports and loaded at the center by a third upper support descending at constant speed until failure. The maximum force applied at the moment of flexural failure is recorded. The flexural strength is then calculated according to the formula prescribed by the standard, depending on the breaking load, the span between lower supports and the cross-sectional dimensions of the specimen.

2.7. Compressive Strength

Following the three-point bending tests, the two resulting prismatic half-tests were subjected to compression tests in accordance with NF EN 196-1 [31]. The compression machine's plate allows for testing of a dimension of 40 mm × 40 mm × 40 mm. An increasing compression load was applied at a constant speed of 2.4 kN/s in accordance with the standard. The maximum force recorded at the moment of crushing failure was used to calculate the compressive strength Rc according to the standardized formula, taking into account the dimensions of the support section. This standardized test, which complements the bending test, provides a complete characterization of the mechanical performance of the different mortar formulations developed.

2.8. Total Shrinkage

Total shrinkage of the various mortar formulations was assessed in accordance with the French standard NF P 15-433 [32], which describes the method for measuring dimensional variations on hardened mortar specimens. Prismatic specimens measuring 40 mm × 40 mm × 160 mm were cast in molds and demolded after 24 h. These specimens were fitted with metal pins at the ends to facilitate accurate measurement of length variations. After demolding, the specimens were stored in an air-conditioned room at 50% relative humidity and 20 °C for curing under controlled conditions.

At different time intervals, the dimensional variations of the specimens were measured using a high-precision length comparator. Total shrinkage was then calculated as the ratio of length variation to initial specimen length.

2.9. Fire Resistance

The fire resistance of the various mortar formulations was assessed on the basis of their residual compressive strength after exposure to high temperatures, using a procedure derived from current investigations [33,34]. For each mix, three 40 mm × 40 mm × 160 mm prismatic specimens were cast and stored under standard conditions for 28 days for complete curing. These specimens were then placed in a preheated muffle furnace and held for 1 h at target temperatures of 200 °C, 400 °C and 900 °C, respectively. The choice of these test temperatures was guided by the results of prior thermogravimetric analyses.

After exposure to the target temperature for 1 h, the samples were removed from the furnace and cooled naturally to room temperature. Once cooled, they were visually inspected for cracks, spalling or other degradation. Their mass was also measured and compared with the initial mass to quantify any losses. Finally, the samples were subjected to a flexion and compression test to determine their strength after thermal aggression.

2.10. Formulation

Four different mortar formulations were prepared, incorporating varying levels of washing fines as a partial replacement for cement. An unadded control formulation (T), as well as three formulations containing, respectively, 10% (L10), 20% (L20) and 30% (L30) washing fines to replace cement were studied. The proportions of each component in each mixture are detailed in Table 3.

Table 3. Mortar incorporating washing fines formulations.

Component	Ref	L10	L20	L30
Cement	450	405	360	315
Washing fine	0	45	90	135
Sand	1350	1350	1350	1350
Water	225	225	225	225

3. Results

3.1. Open Porosity and Density

The physical properties of different mortar formulations incorporating washing fines are presented in Table 4. An increase in porosity is observed with the increasing incorporation of washing fines as a partial replacement for cement. Indeed, the porosity of the reference control mortar is 19.3%, while L10, L20 and L30 formulations have porosities of 23.2%, 24.1% and 25.8%, respectively. This tendency for porosity to increase with the rate of substitution can be explained by the less efficient filling effect of wash fines compared to cement in the cementitious matrix, leaving more voids [35].

Table 4. Physical properties of mortar incorporating washing fine.

Formulation	Porosity [%]	Density [kg/m^3]
Ref	19.3	2008
L10	23.2	1937
L20	24.1	1898
L30	25.8	1835

At the same time, mortar bulk density decreases with the incorporation of washing fines. The reference mortar has a density of 2008 kg/m^3, while the densities of the L10, L20 and L30 formulations are 1937, 1898 and 1835 kg/m^3, respectively. This decrease in density is consistent with the observed increase in porosity, and can also be attributed to the intrinsically lower density of wash fines compared to Portland cement.

These results indicate that the incorporation of washing fines as a partial replacement for cement has a significant impact on the microstructure of mortars, increasing their total porosity and reducing their compactness.

3.2. Water Absorption by Immersion

The water absorption results for the different mortar formulations incorporating washing fines are presented in Table 5.

Table 5. Water absorption of mortars incorporating washing fines.

Formulation	Water Absorption [%]
Ref	4.6
L10	5.9
L20	5.3
L30	10.3

The results indicate an increase in water absorption with a higher incorporation of wash fines. The reference mortar (Ref) showed a water absorption rate of 4.6%. Formulations L10 and L20 showed slightly higher absorption rates of 5.9% and 5.3%, respectively. However, formulation L30, which contained the highest amount of washing fines, showed a significant increase in water absorption to 10.3%. This increase in water absorption with the addition of washing fines can be attributed to several factors. Firstly, higher porosity: as observed in the porosity results, mortars with wash fines have higher porosity. This increased porosity creates more capillary pores capable of absorbing and retaining water, resulting in higher water absorption rates. Secondly, particle compaction density: the less efficient compaction of washing fines compared to cement particles results in a more open microstructure, facilitating water entry because of their irregular shape [29,36]. Thirdly, surface characteristics: washing fines may have different surface characteristics to cement, possibly with higher surface roughness or specific surface, which enhances water absorption. Finally, hydration products: the incorporation of washing fines can alter the type and quantity of hydration products used. This may be due in part to the fact that washing fines contain far fewer active ingredients than cement, as previous studies on this topic have pointed out [29].

3.3. Total Shrinkage

The total shrinkage of the various mortar formulations incorporating wash fines was monitored over time, with measurements taken at 1, 3, 28 and 70 days. Shrinkage values expressed in "µm/m" are shown in Table 6.

Table 6. Total shrinkage of mortar incorporating washing fine.

Total Shrinkage [µm/m]	Ref	L10	L20	L30
1 day	509	1054	1203	1637
3 days	603	1451	1504	2162
28 days	713	1473	1624	2222
70 days	800	1608	1684	2362

It can be seen that the increasing incorporation of washing fines as a partial replacement for cement leads to a significant increase in total shrinkage at all ages. After 70 days, the shrinkage of the reference mortar without addition reaches 800 µm/m. In contrast, the L10, L20 and L30 formulations show significantly higher shrinkages of 1608, 1684 and 2362 µm/m, respectively.

This tendency for shrinkage to increase with the substitution rate can be explained by several factors. Firstly, the greater porosity of mixes with washing fines favors water movement during drying, thus amplifying shrinkage. Secondly, the very nature of wash fines, rich in fine silica and calcium particles, may result in higher water demand for hydration, resulting in a larger pore network after drying.

3.4. Mechanical Strength

The flexural strength test results for the different mortar formulations incorporating wash fines are presented in Table 7. The values correspond to the average strength measured for three specimens at 14 and 28 days, with the associated standard deviation.

Table 7. Flexural strength of mortar incorporating washing fines.

Formulation	Flexural Strength	
	14 Days [MPa]	28 Days [MPa]
Ref	6.4 ± 0.4	6.7 ± 0.2
L10	5.4 ± 0.2	5.8 ± 0.3
L20	4.7 ± 0.3	5.1 ± 0.4
L30	2.5 ± 0.3	2.9 ± 0.2

A decrease in flexural strength is observed as the rate of substitution of cement by washing fines increases. At 28 days, the unadded reference mortar had a flexural strength of 6.7 MPa. The incorporation of 10% washing fines results in a slight drop to 5.8 MPa. This decrease becomes more pronounced for P20 and P30 formulations, with strengths of only 5.1 and 2.9 MPa, respectively.

This downward trend in flexural performance with the incorporation of washing fines can be attributed to several factors. Firstly, the greater porosity of these mixes [35], as highlighted above, weakens the compactness of the cementitious matrix and creates stress concentration sites. Secondly, the very nature of wash fines, composed mainly of fine particles of low reactivity, does not allow for the development of a microstructure as dense and resistant as Portland cement [29,37,38].

On the other hand, Table 8 presents the results of compressive strength tests carried out on the various mortar formulations incorporating the washing fines. The values correspond to the average strength measured for three cubic specimens at 14 and 28 days.

Table 8. Compressive strength of mortar incorporating washing fines.

Formulation	Compressive Strength	
	14 Days [MPa]	28 Days [MPa]
Ref	47 ± 1.8	51.1 ± 1.2
L10	33.3 ± 1.1	36.1 ± 1.2
L20	26.9 ± 1.3	29.3 ± 1.0
L30	17.9 ± 1.2	21.5 ± 1.3

A significant decrease in compressive strength is observed as the rate of substitution of cement by washing fines increases. The reference mortar with no added fines shows a strength of 51 MPa after 28 days of curing. The incorporation of 10% washing fines (L10) results in a moderate decrease to 36 MPa. However, this reduction becomes more marked for formulations L20 and L30, with strengths of just 29 MPa and 21 MPa, respectively, at 28 days.

This drop in compressive performance with an increasing substitution ratio can be attributed to several factors. Firstly, as previously observed, the incorporation of washing fines increases the total porosity of the mortar, thus weakening its compactness and density [35]. This more porous and less dense microstructure does not allow the mortar to develop as high mechanical strength as the reference mortar. Secondly, the very nature of the wash fines, composed mainly of fine particles of low reactivity, does not contribute significantly to the development of a dense, resistant microstructure [29,37,38].

These results of a compressive strength decrease with increasing substitution rate are consistent with those of Wu et al. (2023) [29], who observed a similar trend in their study on waste powders as cement mortar components.

3.5. Fire Resistance

The results of fire resistance tests reveal a worrying trend in the incorporation of washing fines in mortars. All formulations containing washing fines (L10, L20, L30) showed a significant reduction in flexural (Table 9) and compressive strength (Table 10) compared with the reference formulation (Ref) at all temperatures tested (200 °C, 400 °C, 900 °C). This drop in performance became more pronounced as the rate of substitution of cement by washing fines increased. At 900 °C, all formulations, including the reference, show a drastic loss of strength, which is consistent with the typical behavior of cementitious materials at high temperatures. However, the incorporation of wash fines seems to exacerbate this loss of strength.

Table 9. Flexural strength of mortar incorporating washing fines after exposure to fire.

Formulation	Flexural Strength [MPa]		
	200 °C	400 °C	900 °C
Ref	5.8	4.8	0.6
L10	5.1	3.3	0.4
L20	3.7	2.7	0.3
L30	1.1	0.9	0.1

Table 10. Compressive strength of mortar after exposure to fire.

Formulation	Compressive Strength [MPa]		
	200 °C	400 °C	900 °C
Ref	27.7	25.9	4.4
L10	18.7	16.9	3.6
L20	15.6	10.6	2.5
L30	6.2	4.9	1.3

This heat-induced degradation of mechanical properties could be explained by several factors. Firstly, the increased porosity of mortars containing washing fines, as observed in the porosity results, could facilitate heat penetration and accelerate the degradation of the material's internal structure. Secondly, the partial substitution of cement by washing fines could reduce the amount of cement hydrates formed, which are primarily responsible for the mortar's mechanical strength. Finally, potential differences in thermal behavior between washing fines and Portland cement could induce internal stresses and microcracks when exposed to high temperatures, further weakening the mortar's structure.

These results underline the importance of considering fire resistance when using washing fines as a partial cement substitute, particularly in applications where fire safety is paramount. Although the incorporation of washing fines offers environmental and economic benefits, as demonstrated in the previous sections, their negative impact on fire resistance could limit their use in certain contexts. However, it is important to note that the results obtained are preliminary and require more detailed analysis and research to determine the feasibility and effectiveness of these mortars in practical applications. Further studies will be required to comprehensively assess the fire performance of these materials, taking into account various parameters such as processing conditions and variations in composition.

3.6. Embodied Carbon

Cement is the factor with the greatest carbon impact. On the other hand, sand, used as a fine aggregate, although less impactful than cement, also contributes to CO_2 emissions through its extraction and transportation. Wash fines, considered in this study as a partial substitute for cement, are by-products of concrete plant cleaning operations, the use of which reduces the demand for Portland cement and, therefore, the emissions associated with its production.

To assess the carbon impact of mortar manufacturing, several assumptions were made. Firstly, it was assumed that the integrated carbon emissions for each raw material were taken from the ICE (Inventory of Carbon and Energy) database [39]. Secondly, washing powder was dried for 8 hours in a laboratory oven, followed by sieving. As the product is sinterable, no grinding was required. To assess the carbon impact of this process, it is essential to analyze the energy consumption of production stages.

Carbon impact calculations for each kg of powder produced have taken into account two key stages in the process: drying the waste in an oven and sieving. These two operations require specific energy consumption, which varies according to production parameters and depending on the type of energy source used (low-carbon electricity or coal-based electricity).

1. Energy consumption:
    - Oven drying: For a batch of 50 kg of powder, a laboratory oven consuming 1000 W [40] was used for 8 h. The total energy consumption for this stage was 8 kWh for 50 kg, or 0.16 kWh per kg of powder.
    - Sieving: Each kilogram of powder was then sieved using a sieve shaker consuming 360 W [41] for 20 min. The energy consumption for this stage is 0.12 kWh per kg.
    - The total energy consumption to prepare fine wash is 0.28 kWh per kg.
2. CO_2 emissions: CO_2 emissions vary according to the source of electricity used [42]:
    - With low-carbon electricity (e.g., nuclear in France, 50 g CO_2/kWh), the total carbon footprint is 14 g CO_2 per kg of powder.
    - With coal-based electricity (1000 g CO_2/kWh), the total carbon footprint is 280 g CO_2 per kg of powder.
    - As this study was located in France, it is estimated that 1 kg of fine washing causes an emission equivalent to 14 g of CO_2.

These results cover the energy consumed in drying and screening the waste but do not include the transport of raw materials or finished products. With regard to transport distance, it was assumed that all mortar constituents are located within 30 miles of the manufacturing plant, thus minimizing the carbon impact associated with transport. These assumptions make it possible to standardize the calculation conditions and provide a consistent estimate of the carbon footprint of different mortar formulations.

The results in Figure 5 show a significant reduction in carbon footprint with the gradual incorporation of washing fines. Mortar formulations L10, L20 and L30, with 10%, 20% and 30% cement substitution by washing fines, respectively, show carbon footprint reductions of 9%, 19% and 28% compared to the reference formulation (Ref), whose integrated carbon is 436 kg CO_2/m^3. Thus, integrated carbon values decrease to 395 kg CO_2/m^3, 355 kg CO_2/m^3 and 313 kg CO_2/m^3, respectively, for L10, L20 and L30 formulations.

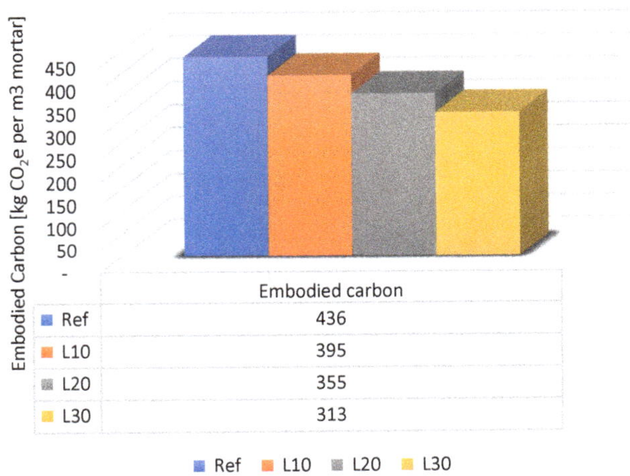

Figure 5. Embodied carbon of mortar incorporating washing fines.

This reduction is due to the lower quantity of cement required, with the cement saved translating into a significant reduction in CO_2 emissions. Consequently, using wash fines as a cement substitute in mortar manufacture is an effective strategy for reducing carbon impact by valorizing an industrial by-product while reducing dependence on Portland cement. This approach not only reduces overall CO_2 emissions but also promotes a more sustainable, environmentally friendly construction industry.

4. Discussion

The incorporation of washing fines into mortars offers significant environmental benefits but also presents technical challenges that need to be analyzed in depth. On the one hand, the results show that the addition of washing fines can considerably reduce the carbon footprint of mortars. In particular, a partial substitution of cement by these fines can reduce CO_2 emissions by up to 28% for a 30% substitution rate, which is in line with the construction industry's current objectives to minimize the environmental impact of construction materials [43]. This reduction is mainly due to a reduction in the use of cement, an energy-intensive material, which supports a circular economy approach by recovering industrial waste.

However, the incorporation of washing fines has negative effects on certain technical properties of the mortar, in particular, an increase in porosity and a decrease in mechanical and heat resistance. Tests have shown that the higher the rate of substitution of cement by washing fines, the greater the increase in porosity, resulting in a loss of density and a reduction in flexural and compressive strengths. For example, at a 30% substitution rate, compressive strength falls by almost 50% compared to the reference mortar without the addition of fines. This behavior is consistent with other studies on the use of secondary cementitious materials (SCM), which show that increased porosity due to poor compaction of fine particles affects mechanical strength [22,44].

Wash fines, mainly composed of silica and calcium particles, have limited chemical reactivity compared to traditional Portland cement, which reduces their ability to form a dense network of hydration products. In addition, increased porosity promotes water absorption, increasing susceptibility to shrinkage and reducing durability [45]. These results suggest that simply replacing cement with washing fines is not enough to guarantee acceptable mechanical performance in demanding applications. It would therefore be appropriate to explore combinations with other SCMs, such as fly ash or blast furnace slag, to offset these negative effects while maintaining the environmental benefits [46,47].

Finally, although the environmental benefits are undeniable, further research is needed to improve the formulation of mortars incorporating washing fines. Pre-treatment of the fines, such as thermal or chemical activation, could increase their reactivity and thus improve their performance. In addition, the addition of specific additives could improve compaction and limit the increase in porosity. Future studies should also assess the long-term impact of these formulations on the durability of the structures, in particular, their fire resistance, which showed a significant drop at high temperatures in this study.

5. Conclusions

This study on the use of washing fines as supplementary cementing materials (SCMs) in mortars offers interesting prospects for the concrete industry while highlighting some of the challenges ahead.

The incorporation of washing fines as a partial replacement for cement offers significant environmental benefits, including a substantial reduction in the carbon footprint of mortars, up to 28% for a substitution rate of 30%. This approach fits in perfectly with a circular economy approach by transforming an industrial waste product into a recoverable resource.

However, the results also highlight trade-offs to be considered. The increase in porosity and the reduction in mechanical and fire resistance with the increasing incorporation of washing fines raise questions about the limits of use of this material in certain structural applications or those subject to specific constraints.

These observations pave the way for future research aimed at optimizing formulations to strike a balance between environmental performance and technical properties. Avenues to be explored could include combining washing fines with other supplementary cementing materials, optimizing fine pre-treatment processes, or developing methods to improve their reactivity.

In conclusion, this study demonstrates the promising potential of wash fines as an alternative SCM, while highlighting the need for further research to overcome the technical challenges identified. It thus makes a significant contribution to the advancement of knowledge in the field of sustainable building materials and paves the way for future innovations in the concrete industry.

Author Contributions: Conceptualization, B.H., F.A. and H.A.; methodology, B.H., F.A., H.A., A.M. and N.S.; formal analysis B.H., F.A., H.A., A.M. and N.S.; writing—original draft preparation, B.H., F.A., H.A. and A.M.; supervision, N.S. All authors have read and agreed to the published version of the manuscript.

Funding: This research received no external funding.

Institutional Review Board Statement: Not applicable.

Informed Consent Statement: Not applicable.

Data Availability Statement: The raw data supporting the conclusions of this article will be made available by the authors upon request.

Acknowledgments: The authors would like to acknowledge the contribution of Master Builders Solution in providing raw materials and technical support.

Conflicts of Interest: Author Farjallah Alassaad was employed by the company CMEG. Author Abdelrahman Mohamad was employed by the company Vicat. The remaining authors declare that the research was conducted in the absence of any commercial or financial relationships that could be construed as potential conflicts of interest.

References

1. Tayebani, B.; Said, A.M.; Memari, A. Less Carbon Producing Sustainable Concrete from Environmental and Performance Perspectives: A Review. *Constr. Build. Mater.* **2023**, *404*, 133234. [CrossRef]
2. Chaudhury, R.; Sharma, U.; Thapliyal, P.C.; Singh, L.P. Low-CO_2 Emission Strategies to Achieve Net Zero Target in Cement Sector. *J. Clean. Prod.* **2023**, *417*, 137466. [CrossRef]
3. de Pedro, J.P.Q.; Lagao, J.A.T.; Ongpeng, J.M.C. Life Cycle Assessment of Concrete Using Copper Slag as a Partial Cement Substitute in Reinforced Concrete Buildings. *Buildings* **2023**, *13*, 746. [CrossRef]
4. Zito, S.V.; Irassar, E.F.; Rahhal, V.F. Recycled Construction and Demolition Waste as Supplementary Cementing Materials in Eco-Friendly Concrete. *Recycling* **2023**, *8*, 54. [CrossRef]
5. Dey, A.; Pani, A.K.; Acharya, P.K. Properties of Mortar Using Supplementary Cementitious Materials. *AIP Conf. Proc.* **2023**, *2740*, 060015. [CrossRef]
6. Naveen, S.; Bhat, G. Effect of Supplementary Cementitious Materials on Compressive Strength of Engineered Cementitious Composites: A Review. In *Recent Advances in Structural Engineering and Construction Management*; Hau, K.K., Gupta, A.K., Chaudhary, S., Gupta, T., Eds.; Springer Nature: Singapore, 2023; pp. 681–691.
7. Fode, T.A.; Chande Jande, Y.A.; Kivevele, T. Effects of Different Supplementary Cementitious Materials on Durability and Mechanical Properties of Cement Composite—Comprehensive Review. *Heliyon* **2023**, *9*, e17924. [CrossRef]
8. Snellings, R.; Suraneni, P.; Skibsted, J. Future and Emerging Supplementary Cementitious Materials. *Cem. Concr. Res.* **2023**, *171*, 107199. [CrossRef]
9. Aquino Rocha, J.H.; Toledo Filho, R.D. The Utilization of Recycled Concrete Powder as Supplementary Cementitious Material in Cement-Based Materials: A Systematic Literature Review. *J. Build. Eng.* **2023**, *76*, 107319. [CrossRef]
10. Zhou, J.; Tian, Q.; Ahmad, A.; Huang, J. Exploring the Resilience of Supplementary Cementitious Materials-Based Concrete to Elevated Temperatures via Modern Computing Techniques. *Mater. Today Commun.* **2024**, *38*, 107894. [CrossRef]
11. Aslam, F.; Shahab, M.Z. Supplementary Cementitious Materials in Blended Cement Concrete: Advancements in Predicting Compressive Strength through Machine Learning. *Mater. Today Commun.* **2024**, *38*, 107725. [CrossRef]
12. Affan, H.; El Haddaji, B.; Ajouguim, S.; Khadraoui, F. A Review—Durability, Mechanical and Hygrothermal Behavior of Building Materials Incorporating Biomass. *Eng* **2024**, *5*, 992–1027. [CrossRef]
13. Haddad, B.; Karaky, L.H.; Boutouil, M.; Sebaibi, N. Investigation Properties of Pervious and Water-Retaining Recycled Concrete to Mitigate Urban Heat Island Phenomena. *Sustainability* **2023**, *15*, 22. [CrossRef]
14. Zahid, C.Z.B.; Rezwan, M.M.; Mohammed, T.U. Service Life Optimization and Life Cycle Assessment of Concrete Using SCMs as Partial Replacement of Cement. *J. Phys. Conf. Ser.* **2023**, *2521*, 012008. [CrossRef]
15. Haddad, B.; Alassaad, F.; Sebaibi, N. Evaluation of Early-Age Compressive Strength in Winter Prefabrication: A Comparative Study. *Appl. Sci.* **2024**, *14*, 3653. [CrossRef]
16. Panesar, D.K. *University of Toronto, Toronto, Ontario, Canada. Developments in the Formulation and Reinforcement of Concrete*, 2nd ed.; Woodhead Publishing: Cambridge, UK, 2019; pp. 55–85.
17. Mehta, A.; Ashish, D.K. Silica Fume and Waste Glass in Cement Concrete Production: A Review. *J. Build. Eng.* **2020**, *29*, 100888. [CrossRef]
18. Chandak, M.A.; Pawade, P.Y. Influence of Metakaolin in Concrete Mixture: A Review. *Int. J. Eng. Sci. (IJES)* **2018**, 37–41.
19. Homayoonmehr, R.; Ramezanianpour, A.A.; Mirdarsoltany, M. Influence of Metakaolin on Fresh Properties, Mechanical Properties and Corrosion Resistance of Concrete and Its Sustainability Issues: A Review. *J. Build. Eng.* **2021**, *44*, 103011. [CrossRef]
20. Ahmad, J.; Kontoleon, K.J.; Majdi, A.; Naqash, M.T.; Deifalla, A.F.; Ben Kahla, N.; Isleem, H.F.; Qaidi, S.M.A. A Comprehensive Review on the Ground Granulated Blast Furnace Slag (GGBS) in Concrete Production. *Sustainability* **2022**, *14*, 8783. [CrossRef]
21. Özbay, E.; Erdemir, M.; Durmuş, H.İ. Utilization and Efficiency of Ground Granulated Blast Furnace Slag on Concrete Properties—A Review. *Constr. Build. Mater.* **2016**, *105*, 423–434. [CrossRef]
22. Nath, P.; Sarker, P.K.; Biswas, W.K. Effect of Fly Ash on the Service Life, Carbon Footprint and Embodied Energy of High Strength Concrete in the Marine Environment. *Energy Build.* **2018**, *158*, 1694–1702. [CrossRef]
23. Hindu, A.; Bheel, N.; Ahmed, I.; Rizvi, S.; Kumar, R.; Jhatial, A. Effect of Silica Fume and Fly Ash as Cementitious Material on Hardened Properties and Embodied Carbon of Roller Compacted Concrete. *Environ. Sci. Pollut. Res.* **2022**, *29*, 1210–1222. [CrossRef]
24. Keerio, M.; Saand, A.; Hindu, A.; Bheel, N.; Ali, K. Effect of Local Metakaolin Developed from Natural Material Soorh and Coal Bottom Ash on Fresh, Hardened Properties and Embodied Carbon of Self-Compacting Concrete. *Environ. Sci. Pollut. Res.* **2021**, *28*, 60000–60018. [CrossRef] [PubMed]
25. Yahyaee, T.; Elize, H.S.; Yahyaee, T.; Elize, H.S. A Comprehensive Study on Mechanical Properties, Durability, and Environmental Impact of Fiber-Reinforced Concrete Incorporating Ground Granulated Blast Furnace Slag. *Case Stud. Constr. Mater.* **2024**, *20*, E03190. [CrossRef]

26. Silva, D.O.F.; Quattrone, M.; Romano, R.C.O.; Angulo, S.C. Reuse of Fines from Ready-Mix Concrete Washing Slurries. *Resour. Conserv. Recycl.* **2020**, *155*, 104653. [CrossRef]
27. Othmane, B.; Rabah, H.; Céline, F.; Laure, R. Valorisation of Fines Recycled Concrete Obtained by Mechanosynthesis for Construction Materials Production. *Mater. Today Proc.* **2023**. [CrossRef]
28. Kathi, S.; Singh, S.; Yadav, R.; Singh, A.N.; Mahmoud, A.E.D. Wastewater and Sludge Valorisation: A Novel Approach for Treatment and Resource Recovery to Achieve Circular Economy Concept. *Front. Chem. Eng.* **2023**, *5*, 1129783. [CrossRef]
29. Wu, C.-H.; Wang, W.-C.; Jung, C.H. Potential Pozzolanic Reactivity of Recycled Aggregates and Waste Powders as Cement Mortar Component. *J. Chin. Inst. Eng.* **2023**, *46*, 399–408. [CrossRef]
30. *NF P18-459*; Concrete—Testing Hardened Concrete—Testing Porosity and Density. AFNOR: Paris, France, 2022.
31. *NF EN 196-1*; Methods of Testing Cement—Part 1: Determination of Strength. AFNOR: Paris, France, 2016.
32. *NF P15-433*; Hydraulic Binders—Methods of Testing Cement—Determination of Shrinkage and Swelling. AFNOR: Paris, France, 2023.
33. Yoo, D.H.; Jeon, I.K.; Kim, H.G.; Lee, J.S.; Ryou, J.-S. Experimental Evaluation of Fire Resistance Performance of Cement Mortar with PCM/$Mg(OH)_2$-Based Composite Fine Aggregate. *Constr. Build. Mater.* **2021**, *287*, 123018. [CrossRef]
34. Shankar, S.V.; Mohan, K.S.R. Durability Studies on Cement Mortar with Granite Powder as A Partial Replacement of Cement. *Int J. ChemTech Res.* **2015**, *8*, 1417–1422.
35. Cantero, B.; Bravo, M.; de Brito, J.; del Bosque, I.F.S.; Medina, C. The Influence of Fly Ash on the Mechanical Performance of Cementitious Materials Produced with Recycled Cement. *Appl. Sci.* **2022**, *12*, 2257. [CrossRef]
36. Kim, J.; Nciri, N.; Sicakova, A.; Kim, N. Characteristics of Waste Concrete Powders from Multi-Recycled Coarse Aggregate Concrete and Their Effects as Cement Replacements. *Constr. Build. Mater.* **2023**, *398*, 132525. [CrossRef]
37. Villagrán-Zaccardi, Y.; Broodcoorens, L.; Van den Heede, P.; De Belie, N. Fine Recycled Concrete Aggregates Treated by Means of Wastewater and Carbonation Pretreatment. *Sustainability* **2023**, *15*, 6386. [CrossRef]
38. Mahajan, L.S.; Bhagat, S.R. Utilization of Pozzolanic Material and Waste Glass Powder in Concrete. In *Recent Trends in Construction Technology and Management*; Ranadive, M.S., Das, B.B., Mehta, Y.A., Gupta, R., Eds.; Springer Nature: Singapore, 2023; pp. 201–206.
39. Concrete Embodied Carbon Footprint Calculator. *Circ. Ecol.* Available online: https://circularecology.com/concrete-embodied-carbon-footprint-calculator.html (accessed on 14 July 2024).
40. Étuves Universelles Ventilées. *VIALAB*. Available online: https://www.vialab.fr/fr/produits/accueil/equipement-de-laboratoire-routier/equipement-general/etuves/ (accessed on 14 July 2024).
41. Tamiseuse pour Tamis ø 200, ø 300 et ø 400 mm. *3R—Rech. Réalisations Rémy*. Available online: https://3r-labo.com/produit/tamiseuse-pour-tamis-o-200-o-300-et-o-400-mm (accessed on 14 July 2024).
42. Renewable Energy Sources and Climate Change Mitigation—IPCC. Available online: https://www.ipcc.ch/report/renewable-energy-sources-and-climate-change-mitigation (accessed on 14 July 2024).
43. Mehta, P.K.; Monteiro, P.J.M. *Concrete: Microstructure, Properties, and Materials*, 4th ed.; McGraw-Hill Education: New York, NY, USA, 2014; ISBN 978-0-07-179787-0.
44. De Brito, J.; Saikia, N. Recycled Aggregate in Concrete: Use of Industrial, Construction and Demolition Waste. In *Green Energy and Technology*; Springer: London, UK, 2013; ISBN 978-1-4471-4539-4.
45. Gjrv, O.E. Durability Design of Concrete Structures in Severe Environments, 2nd ed.CRC Press: Boca Raton, FL, USA, 2014.
46. Schneider, M.; Romer, M.; Tschudin, M.; Bolio, H. Sustainable Cement Production—Present and Future. *Cem. Concr. Res.* **2011**, *41*, 642–650. [CrossRef]
47. Juenger, M.C.G.; Siddique, R. Recent Advances in Understanding the Role of Supplementary Cementitious Materials in Concrete. *Cem. Concr. Res.* **2015**, *78*, 71–80. [CrossRef]

applied sciences

Article

Utilizing Graphite Waste from the Acheson Furnace as Anode Material in Lithium-Ion Batteries

Inchan Yang [1], Seonhui Choi [1], Sang-Wook Kim [2], Man Youl Ha [2], Sei-Min Park [1,*] and Jung-Chul An [1,*]

1 Carbon Materials Research Cell, Research Institute of Industrial Science & Technology (RIST), Pohang 37673, Republic of Korea; iyang@rist.re.kr (I.Y.); hes02427@rist.re.kr (S.C.)
2 Technical R&D Center, Dong-Suh Chemical, Pohang 37860, Republic of Korea; airsendo@naver.com (S.-W.K.); w1www@naver.com (M.Y.H.)
* Correspondence: smpark1@rist.re.kr (S.-M.P.); jcan@rist.re.kr (J.-C.A.)

Featured Application: Lithium-ion battery.

Abstract: This study investigates the potential of graphite waste (GW) from the Acheson furnace as a sustainable and cost-effective anode material for lithium-ion batteries (LIBs). Conventional anode materials face challenges such as energy-intensive production processes and reliance on virgin graphite resources, leading to high costs and environmental concerns. GW from the Acheson furnace, which already possesses high carbon purity (98.5%–99.9%) and crystallinity (93.5%), offers a promising alternative by eliminating the need for graphitization and extensive purification. Through spheronization and carbon coating, GW was successfully optimized to achieve electrochemical properties comparable to commercial anode materials (CAM), including an initial Coulombic efficiency of 85.1% and a specific capacity of 348.9 mAh/g. These findings suggest that GW from the Acheson furnace represents a viable pathway toward cost-effective and environmentally friendly LIB anodes.

Keywords: graphite waste; Acheson furnace; spheronization; carbon coating; lithium-ion battery; anode

1. Introduction

In modern technologies, including electric vehicles (EVs) and energy storage systems (ESS), the lithium-ion battery (LIB) market has been pivotal [1–6]. Over the past decade, the LIB market has experienced steady growth due to the increasing demand for energy storage solutions. However, recent projections indicate that LIB market expansion is slower than initially anticipated, raising concerns among manufacturers and stakeholders. This slowdown has intensified competition among battery material manufacturers, particularly as the market becomes saturated with cost-sensitive applications [6].

The anode material, one of the four key components of LIBs, is primarily composed of graphite [7–11]. Graphite-based anodes are widely favored for their stability, conductivity, and compatibility with existing LIB technologies. Although silicon-based anode materials have recently gained attention for their high-capacity characteristics, challenges such as low initial Coulombic efficiency (ICE) and severe volume expansion during the charge–discharge process limit their adoption to less than 10% of the anode material composition when mixed with graphite [12,13]. This underscores the continued reliance on graphite as the dominant anode material. Consequently, the price of graphite plays a significant role in determining the overall cost of anode materials, creating additional pressure on manufacturers to optimize costs without compromising quality. In 2024, artificial graphite anodes produced in China were priced at approximately USD 5 per kilogram. This competitive pricing environment has forced manufacturers to seek innovative ways to reduce raw material costs while maintaining or improving product performance.

In response to these market pressures, the use of alternative and cost-effective raw materials has emerged as a key strategy [14–18]. One promising approach involves utilizing industrial waste products, particularly graphite waste (GW) [14–22]. For instance, insulation material composed of granulated coke used for graphitization in an Acheson furnace (Figure 1), which has been used for over a century for the synthesis of artificial graphite, could be considered a representative example of GW. Due to its existing graphitization, this by-product offers unique potential for reuse, presenting an opportunity to reduce production costs and mitigate environmental impacts. Yet, it is often discarded or underutilized, reflecting a missed opportunity for sustainable material recovery.

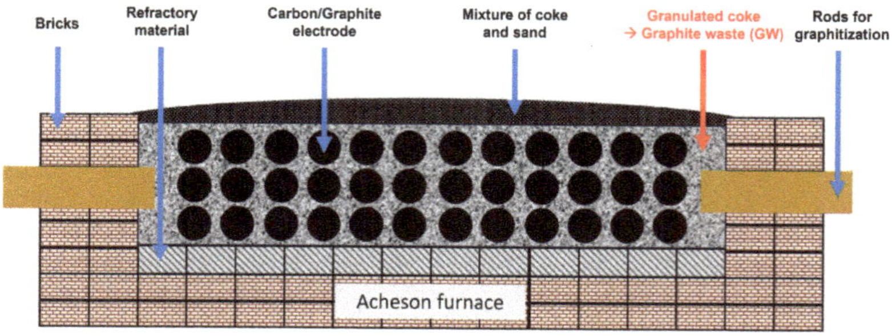

Figure 1. Acheson furnace for graphite manufacturing.

GW from the Acheson furnace holds great potential as a raw material for battery anode production because it is already graphitized. Graphitization is a crucial and energy-intensive process in the production of artificial graphite anodes. Typically, raw carbonaceous materials undergo heat treatment at temperatures exceeding 2800 °C for several days to convert them into graphite, a process that consumes substantial amounts of energy and time [23,24]. This high-temperature process not only contributes to significant energy consumption but also results in high carbon emissions, making it a target for environmental sustainability efforts. By contrast, GW has already undergone this transformation, providing manufacturers with a resource that requires minimal additional processing. By utilizing GW, manufacturers can bypass the graphitization process, significantly reducing energy consumption, production time, and overall costs. Additionally, the use of GW aligns with global sustainability initiatives aimed at promoting the circular economy. The reuse of industrial waste materials not only reduces dependency on virgin graphite resources but also minimizes waste generation, addressing key environmental and economic challenges.

Despite these advantages, challenges remain in adopting GW for LIB anodes. One key concern is the purity of GW. While GW is graphitized, it may still contain impurities or exhibit non-uniform crystalline structures that could negatively impact the performance of the resulting anode material. Therefore, a thorough characterization of GW properties is essential to ensure it meets the performance standards required for LIBs. Previous studies have shown that impurities in graphite can reduce conductivity and increase resistance, thereby diminishing the electrochemical performance of LIBs [25–28]. Another concern is the physical and morphological properties of GW, such as particle size, specific surface area (SSA), and tap density (TD), which must be optimized to align with commercial anode material benchmarks. These parameters directly influence the material's ability to perform under the rigorous demands of LIB applications. For instance, excessively high SSA can result in lower initial Coulombic efficiency (ICE), while low TD can reduce the energy density of the anode.

In this study, granulated coke, a waste discarded after the Acheson process—a core process for graphite production—was repurposed as a starting material for LIB anodes. This approach seeks to reduce the production cost of anode materials while simultaneously improving the performance of granulated coke as a viable anode material through further processing. We focus on evaluating the structural and electrochemical characteristics of

GW to determine its compatibility with commercial anode materials (CAM). This includes optimizing key parameters such as particle size, SSA, and TD, as well as applying advanced processing techniques such as spheronization and carbon coating to improve its performance [29,30]. The successful adoption of GW for anode production could provide significant economic and environmental benefits, contributing to a more competitive and sustainable battery industry. By reducing reliance on virgin graphite and minimizing the energy-intensive graphitization process, this approach offers a pathway toward more cost-effective and environmentally friendly LIBs.

2. Materials and Methods

2.1. Materials

Graphite waste (GW) was obtained from POSCO Future M (Pohang, Republic of Korea). Jet-milled GW (JMGW) was prepared by jet-milling GW using s-JET (Netzsch, Selb, Germany). Refined coal-tar (RCT, Dong-Suh Chemical, Pohang, Republic of Korea) was prepared by removing quinoline-insoluble (QI) components from coal-tar (CT, POSCO, Pohang, Republic of Korea). A commercial anode material (CAM) was supplied by Sunstone Development (Linyi, China). All other chemicals and reagents were of analytical grade and used without further purification unless otherwise specified.

2.2. Proximate Analysis (PA), X-Ray Fluorescence (XRF) Analysis

The carbon content was analyzed through proximate analysis (PA) to determine the fixed carbon, volatile matter, and ash content. The analysis was conducted using a TGA701 (LECO, St. Joseph, MI, USA). XRF analysis was performed to determine the elemental composition and carbon purity of the JMGW. The analysis was carried out using a ZSX Primus II (Rigaku, Tokyo, Japan). Calibration was performed using certified standards.

2.3. X-Ray Diffraction (XRD)

The crystallinity of JMGW and its spheronized derivatives was assessed using XRD analysis. The samples were analyzed on a D-max 2500 VPC (Rigaku, Tokyo, Japan) diffractometer with Cu Kα radiation (λ = 1.5406 Å). The scans were performed over a 2θ range of 10° to 90° at a scanning rate of 0.5°/min. For normalization purposes, 20 wt.% of pure silicon (Kojundo Chemical, Sakado, Japan) relative to the target material was introduced as a standard reference. The interlayer spacing of (002) plane (d_{002}) and graphitization degree (GD) were calculated using Bragg's law and Franklin Equation (Equations (1)–(3)) [31–33]. The value of d_{002} is a fundamental parameter that describes the distance between adjacent graphene layers in graphite materials. A smaller d_{002} value signifies higher crystallinity, which is crucial for enhancing the electrical conductivity and lithium-ion intercalation properties of anode materials in LIBs. The graphitization degree (GD) is another important parameter that quantifies the structural order of graphite. It is calculated by comparing the measured d_{002} value with theoretical benchmarks for ideal graphite (0.3354 nm) and disordered carbon (0.3440 nm). A higher GD reflects superior crystallinity, which directly contributes to improved electrochemical performance by reducing internal resistance and increasing electrical conductivity.

2.4. Particle Size (PS), Specific Surface Area (SSA), and Tap Density (TD) Measurement

The particle size distribution of JMGW and spheronized JMGW (SGWX, X = spheronization time (min)) was measured using a PSA1090 (Anton Paar, Graz, Austria) laser diffraction particle size analyzer. The specific surface area was determined via nitrogen adsorption using the Brunauer–Emmett–Teller (BET) method [34] on a Tri-Star (Micromeritics, Norcross, GA, USA). Tap density was measured using a Dual Autotap (Anton Paar, Graz, Austria).

2.5. Spheronization Process

JMGW was subjected to a spheronization process to improve its particle size distribution, specific surface area, and tap density. The spheronization was performed using a

F-LABX (Hosokawa Micron, Hirakata, Japan) under varying spheronization times (X = 3, 6, or 9 min). The rotational speed was fixed to 8000 rpm.

2.6. Carbon Coating and Granulation

To enhance the electrochemical properties of the SGW9, the material was coated with RCT. The coating and granulation processes were conducted in a box-type furnace (CM furnaces, Bloomfield, NJ, USA). SGW9 was mixed with RCT at a mass ratio of 10:1, and carbonization was carried out at 1200 °C for 2 h under a nitrogen atmosphere. The obtained product was labeled SGW9@RCT.

2.7. Scanning Electron Microscopy (SEM) and Transmission Electron Microscopy (TEM)

The morphology of JMGW, SGWXs, and CAM was observed using a scanning electron microscope (Gemini500, ZEISS, Oberkochen, Germany). Samples were prepared by sputter-coating with gold to ensure conductivity. Images were taken at an accelerating voltage of 2 kV to study the surface and morphological changes induced by spheronization and carbon coating.

High-resolution TEM was used to examine the carbon coating layer on SGW9@RCT. TEM analysis was conducted using ARM200 (JEOL, Tokyo, Japan). The samples were prepared by dispersing the powder in high-purity ethanol and depositing it onto a copper grid with carbon film.

2.8. Fabrication of Lithium-Ion Batteries Cells and Electrochemical Characterization

Coin-type LIB half-cells were fabricated using SGW9, SGW9@RCT, and CAM as the anode materials. The electrodes were prepared by mixing the active material, conductive additive, and binder in a ratio of 96:1:3. The electrode slurry was cast onto copper foil with 18 μm and dried at 100 °C for 0.5 h. The coin cells were assembled in dry room with lithium metal as the counter electrode. The electrolyte used was 1 M $LiPF_6$ in EC/EMC = 30:70, and a PE-type was used to separate the electrodes.

The cells were tested in half-cell configuration using a VMP-3 potentiostat (BioLogic Science Instruments, Claix, France). Galvanostatic charge–discharge cycling was performed at a C-rate of 0.1 C, and the voltage range was set to 0.005 V to 3.0 V. Initial Coulombic efficiency (ICE) was determined from the capacity ratio between 1st charge and discharge (Equation (4)). Specific capacity was calculated based on the 3rd discharge data.

We have illustrated the entire experimental process in Figure 2.

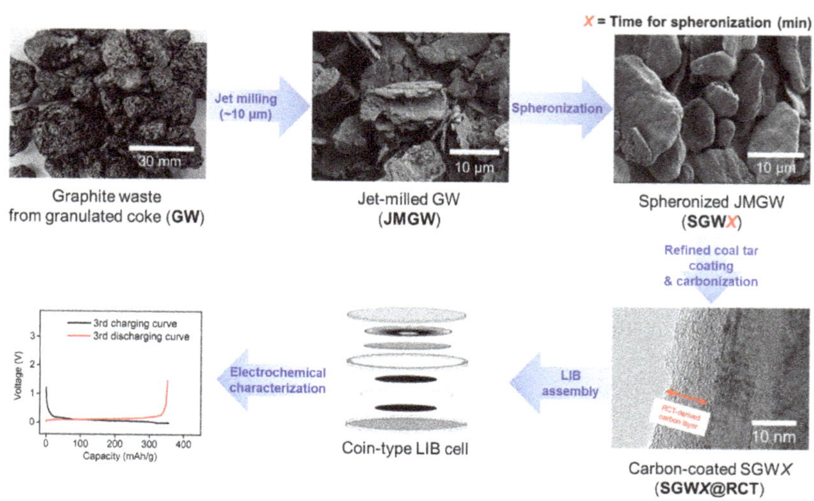

Figure 2. The experimental processes for sample preparation, cell fabrication, and evaluation.

3. Results and Discussion

3.1. Characterization of Jet-Milled Graphite Waste

The carbon purity of jet-milled graphite waste (JMGW) was thoroughly analyzed using both proximate analysis (PA) and X-ray fluorescence (XRF). The PA results indicate that JMGW consists of 98.5% fixed carbon, with a minimal presence of volatile matter (0.8%) and ash (0.7%) (Table S1). The high level of fixed carbon is a clear indicator of JMGW's suitability for applications requiring high-purity carbon. Furthermore, the XRF analysis confirmed that the carbon purity of JMGW is as high as 99.9% (Table S2). These complementary results collectively demonstrate that JMGW boasts excellent carbon purity, effectively removing the need for additional purification processes. Given this level of purity, JMGW could be directly utilized in various industrial applications, including electrode materials for energy storage systems and carbon-based composites. The direct application of JMGW could reduce overall production costs, energy consumption, and carbon footprint, as fewer resources would be required for further refinement or purification. Such features are especially appealing for industries aiming to implement more sustainable production practices.

To further explore the crystalline properties of JMGW, X-ray diffraction (XRD) analysis was performed (Figure 3 and Table 1). The XRD results indicate that JMGW possesses high crystallinity, evidenced by a d-spacing value of 0.3360 nm for the (002) plane (d_{002}) and a graphitization degree (GD) of 93.5%. In comparison, commercial anode materials (CAM) exhibit a d_{002} value of 0.3359 nm and a GD of 94.2%. This close proximity in crystallinity between JMGW and CAM suggests that JMGW is highly compatible with commercial graphite anode materials, making it a strong candidate for integration into lithium-ion battery (LIB) applications (Figure S1 and Table S3). The ability of JMGW to meet the crystallinity standards required for LIBs underscores its potential for use in large-scale energy storage solutions.

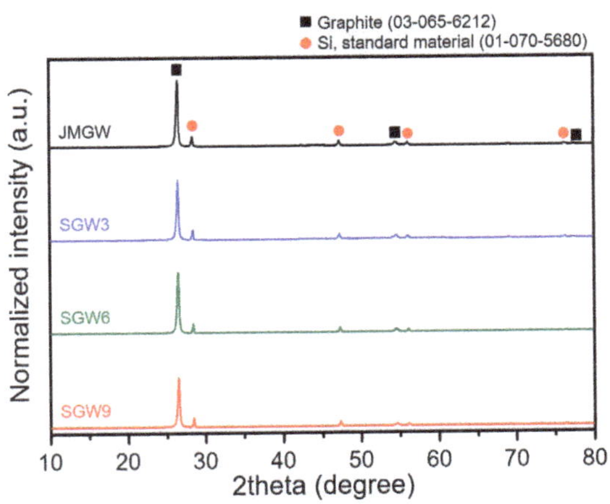

Figure 3. XRD patterns of JMGW and SGWXs.

Table 1. d-spacing value of the (002) plane and graphitization degree of JMGW and SGWXs.

Sample	d_{002} [1] (nm)	GD [2] (%)
JMGW	0.3360	93.5
SGW3	0.3360	94.2
SGW6	0.3360	92.5
SGW9	0.3361	93.2

[1] d-spacing value of the (002) plane; [2] graphitization degree determined by Franklin's Equation.

In addition to crystallinity, the particle size (PS), specific surface area (SSA), and tap density (TD) of JMGW were measured and presented in Table 2. The results show that JMGW has a PS of 9.3 μm, a span of 1.16, an SSA of 14.8 m^2/g, and a TD of 0.27 g/cc. These values highlight the material's potential but also suggest that improvements are necessary, particularly in SSA and TD, to meet the performance benchmarks set for LIB anode materials. To address these issues and improve JMGW's viability as a LIB anode material, a spheronization process was employed. Spheronization is widely recognized for its ability to modify particle morphology, thereby enhancing physical properties such as PS, SSA, and TD.

Table 2. Average particle size, span, specific surface area, and tap density of JMGW and SGWXs.

Sample	D50 [1] (μm)	Span (-)	SSA [2] (m^2/g)	TD [3] (g/cc)
JMGW	9.3	1.16	14.8	0.27
SGW3	9.1	1.06	14.2	0.52
SGW6	9.0	1.00	14.0	0.58
SGW9	8.7	0.97	13.2	0.65

[1] Average particle size; [2] specific surface area calculated using the Brunauer–Emmett–Teller (BET) equation; [3] tap density.

3.2. *Characterization of Spheronized JMGW*

The morphology, PS, SSA, and TD of spheronized JMGW, denoted as SGW*X* (*X* = spheronization time (min)), were characterized. The results are presented in Figures 4 and S2, and Table 2. The SEM images of JMGW and SGWXs (Figures 4 and S2) clearly demonstrate that the spheronization process induced a morphological change in JMGW. Based on these changes, it was predicted that the spheronization process would influence the PS, SSA, and TD of JMGW. However, as illustrated in Figure 4 and Table 2, PS and SSA exhibited only slight variations during the spheronization process, while TD gradually increased. The observed change in tap density (TD) can be associated with the reduction in span after the spheronization process. While the mean particle size of JMGW showed negligible variation, the reduced span indicates a change in the distribution of coarse and fine particles. This change in particle size distribution is identified as a key factor contributing to the increase in TD. This suggests that although the applied spheronization process had minimal effect on the PS and SSA of JMGW, it significantly improved TD, which is a favorable outcome. Furthermore, it was confirmed that the spheronization process did not affect the crystallinity of GW (Figure 3 and Table 1).

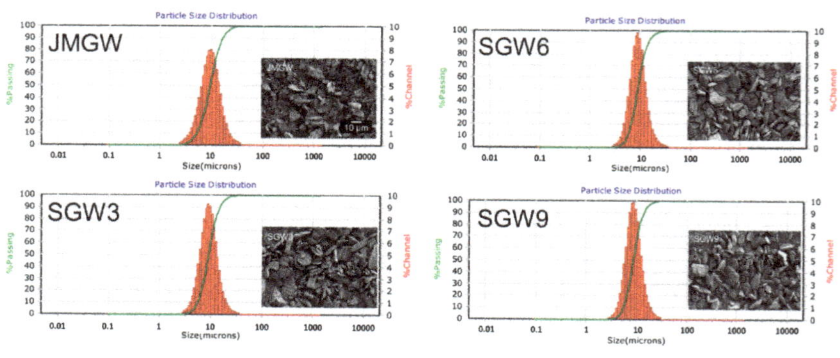

Figure 4. SEM images and particle size distribution of JMGW and SGWXs.

3.3. *Characterization of the Carbon-Coated SGW9*

Considering the characteristics of CAM, further improvement in PS, SSA, and TD is required to reach commercially viable levels (Table S4). Therefore, refined coal-tar (RCT), which is coal-tar (CT) with the quinoline insoluble (QI) removed (Table S5), was utilized for the carbon coating and granulation of SGW9 (Figure 5). In other words, to improve the PS, SSA, and TD of SGW9 to a level comparable to commercial products, a method involving mixing with RCT followed by carbonization was employed. Figure 5a shows the manufacturing flow for RCT, while Figure 5b illustrates the carbon coating and granulation processes through RCT coating and carbonization. The RCT-derived carbon-coated SGW9 was labeled as SGW9@RCT.

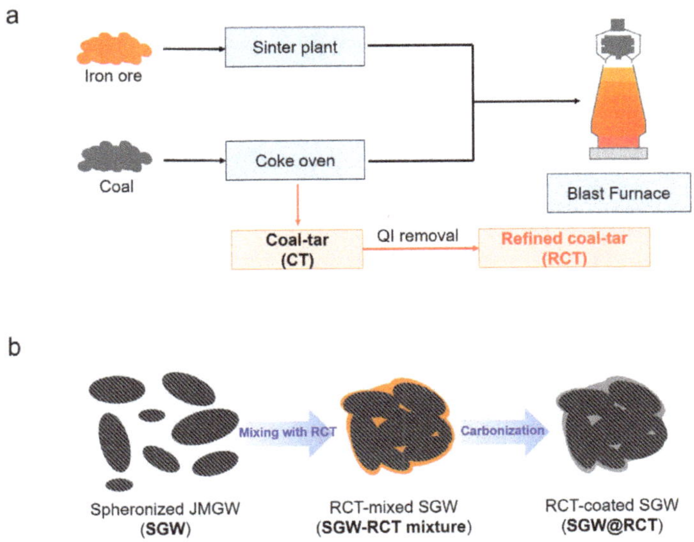

Figure 5. (**a**) Manufacturing flow for RCT and (**b**) carbon coating processes.

From the TEM images of SGW9 and SGW9@RCT (Figure 6), we directly observed a carbon-coated layer on SGW9. This confirms that the applied carbon coating process was successfully implemented. Subsequently, we measured the PS, SSA, and TD of SGW9@RCT (Table 3). Compared to SGW9, both PS and TD increased, while SSA significantly decreased through the applied carbon coating process. All of these trends represent positive changes from the perspective of anode material performance. Moreover, the properties of SGW9@RCT were found to be at a level comparable to commercial anode materials (CAM). Consequently, we successfully converted GW into SGW9@RCT, an anode material that is competitive with commercial products.

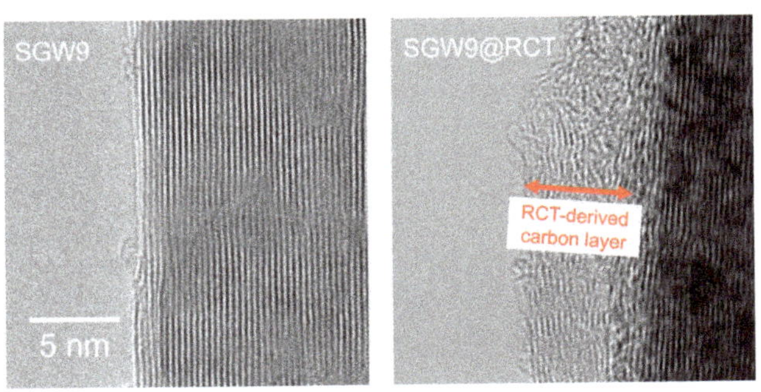

Figure 6. TEM images of SGW9 and SGW9@RCT.

Table 3. Average particle size, span, specific surface area, and tap density of JMGW and SGWXs.

Sample	D50 [1] (μm)	Span (-)	SSA [2] (m^2/g)	TD [3] (g/cc)
SGW9@RCT	9.6	1.00	2.4	0.83

[1] Average particle size; [2] specific surface area calculated using the Brunauer–Emmett–Teller (BET) equation; [3] tap density.

3.4. Characterization of SGW9, SGW9@RCT, and CAM as Anode Materials for LIBs

To confirm the electrochemical performance of SGW9, SGW9@RCT, and CAM, we prepared coin-type lithium-ion battery (LIB) cells using SGW9, SGW9@RCT, or CAM as anode materials (Figure 7). For objective comparison, the mass loading and electrode density were controlled (Table S6). The mass loading of the anode materials was controlled within the range of 10.2 to 10.6 mg/cm^2, and the electrode density was maintained between 1.53 and 1.55 g/cc.

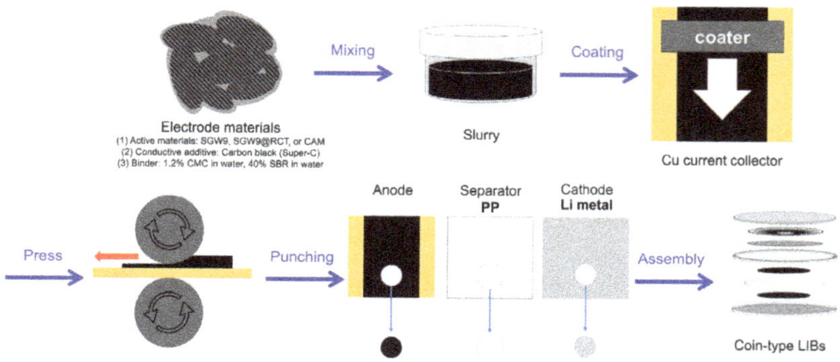

Figure 7. Preparation of LIBs using SGW9, SGW9@RCT, or CAM as anode materials.

As shown in Figure 8 and Table 4, the prepared LIB cells exhibited typical charge–discharge behavior during the half-cell tests. Specifically, the open-circuit voltage (OCV) of all cells exceeded 3.0 V, which is higher than the standard reduction potential of lithium. This confirms the successful fabrication of all LIB cells. We then calculated the ICE and capacity of the LIB cells. For ICE, SGW9, SGW9@RCT, and CAM exhibited values of 82.1%, 85.1%, and 93.9%, respectively. These results demonstrate that the applied carbon coating process improved the ICE of GW, although it remained lower than that of CAM. Given the inverse relationship between ICE and SSA, the higher SSA of SGW9 and SGW9@RCT resulted in lower ICE compared to CAM. However, with respect to capacity, the samples derived from GW outperformed CAM. The capacities of SGW9, SGW9@RCT, and CAM were 352.1 mAh/g, 348.9 mAh/g, and 348.3 mAh/g, respectively. This superior capacity can be attributed to the inherent high crystallinity of the starting material, GW. In other words, the developed SWG9@RCT demonstrates a specific capacity comparable to that of CAM, a high-end grade commercial anode material, while exhibiting an ICE of 85.1%. These findings highlight the practical applicability and mid-to-high-quality performance of SWG9@RCT as a sustainable and cost-effective alternative for LIB anodes. Consequently, while the anode samples derived from GW may face challenges in achieving high efficiency, they exhibit significant potential as high-capacity anode materials. It was concluded that GW from the Acheson furnace holds promise for conversion into commercially viable high-capacity anode materials.

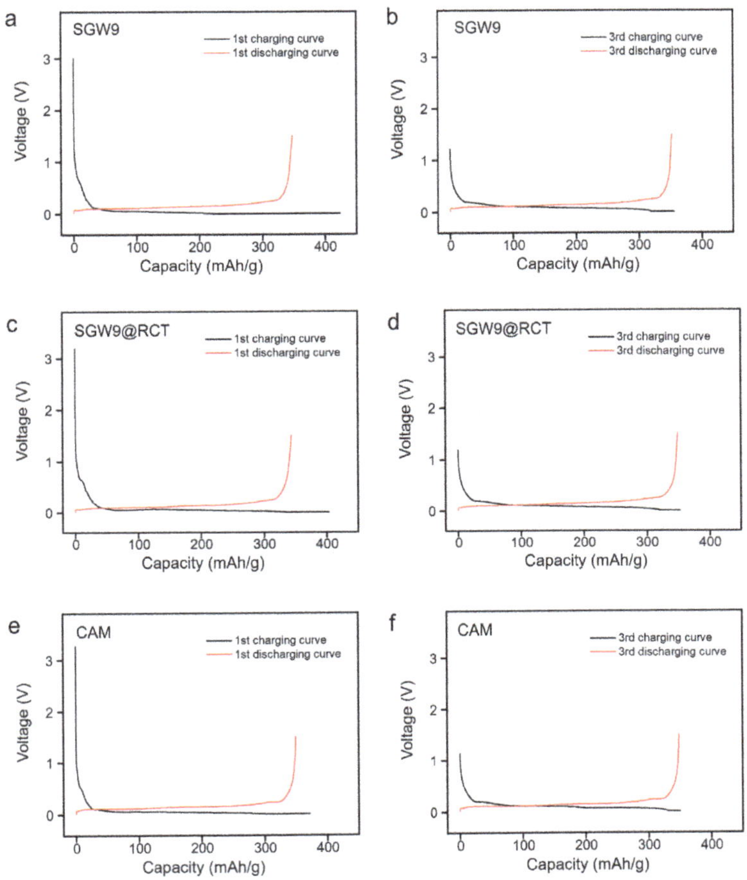

Figure 8. First and third charge–discharge curves of prepared LIB cells: (**a**,**b**) SGW9, (**c**,**d**) SGW9@RCT, (**e**,**f**) CAM.

Table 4. OCV, ICE, and capacity results for LIB cells.

Sample	OCV [1] (V)	ICE [2] (%)	Capacity (mAh/g)		
			1st Charge	1st Discharge	3rd Discharge
SGW9	3.26	82.1	423.5	347.6	352.1
SGW9@RCT	3.20	85.1	403.6	343.5	348.9
CAM	3.29	93.9	372.1	349.3	348.3

[1] Open circuit voltage; [2] initial Coulombic efficiency.

3.5. Formatting of Mathematical Components

$$2 \times d_{002} \sin \theta = n\lambda, \quad (1)$$

d_{002} refers to the distance between adjacent graphene layers in graphite, which is calculated using Bragg's Law. This parameter is a critical indicator of the material's crystallinity and graphitization degree. The diffraction angle (θ) is the angle at which X-rays are scattered by the crystal lattice, and it is used alongside the X-ray wavelength (λ) to determine d_{002}. The wavelength is a fixed property of the X-ray source, and λ is typically provided by the XRD instrument.

$$d_{002} = (n\lambda)/(2 \sin \theta), \quad (2)$$

$$GD = (0.3440 - d_{002})/(0.3440 - 0.3354) \times 100(\%), \quad (3)$$

The order of diffraction (n) indicates the diffraction plane being measured, with n = 1 commonly used for simplicity in XRD analysis. The graphitization degree (GD) quantifies the structural order of the graphite relative to ideal graphite and disordered carbon. This parameter is essential for understanding the material's suitability as an anode in LIBs, as higher graphitization correlates with better electrochemical performance.

$$ICE = (\text{capacity @1st discharge})/(\text{capacity @1st charge}) \times 100(\%), \quad (4)$$

The initial Coulombic efficiency (ICE) represents the ratio of the discharge capacity to the charge capacity during the first charge–discharge cycle. Expressed as a percentage, this metric evaluates how efficiently lithium ions are stored and released, which is a critical measure of the anode material's practical performance. These parameters collectively provide a comprehensive understanding of the material's structural and functional characteristics.

4. Conclusions

In this study, we demonstrated the feasibility of utilizing graphite waste (GW) from the Acheson furnace as a cost-effective and sustainable anode material for lithium-ion batteries (LIBs). By leveraging the high carbon purity (98.5%–99.9%) and crystallinity (93.5%) of GW, we successfully eliminated the need for the energy-intensive graphitization process, thereby significantly reducing production costs and environmental impact. The structural and electrochemical characterization revealed that GW-derived anodes, after spheronization and carbon coating, exhibit physical and electrochemical properties comparable to commercial anode materials. Moreover, the findings underscore the importance of optimizing processing parameters, such as tap density (TD) and specific surface area (SSA), which were found to significantly influence the material's electrochemical performance. The study highlights that achieving a balance between TD and SSA is critical for ensuring competitive performance in LIB anode applications. These results validate the potential of GW as a scalable and practical alternative to conventional graphite-based anodes. While previous studies on recycled carbon anodes have often been limited by low production yields, complex manufacturing processes, and challenges in industrial scalability, this study highlights a simpler and more cost-efficient approach by directly using industrial graphite waste. Additionally, the elimination of energy-intensive processes, such as high-temperature graphitization, demonstrates a clear pathway toward reducing the carbon footprint of LIB manufacturing. The reuse of GW not only reduces dependency on virgin graphite resources but also addresses key environmental challenges by minimizing waste generation and promoting sustainable manufacturing practices. This research provides a foundation for future work to explore long-term cycling performance and further optimize the industrial-scale application of GW-based anodes. With continued development, GW-derived anodes could offer a transformative solution for the production of environmentally friendly and economically viable LIBs, contributing to the advancement of sustainable energy storage technologies.

Supplementary Materials: The following supporting information can be downloaded at: https://www.mdpi.com/article/10.3390/app142311353/s1, Figure S1: XRD pattern and SEM image of CAM; Figure S2: SEM images of JMGW and SGWX samples; Table S1: Result of proximate analysis for GW; Table S2: Result of X-ray fluorescence analysis for GW; Table S3: d-spacing value of the (002) plane and graphitization degree of CAM; Table S4: Average particle size, span, specific surface area, and tap density of CAM; Table S5: Proximate analysis result and QI value for CT and RCT; Table S6: Mass loading and electrode density of the prepared electrodes.

Author Contributions: Conceptualization, I.Y.; methodology, I.Y., S.-M.P. and J.-C.A.; validation, S.-W.K., M.Y.H. and J.-C.A.; formal analysis, I.Y., S.C. and S.-W.K.; investigation, I.Y., S.-M.P. and J.-C.A.; resources, S.-W.K., M.Y.H. and J.-C.A.; data curation, I.Y., S.-M.P. and J.-C.A.; writing—original draft preparation, I.Y.; writing—review and editing, I.Y., S.-M.P. and J.-C.A.; visualization, I.Y. and S.C.; supervision, S.-M.P. and J.-C.A.; project administration, M.Y.H. and J.-C.A. All authors have read and agreed to the published version of the manuscript.

Funding: This work was supported by the Technology Innovation Program (20020300, Development of waste carbon resource-based anode material manufacturing technology for secondary battery) funded By the Ministry of Trade, Industry & Energy (MOTIE, Korea). This work was supported by the Technology Innovation Program (RS-2024-00422895), Development of ultra high-quality precursors for super premium grade graphite block manufacturing) funded by the Ministry of Trade Industry & Energy (MOTIE, Korea).

Institutional Review Board Statement: Not applicable.

Informed Consent Statement: Not applicable.

Data Availability Statement: The original contributions presented in the study are included in the article and Supplementary Materials, further inquiries can be directed to the corresponding author.

Conflicts of Interest: Authors Sang-Wook Kim and Man Youl Ha were employed by the company Dong-Suh Chemical. The remaining authors declare that the research was conducted in the absence of any commercial or financial relationships that could be construed as a potential conflict of interest.

References

1. Fernandez, V.; Pérez, V. Optimization of Electric Vehicle Charging Control in a Demand-Side Management Context: A Model Predictive Control Approach. *Appl. Sci.* **2024**, *14*, 8736. [CrossRef]
2. Khan, F.M.N.U.; Rosul, M.G.; Sayem, A.S.M.; Mandal, F. Maximizing Energy Density of Lithium-Ion Batteries for Electric Vehicles: A Critical Review. *Energy Rep.* **2023**, *9*, 11–21. [CrossRef]
3. Khaleel, M.; Nassar, Y.; Hala, E.-K.; Elmnif, M.; Rajab, Z.; Yaghoubi, E.; Yaghoubi, E. Electric Vehicles in China, Europe, and the United States: Current Trend and Market Comparison. Electric Vehicles in China, Europe, and the United States: Current Trend and Market Comparison. *Int. J. Electr. Eng. Sustain.* **2024**, *2*, 1–20.
4. Sun, X.; Li, Z.; Wang, X.; Li, C. Technology Development of Electric Vehicles: A Review. *Energies* **2020**, *13*, 90. [CrossRef]
5. Qizo, Q.; Zhao, F.; Liu, Z.; He, X.; Hao, H. Life cycle greenhouse gas emissions of Electric Vehicles in China: Combining the vehicle cycle and fuel cycle. *Energy* **2019**, *177*, 222–233.
6. Figgener, J.; Stenzel, P.; Kairies, K.-P.; Linßen, J.; Haberschusz, D.; Wessels, O.; Robinius, M.; Stolten, D.; Sauer, D.U. The Development of Stationary Battery Storage Systems in Germany—Status 2020. *J. Energy Storage* **2021**, *33*, 101982. [CrossRef]
7. Liu, Y.; Shi, H.; Wu, Z.-S. Recent status, Key Strategies and Challenging Perspectives of Fast-Charging Graphite Anodes for Lithium-Ion Batteries. *Energy Environ. Sci.* **2023**, *16*, 4834–4871. [CrossRef]
8. Ma, C.; Zhao, Y.; Li, J.; Song, Y.; Shi, J.; Guo, Q.; Liu, L. Synthesis and Electrochemical Properties of Artificial Graphite as an Anode for High-Performance Lithium-Ion Batteries. *Carbon* **2013**, *64*, 553–556. [CrossRef]
9. Lee, S.H.; Park, C.R. Correlation Between Graphite Morphology and Interparticle Space Affecting Rate Performance in Li-Ion Battery Anodes. *Carbon Lett.* **2024**, *34*, 627–636. [CrossRef]
10. Qin, H.; Mo, Z.; Lu, J.; Sui, X.; Song, Z.; Chen, B.; Zhang, Y.; Zhang, Z.; Lei, X.; Lu, A.; et al. Ultrafast Transformation of Natural Graphite into Self-Supporting Graphene as Superior Anode Materials for Lithium-Ion Batteries. *Carbon* **2024**, *216*, 118559. [CrossRef]
11. Xu, L.; Wang, J.; Su, C. Preparation and Electrochemical Performances for Silicon-Carbon Ternary Anode Materials with Artificial Graphite as Conductive Skeleton. *Carbon Lett.* **2024**, *28*, 2445–2454. [CrossRef]
12. Cho, C.-H.; Jeong, W.U.; Lim, C.-H.; Yoon, J.-K.; Bon, C.Y.; Yang, K.-S.; Park, M.-S. Facile Synthesis of Nano-Si/Graphite-Carbon Anode Through Microwave-Induced Carbothermal Shock for Lithium-Ion Batteries. *Carbon* **2024**, *229*, 119542. [CrossRef]
13. Qiao, Y.; Zhao, H.; Shen, Y.; Li, L.; Rao, Z.; Shao, G.; Lei, Y. Recycling of graphite anode from spent lithium-ion batteries: Advances and perspectives. *EcoMat* **2023**, *5*, e12321. [CrossRef]
14. Choi, J.; Yang, I.; Kim, S.-S.; Cho, S.Y.; Lee, S. Upcycling Plastic Waste into High Value-Added Carbonaceous Materials. *Macromol. Rapid Commun.* **2022**, *43*, 2100467. [CrossRef]
15. Villagómez-Salas, S.; Manikandan, P.; Guzmán, S.F.A.; Pol, V.G. Amorphous Carbon Chips Li-Ion Battery Anodes Produced through Polyethylene Waste Upcycling. *ACS Omega* **2018**, *3*, 17520. [CrossRef]
16. Naskar, A.K.; Bi, Z.; Li, Y.; Akato, S.K.; Saha, D.; Chi, M.; Bridges, C.A.; Paranthaman, M.P. Tailored Recovery of Carbons from Waste Tires for Enhanced Performance as Anodes in Lithium-Ion Batteries. *RSC Adv.* **2014**, *4*, 38213. [CrossRef]
17. Kim, K.S.; Hwang, J.U.; Im, J.S.; Lee, J.D.; Kim, J.H.; Kim, M.I. The Effect of Waste PET Addition on PFO-Based Anode Materials for Improving the Electric Capacity in Lithium-Ion Battery. *Carbon Lett.* **2020**, *30*, 545. [CrossRef]
18. Feng, J.; Zhang, B.; Du, P.; Yuan, Y.; Li, M.; Chen, X.; Guo, Y.; Xie, H.; Yin, H. Recovery of $LiCoO_2$ and Graphite from Spent Lithium-Ion Batteries by Molten-Salt Electrolysis. *iScience* **2023**, *26*, 108097. [CrossRef]

19. Chernyaev, A.; Kobets, A.; Liivand, K.; Tesfaye, F.; Hannula, P.-M.; Kallio, T.; Hupa, L.; Lundström, M. Graphite Recovery from Waste Li-Ion Battery Black Mass for Direct Re-Use. *Miner. Eng.* **2024**, *208*, 108587. [CrossRef]
20. Tian, H.; Graczyk-Zajac, M.; Kessler, A.; Weidenkaff, A.; Riedel, R. Recycling and Reusing of Graphite from Retired Lithium-ion Batteries: A Review. *Adv. Mater.* **2023**, *36*, 2308494. [CrossRef]
21. Moradi, B.; Botte, G.G. Recycling of Graphite Anodes for the Next Generation of Lithium Ion Batteries. *J. Appl. Electrochem.* **2016**, *46*, 123–148. [CrossRef]
22. Gao, S.; Yan, L.; Qin, J.; Liu, R.; Liang, B.; Wang, Q.; Geng, M.; Wang, B. Through-Hole Graphite Made from Waste Graphite for High-Rate Lithium-Ion Battery Anode. *J. Mater. Chem. A* **2023**, *11*, 4729–4738. [CrossRef]
23. Acheson, E.G. *A Pathfinder: Inventor, Scientist, Industrialist*; Acheson Industries: Port Huron, MI, USA, 1965.
24. Ragan, S. Science and Technology of Graphite Manufacture. *J. Mater. Sci.* **1983**, *18*, 3161–3176. [CrossRef]
25. Beak, M.; Park, J.; Park, S.; Jeong, S.; Kang, J.; Choi, W.; Yoon, W.-S.; Kwon, K. Understanding the Effect of Nonmetallic Impurities in Regeneratred Cathode Materials for Lithium-Ion Battery Recycling by Tracking Down Impurity Elements. *J. Hazard. Mater.* **2022**, *425*, 127907. [CrossRef] [PubMed]
26. Klaehn, J.R.; Shi, M.; Diza, L.A.; Molina, D.E.; Reich, S.M.; Palasyuk, O.; Repukaiti, R.; Lister, T.E. Removal of Impurity Metals as Phosphates from Lithium-Ion Battery Leachates. *Hydrometallurygy* **2023**, *217*, 106041. [CrossRef]
27. Gao, Y.; Zhang, J.; Chen, Y.; Wang, L.; Wang, C. Graphite Regenerating from Retired (LFP) Lithium-Ion Battery: Phase Transformation Mechanism of Impurities in Low-Temperature Sulfation Roasting Process. *Renew. Energy* **2023**, *204*, 290–299. [CrossRef]
28. Liu, Y.; Su, Z.; Wang, Y.; Shui, J.; Jin, Z.; Bai, B.; Qiu, L.; Du, P. Fabrication of High-Performance Silicon Anode Materials for Lithium-Ion Batteries by the Impurity Compensation Doping Method. *J. Solid State Electrochem.* **2023**, *27*, 969–976. [CrossRef]
29. Yan, X.; Jiao, J.; Ren, J.; Luo, W.; Mai, L. Fast-Charging Graphite Anode for Lithium-Ion Batteries: Fundamentals, Strategies, and Outlooks. *Appl. Phys. Lett.* **2024**, *124*, 040501. [CrossRef]
30. Wang, C.; Zhao, H.; Wang, J.; Wang, J.; Lv, P. Electrochemical Performance of Modified Artificial Graphite as Anode Material for Lithium Ion Batteries. *Ionics* **2012**, *19*, 221–226. [CrossRef]
31. Franklin, R.E. Crystallite Growth in Graphitizing and Non-Graphitizing Carbons. *Proc. R. Soc. A* **1951**, *209*, 196–218.
32. Franklin, R.E. The Structure of Graphite Carbons. *Acta Crystal.* **1951**, *4*, 253–261. [CrossRef]
33. Vlahov, A. XRD Graphitization Degree: A Review of the Published Data and New Calculations, Correlations, and Applications. *Geol. Balc.* **2021**, *50*, 11–35.
34. Brunauer, S.; Emmett, P.H.; Teller, E. Adsorption of Gases in Multimolecular Layers. *J. Am. Chem. Soc.* **1938**, *60*, 309–319. [CrossRef]

applied sciences

MDPI

Review

Recent Advances in Hydrothermal Oxidation Technology for Sludge Treatment

Hang Yu [1,*], Yuanyuan Liu [1], Nana Guo [1], Weiling Piao [2], Zonglin Pan [1], Bin Zhu [1], Yimin Zhu [1], Libo Wu [3,4], Jinling Wan [3,4] and Huangzhao Wei [2,*]

1 Collaborative Innovation Center for Vessel Pollution Monitoring and Control, Dalian Maritime University, Dalian 116026, China; lyy17838797206@163.com (Y.L.); guonana0921@163.com (N.G.); zonglinpan@dlmu.edu.cn (Z.P.); binzhu@dlmu.edu.cn (B.Z.); ntp@dlmu.edu.cn (Y.Z.)

2 Dalian Institute of Chemical Physics, Chinese Academy of Sciences, Dalian 116026, China; piaoweiling@dicp.ac.cn

3 Research Center of Supercritical and Subcritical Fluid, Zhejiang University of Technology, Hangzhou 310014, China; libo.wu@sr-water.cn (L.W.); jinling.wan@sr-water.cn (J.W.)

4 Hangzhou Sunrise Environment Co., Ltd., Hangzhou 310000, China

* Correspondence: yuhang891@126.com or yuhang@dlmu.edu.cn (H.Y.); whzhdicpwtg@dicp.ac.cn (H.W.); Tel./Fax: +86-411-8472-4357 (H.Y.)

Abstract: With the rapid development of urbanization and the widespread adoption of wastewater treatment facilities, the volume of sludge produced has steadily increased. Hydrothermal oxidation (HTO) technology offers an effective solution for sludge reduction, harmless disposal, and resource recovery, making it a highly promising method for sludge treatment. In recent years, HTO has attracted significant attention due to its efficiency and environmental benefits. This paper provides a detailed explanation of the fundamental principles of HTO in sludge treatment, with a focus on the removal of organic pollutants, nitrogen transformation, and phosphorus recovery. The influence of key operational parameters, such as reaction temperature, time, initial oxygen pressure, and pH, on the performance of HTO treatment is also explored. In addition, the research status of HTO sludge treatment and an example of product recovery after treatment are also discussed. It examines the challenges associated with scaling up HTO for large-scale sludge treatment, along with potential research directions for future work. Special attention is given to the innovation of catalysts, with the goal of achieving self-catalysis in sludge treatment. Moreover, considering that ammonia nitrogen (NH_3-N) is a major intermediate product in HTO, its removal, as well as the prediction and planning of other unintended products, remains a key issue. Further areas of interest include improving sludge dewatering performance and enhancing the production of valuable single carboxylic acids, which can boost resource recovery efficiency. This paper also highlights the diversification of sludge applications after HTO treatment. By providing insights into future development trends, this review offers valuable references for further research and practical applications. The ultimate goal is to support the development of HTO as a sustainable and efficient solution for sludge treatment, addressing environmental concerns while maximizing resource recovery opportunities.

Keywords: sludge; hydrothermal oxidation technology; principle; influencing factors; optimization scheme; resource utilization

1. Introduction

With the rapid development of industry, the number of wastewater treatment plants has increased significantly. Currently, the main wastewater treatment methods include biological treatment [1]; adsorption [2]; advanced oxidation processes such as UV/H_2O_2, Fenton, and photo-Fenton [3]; UV/$S_2O_8^{2-}$ and UV/HSO_5^- [4]; plasma degradation processes [5,6]; photocatalysis [7]; coagulation–flocculation [8]; electrochemical treatment [9];

and membrane filtration [10]. Although these methods can effectively treat wastewater, another problem that arises after treatment is the residual sludge. Wastewater contains many suspended solids, and most of these solids form sludge through adsorption or sedimentation. Sludge is the main byproduct of wastewater treatment plants, and its production has been increasing year by year. According to the Ministry of Housing and Urban–Rural Development of China, the sludge produced by urban areas and wastewater treatment facilities exceeded 60 million tons in 2022 (based on an 80% moisture content), showing a rapid growth trend compared to 50 million tons in 2020. It is estimated that by 2025, the sludge discharge will reach 90 million tons. Therefore, sludge treatment and disposal have become the primary potential source of secondary pollution in the wastewater treatment process [11]. Research indicates that most sludge consists of microorganisms, organic and inorganic components, water-soluble substances, proteins, pathogens, polysaccharides, difficult-to-degrade organic pollutants, heavy metals, and salts [12]. Consequently, sludge generally has a high organic content, complex composition, and high moisture and ash content. Improper management of harmful substances in sludge can pose a serious environmental threat.

To effectively manage sludge generated from wastewater treatment plants and other sources, researchers have made considerable efforts and conducted in-depth studies. Currently, there are four main traditional sludge treatment methods, as shown in Table 1: landfilling, aerobic composting, anaerobic digestion, and sludge drying and incineration [13]. The results in the Table 1 clearly show that the frequency of use of sludge treatment methods in China is landfill > anaerobic digestion > dry incineration > aerobic composting; The frequency of use in other developed countries is sorted as follows: anaerobic digestion > dry incineration > aerobic composting > landfill. It can be seen that the domestic method of landfill sludge treatment is still extensive, while foreign countries focus on anaerobic digestion due to environmental restrictions. A summary of the costs and advantages/disadvantages of these methods reveals that landfilling is simple and cost-effective, making it widely used in sludge management. However, it occupies substantial land and poses risks of leaching toxic substances due to high moisture content and heavy metal concentrations, potentially contaminating soil and groundwater. Biochemical techniques like anaerobic digestion and aerobic composting exhibit selectivity toward harmful substances but have limited applicability. On the other hand, drying and incineration face constraints due to lengthy processes and high costs, and they may generate secondary pollutants like dioxins [13–15]. In summary, traditional sludge treatment methods face multiple challenges, including land use, high costs, and the potential for secondary pollution.

Table 1. Comparison of technical investment, operating cost, advantages and disadvantages, and development trends of sludge treatment at this stage.

Treatment Method	Investment, Operating COST	Advantages	Disadvantages	Development Trend	
				China	Other Developed Countries
Landfilling	Low, Low	Low treatment cost, simple operation	Occupies large land area, potential groundwater contamination, not sustainable in the long term	Usage is high, declining in recent years, but still common (small urban and rural)	Restricted by environmental laws, the use of the European Union and other countries has gradually declined
Aerobic Composting	Medium, Medium	Enables resource recovery, produces fertilizers	Requires large space, long treatment cycle, specific requirements for sludge properties	Mostly used in urban sewage treatment plants	It is one of the main treatment methods used by sewage treatment plants in developed countries such as Europe

Table 1. *Cont.*

Treatment Method	Investment, Operating COST	Advantages	Disadvantages	Development Trend	
				China	Other Developed Countries
Anaerobic Digestion	High, Medium	Generates biogas as energy, reduces volume	High investment cost, requires rigorous management, specific requirements for sludge solid content	Mostly used in agricultural areas for small-scale treatment	Mainly used in some agricultural developed countries, such as the United States, Canada, France, etc.
Drying and Incineration	High, High	Thorough treatment, high degree of volume reduction	High investment and operating costs, high energy consumption, potential for secondary pollution	Use less frequently than landfills and anaerobic digestion	Mostly used in some countries with tight land resources (such as the Netherlands, Germany and other countries with high incineration rates)

To overcome these drawbacks, hydrothermal oxidation technology (HTO) has garnered increasing attention. As a novel sludge treatment method, as shown in Figure 1, HTO offers high treatment efficiency, effective detoxification, and potential resource recovery. It is well known that "hydrothermal (HT)" refers to homogeneous or heterogeneous chemical reactions occurring in aqueous solutions within closed systems at temperatures above 100 °C and pressures exceeding 1 bar (1 bar = 0.1 MPa). Hydrothermal oxidation, as a subset of hydrothermal technologies, involves thermochemical reactions in the presence of an oxidizing agent during the hydrothermal process. HTO can be classified based on the state of water into subcritical water oxidation (120–370 °C, 2–20 MPa) and supercritical water oxidation (above 574.2 °C and 22.1 MPa). Under subcritical conditions, such as in wet oxidation (WO) or catalytic wet oxidation (CWO), organic matter is partially oxidized, producing small organic molecules like acetic acid as the main byproducts. In contrast, during supercritical hydrothermal oxidation, organic matter is typically mineralized into carbon dioxide and water, with a mineralization efficiency that can reach as high as 99.999% [11,14,15].

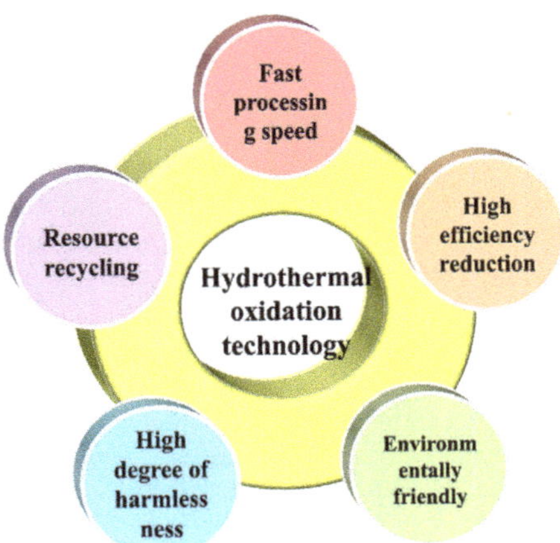

Figure 1. Advantages of hydrothermal oxidation technology for sludge treatment.

Researchers both domestically and internationally have conducted extensive studies on hydrothermal oxidation for sludge treatment. For example, Chu et al. designed a Cu-Ce/γ-Al_2O_3 catalyst to assist in wet oxidation of antibiotic-laden sludge, achieving a chemical oxygen demand (COD) removal rate of 81.2% and a volatile suspended solids (VSS) removal rate of 93.8%. The resulting carboxylic acids included acetic and propionic acids, which can serve as carbon sources [15]. Similarly, Fang et al. applied wet oxidation technology to pharmaceutical sludge and found that the concentration of volatile fatty acids (including acetic, propionic, isobutyric, and isoamyl acids) increased to 4819 mg/L (mg/L in this paper is equivalent to mg/dm^3 in the SI standard) at a pH of 12.56. This suggests that optimizing wet oxidation can enhance the quality of the sludge solution as a potential carbon source [16]. These studies strongly indicate that HTO is a promising sludge treatment method characterized by its high efficiency, effective detoxification, and ability for resource recovery.

In summary, HTO, as a novel sludge treatment method, has significant development potential. A thorough exploration of research advancements in hydrothermal oxidation for sludge treatment is crucial for promoting innovation and development in sludge management and achieving detoxification, volume reduction, and resource recovery. This study focuses on the following key aspects: firstly, the principles of HTO for sludge treatment; secondly, the control of temperature, time, and pressure in the HTO process, which has an impact on sludge treatment; thirdly, optimization strategies to improve the efficiency of HTO; fourthly, the optimization of reactors used in HTO; and lastly, the products generated from sludge after HTO and their related applications. The study summarizes these aspects and addresses the challenges faced in this field, as well as future research directions. This review provides a theoretical foundation for utilizing hydrothermal oxidation technology in sludge treatment, with the hope of achieving sustainable development in this area.

2. Results and Discussion

2.1. Introduction of Sludge

As is shown in Figure 2, sludge, as an inevitable byproduct of domestic wastewater and industrial effluent treatment processes, is a viscous substance composed of solid impurities, suspended solids, and colloidal materials [17]. Essentially, it represents the solid fraction of wastewater. Depending on its source, sludge can mainly be categorized into two types:

(1) Municipal sludge: This type is derived from urban wastewater treatment plants and primarily generated from domestic sewage. It contains a certain amount of organic matter, nitrogen, phosphorus, and other nutrients, which can potentially be utilized as resources under specific conditions. However, municipal sludge also inevitably contains harmful substances such as pathogens, which, if not properly treated, can pose significant risks to the environment and human health [17,18].

(2) Industrial sludge: This sludge is produced during various industrial processes, including those in the chemical, pharmaceutical, dyeing, papermaking, and metallurgy sectors. Due to the diverse raw materials and processes used in different industries, industrial sludge has a highly complex composition. It may contain heavy metals and toxic substances, significantly increasing the challenges associated with its treatment [18].

In-depth research reveals that sludge possesses several key characteristics [19–22]: first, sludge typically has an extremely high moisture content, often exceeding 95%. Such a high water content significantly increases the sludge volume, posing substantial challenges for subsequent treatment and escalating transportation costs and complexity. The water in sludge exists in four forms: free water, surface water, interstitial water, and bound water. Among these, the non-free water (surface, interstitial, and bound water) is tightly bound to sludge particles, making it difficult to separate using conventional methods. This further contributes to the large volume of sludge, complicating treatment and dewatering efforts. Second, sludge is rich in organic matter, containing a large number of microorganisms, bacteria, parasite eggs, and heavy metal ions. The presence of these components makes sludge prone to decomposition and odor, which can negatively impact the surrounding environment if not properly managed. Third, the particles in sludge are very small, resulting in a large specific surface area and strong adsorptive properties. This characteristic complicates the treatment process, as sludge can easily adsorb other substances. Finally, sludge has a relatively low specific gravity, generally less than that of water, which makes it readily suspend in water, further complicating treatment efforts.

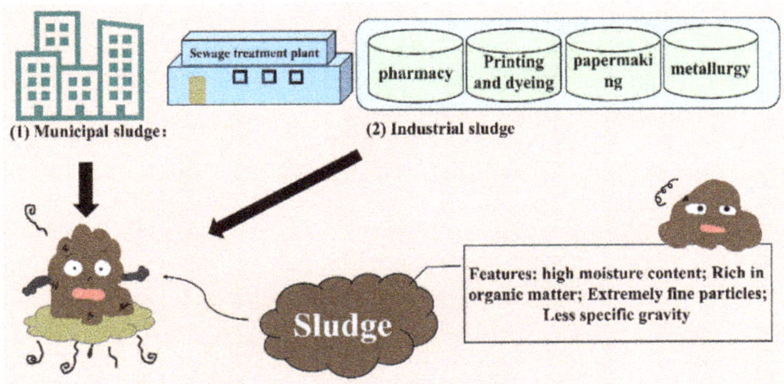

Figure 2. Source and characteristics of sludge.

Given the composition and characteristics of sludge, several key aspects should be emphasized during treatment. On one hand, research should focus on volume reduction, particularly addressing the significant volume caused by the high moisture content. On the other hand, it is essential to remove residual organic pollutants and heavy metals from the sludge to achieve environmentally friendly resource recovery. Therefore, the search for efficient and sustainable sludge treatment technologies is of critical practical significance.

2.2. The Application and Disadvantages of HT in Sludge

To achieve the goals of sludge reduction, harmlessness, and resource recovery, researchers have conducted extensive studies. Findings suggest that hydrothermal technology (HT), as an emerging sludge treatment method, has attracted significant attention in recent years. This technology operates under specific temperature and pressure conditions to treat sludge. Through HT reactions, the organic matter in sludge can be decomposed and transformed, thereby reducing its organic content and overall volume [23]. Additionally, HT technology promotes the stabilization of heavy metals in sludge, mitigating potential environmental hazards. However, literature reviews indicate that the effectiveness of standalone hydrothermal treatments is limited. For example, research by Chen et al. found that the addition of tannic acid (TA) could assist in HT treatment by lowering the effective HT temperature for deep dewatering of sludge proteins. Nonetheless, this approach faces significant economic challenges and has limited effectiveness in removing antibiotics and protein binding [24].

As the development of technology, HT technology has been continuously optimized, and the three commonly used methods at this stage mainly include hydrothermal carbonization (HTC), hydrothermal liquefaction (HTL), and hydrothermal gasification (HTG). Despite their respective advantages, these technologies still face challenges in practical applications, as shown in Table 2. Among the three commonly used technologies, HTC is the most widely applied. It operates at temperatures between 180 °C and 250 °C, producing a solid byproduct that can be used as biochar or a soil amendment. However, the treated sludge may still contain high levels of organic matter, particularly large organic molecules, and requires further dewatering. In addition, HTC has limited effectiveness in removing heavy metals from the sludge, which constrains its potential for safe disposal and resource recovery. As noted by Kossinska et al., the hydrochar produced from HTC contains higher concentrations of heavy metals and has low energy value, limiting its use in agriculture and energy applications. To address this, their study optimized the process by using co-hydrothermal carbonization to improve the performance in these areas [25]. HTL, on the other hand, operates at higher temperatures (250 °C–350 °C) and produces liquid fuels such as bio-oil. However, the efficiency of HTL varies significantly depending on the types of organic materials present in the sludge, and the resulting liquid phase often requires further treatment to meet discharge standards. For example, Thomsen et al. found that the liquid phase from HTL-treated sludge did not meet regulatory standards, necessitating additional treatment using wet oxidation technology [26]. Moreover, persistent organic pollutants and heavy metals may still remain in the treated sludge, affecting the quality and usability of the final product. Compared to HTC and HTL, HTG is less frequently applied. HTG converts sludge into gaseous products, such as methane, at higher temperatures (350 °C–600 °C), achieving significant reduction in sludge volume. However, this process is energy-intensive and requires complex equipment. As noted by Qiu et al., supercritical water gasification (SCWG) and supercritical water oxidation (SCWO) still face technical challenges in achieving high efficiency. For instance, initiating and sustaining the reaction requires preheating the feedstock to supercritical temperatures, typically using direct heating via fuel combustion or electric heaters (EH). Furthermore, coking and salt precipitation during preheating reduce the efficiency of the EH and heat exchangers (HX), potentially leading to blockages and system shutdowns [27].

In summary, while these common hydrothermal technologies have achieved varying degrees of success in sludge treatment, they still face challenges such as difficulty in controlling reaction conditions, inconsistent product quality, and limited removal efficiency of organic pollutants and heavy metals. Therefore, further optimization of hydrothermal technology is necessary to enhance its effectiveness in sludge treatment.

2.3. *Hydrothermal Oxidation Technology*

2.3.1. Comparison with Common Hydrothermal Techniques

As shown in Table 2, the summary and discussion of traditional hydrothermal technologies indicate that they mainly focus on promoting the physical and chemical transformation of sludge components by adjusting reaction conditions. However, a significant amount of macromolecular organic substances may still remain in the treated sludge [24]. To address this issue, researchers have introduced hydrothermal oxidation, a process that incorporates oxygen or oxidants into the hydrothermal reaction to further oxidize organic matter. This technology breaks down macromolecular organic compounds into smaller molecules such as carbon dioxide and water, significantly improving the removal efficiency of organic pollutants [26,28,29]. Therefore, HTO demonstrates clear advantages over traditional HT methods, particularly in terms of sludge detoxification and reduction. For example, a study by Pola et al. compared the effectiveness of hydrothermal and hydrothermal oxidation methods in removing heavy metals from sludge. The results showed that, during hydrothermal treatment, the presence of oxygen had a significant impact on most heavy metals (Hg, Cr, Ni, Cu, Zn, and Fe), and oxidation promoted greater dissolution of these metals. Notably, the most significant differences were observed in the behavior of Hg, Ni, and Cu. For Hg, in the presence of oxygen, the percentage of total Hg in the liquid phase decreased from 40% to 21% after 210 min at 200 °C. In contrast, under anoxic conditions, the concentration of Hg could only be reduced to 30%. The study also found that Hg was almost completely removed in less than 30 min, but after that, at the lowest temperature (160 °C), Hg began to re-dissolve. After further treatment, its concentration could only be reduced to a minimum of 30%. For Ni and Cu, under WO conditions, low temperatures were sufficient for their removal. At 160 – 170 °C, the content of Ni and Cu in the liquid phase accounted for only 1.5% (0.42 ± 0.07 mg/kg DM) and 0.2% (0.34 ± 0.01 mg/kg DM) of the total Ni and total Cu, respectively. In contrast, during the HT process, these two metals were almost entirely retained in the solid phase, which was the opposite trend to that observed during wet oxidation [30]. This advantage positions hydrothermal oxidation as a promising method for sludge treatment, offering efficient detoxification and effectively addressing the challenges of resource recovery. By comparing HTC, HTL, and HTG with hydrothermal oxidation, it becomes evident that HTO provides distinct benefits in sludge treatment. It not only achieves more thorough decomposition of organics but also efficiently removes heavy metals, reducing the environmental risks associated with sludge.

Table 2. Comparison of commonly used hydrothermal technology and hydrothermal oxidation technology for sludge treatment [31–33].

Comparison Item	Hydrothermal Carbonization (HTC)	Hydrothermal Liquefaction (HTL)	Hydrothermal Gasification (HTG)	Hydrothermal Oxidation (HTO)
Temperature Conditions	180 °C–250 °C	250 °C–400 °C	350 °C–600 °C	125 °C–320 °C
Pressure Conditions	2–6 MPa	10–25 MPa	20–30 MPa	0.5–10 MPa
Decomposition of Organic Matter	Limited decomposition: large molecules may remain (the removal rate of COD: 68–75%)	Moderate decomposition, mainly producing bio-oil (the removal rate of COD: 50–85%)	High decomposition into gases, leaving little organic residue	Very efficient organic matter decomposition into small molecules (CO_2, H_2O) (the removal rate of COD: more than 78%)
Dehydration Effect	Moderate dehydration, solid product still retains moisture (reduction around 50–70%)	Produces liquid products, so dehydration is moderate (reduction around 60–80%)	High dehydration, little solid residue	High dehydration efficiency, leading to effective sludge volume reduction (reduction around 80–90%)
Resource Recovery Potential	Produces solid hydrochar (HHV is 24.0–31.5 MJ/kg) for fuel or soil improvement	Produces bio-oil for energy, some nutrient recovery possible (HHV is 30–40 MJ/kg bio-oil)	Gaseous products like methane can be used as energy sources	High potential for heat recovery, metals and other resources are more easily recoverable
Advantages	Low temperature, easy to operate, solid product for energy use	Produces valuable liquid fuels, higher energy recovery potential	High energy yield in the form of gases, efficient organic conversion	Significant reduction and harmless treatment effects, high pollutant decomposition efficiency, and great resource recovery potential
Disadvantages	Large molecules remain, requires post-processing	Complex system, high pressure and temperature needed	Expensive setup, complex reaction control	Overall treatment efficiency: high, but improvements are needed in cost efficiency

2.3.2. Principle of HTO Technology

The principle of HTO in treating sludge primarily relies on the degradation or transformation of organic pollutants and harmful substances through oxidation reactions under high-temperature and high-pressure conditions. This technology takes advantage of the unique physical and chemical properties of water in subcritical or supercritical states, particularly its solubility and the accelerated rate of oxidation reactions, to achieve sludge reduction, detoxification, and resource recovery. Both subcritical (WO, CWO) and supercritical conditions in hydrothermal oxidation share common features, where the process involves two main steps: pyrolysis and oxidation.

In the pyrolysis step, the floc structure of the sludge is disrupted, releasing macromolecular organic compounds into the liquid phase. Subsequently, during the oxidation step, these macromolecules are oxidized into smaller organic molecules, with organic nitrogen and phosphorus being converted into inorganic nitrogen (e.g., N_2, NH_3, NO_x) and inorganic phosphorus (e.g., phosphates). The key difference between the two conditions lies in the oxidation mechanism. In subcritical conditions (WO, CWO), the oxidation process follows a traditional free-radical chain reaction, which may leave behind small molecular compounds like acetic acid in the final products [26,28,29]. In contrast, the SCWO process exhibits a higher efficiency in organic matter mineralization. Due to the extremely high temperature and pressure, oxidation is not limited to the conventional free-radical chain reactions; instead, direct oxidation between oxygen molecules and organic compounds occurs in the supercritical state, leading to complete mineralization into carbon dioxide and water. Therefore, the reaction mechanisms involved in the oxidation process include the oxidation and mineralization of organic matter, the transformation of organic nitrogen, and the transformation of organic phosphorus, as shown in Figure 3.

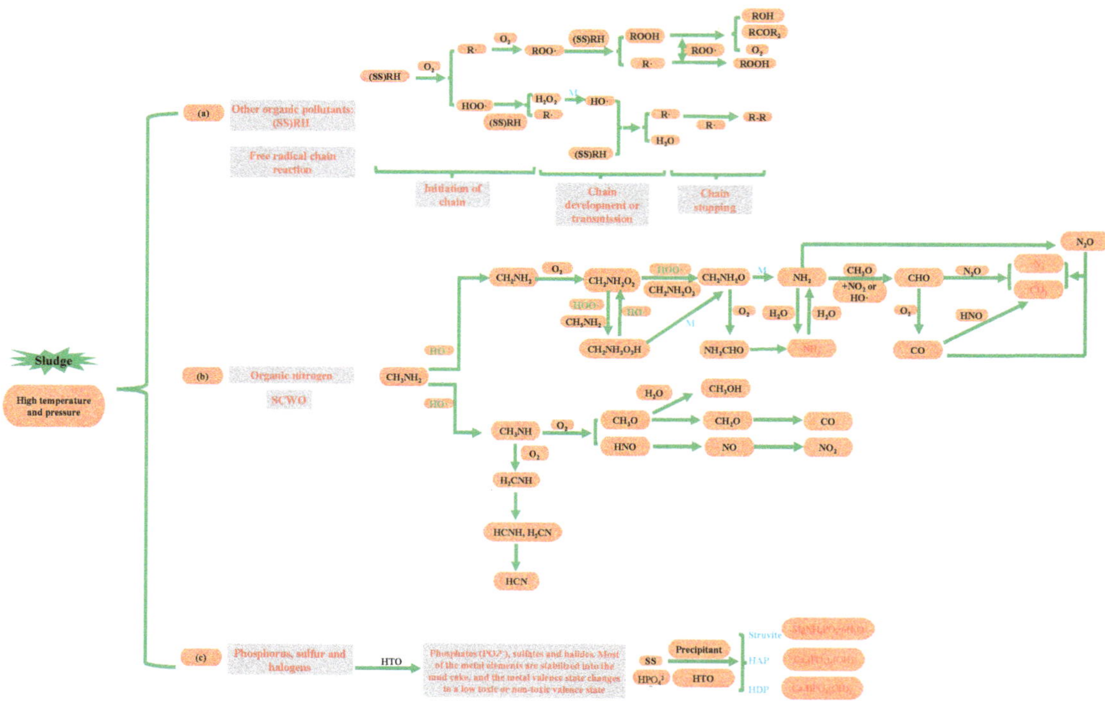

Figure 3. Reaction mechanism of sludge treatment by hydrothermal oxidation technology (including (**a**) free radical reaction, (**b**) organic nitrogen conversion, (**c**) phosphorus conversion.

First is oxidation and mineralization of organic matter: Under subcritical conditions in HTO, the process mainly involves the traditional free-radical chain reactions, as shown in Figure 3a, which can be divided into three stages: chain initiation, chain propagation and development, and chain termination [34,35].

(1) Chain initiation: the process by which reactant molecules generate free radicals. In this process, the organic matter in the sludge will produce HO· through thermal reaction with oxygen, as follows (1)–(3):

$$(SS)RH + O_2 \rightarrow R\cdot + HOO\cdot \ \text{((SS)RH is organic matter in sludge)}, \tag{1}$$

$$(SS)RH + HOO\cdot \rightarrow R\cdot + H_2O_2, \tag{2}$$

$$H_2O_2 + M \rightarrow 2OH\cdot \ \text{(M as catalyst)}, \tag{3}$$

(2) Chain development or transfer: Free radicals interact with molecules, alternately making the number of free radicals increase rapidly. As shown in the following Formulas (4)–(6),

due to its high electron affinity (568 kJ), HO· can oxidize all hydrogen-containing organic compound RH in the sludge to generate organic free radical R·, while organic free radical R· reacts with O_2 to form organic peroxyradical ROO·, which further removes hydrogen atoms from RH. The organic hydroperoxide ROOH and another organic radical R· are formed.

$$(SS)RH + HO\cdot \rightarrow R\cdot + H_2O, \tag{4}$$

$$(SS)RH + O_2 \rightarrow ROO\cdot, \tag{5}$$

$$ROO\cdot + (SS)RH \rightarrow ROOH + R\cdot, \tag{6}$$

(3) The termination of the chain (7)–(9): Free radicals collide with each other to generate stable molecules, and the growth process of the chain is interrupted.

$$R\cdot + R\cdot \rightarrow R - R, \tag{7}$$

$$ROO\cdot + R\cdot \rightarrow ROOR, \tag{8}$$

$$ROO\cdot + ROO\cdot + H_2O \rightarrow ROOH + ROH + O_2, \tag{9}$$

From the above reaction mechanism, it can be seen that the concentration of free radicals (HO·, HOO·, R·, ROO·), as the intermediate transfer of the whole chain reaction, plays a decisive role in enhancing the efficiency of the hydrothermal oxidation technology. In order to improve the application of hydrothermal oxidation technology in practice, it is necessary to comprehensively consider and optimize according to the characteristics of sludge, treatment requirements, and economic costs. In the supercritical state, the mineralization process is a simple direct oxidation of oxygen molecules and organic matter to produce carbon dioxide and water [36].

Second is transformation of organic nitrogen [36]. A comparison between subcritical and supercritical conditions reveals differences in the transformation of organic nitrogen. In subcritical conditions, due to the relatively mild reaction environment, the conversion of organic nitrogen is often incomplete, and the main products include ammonia (NH_3), nitrites (NO_2^-), and nitrates (NO_3^-). In contrast, under supercritical conditions, the conversion efficiency is significantly faster, and the residual nitrogen oxides (NO_x) are minimal. Ammonia nitrogen (NH_3-N) can be further oxidized to small molecules like N_2 under appropriate conditions. As illustrated in Figure 3b, Li et al. summarized the transformation of organic nitrogen in the SCWO process, finding that the primary products are formamide, NH_3-N, CO_2, and formic acid, with NH_3-N primarily originating from the hydrolysis of formamide. Furthermore, NH_3-N can be oxidized into N_2O and N_2 under suitable conditions. The yield of NH_3-N increases with rising temperature. When dimethylamine and trimethylamine are used as reactants, NH_3-N primarily originates from the intermediate dimethylamine, while the concentrations of NO_x, N_2O, and N_2 remain relatively low. Similar conclusions apply to aromatic organic compounds containing amino groups. For instance, aniline, under the action of hydroxyl radicals (HO·), generates intermediates such as phenol and NH_3-N. Under the influence of metal catalysts, phenol and NH_3-N are further converted into final products like N_2, N_2O, and NO_3^-, as reaction temperature, oxygen availability, and retention time increase.

Third is transformation of organic phosphorus [37]. Phosphorus in sludge is typically converted from organic phosphorus compounds (R-PO_4) to inorganic phosphate (PO_4^{3-}) during the oxidation process. Following this, dissolved phosphorus can be recovered by adding appropriate precipitating agents, usually magnesium (Mg^{2+}) or calcium (Ca^{2+}), along with available ammonium (NH_4^+). Magnesium phosphate precipitates primarily in the form of magnesium ammonium phosphate (struvite), while calcium phosphate (CaP) can form a broader range of compounds. The most common forms are hydroxyapatite (HAP) and dicalcium phosphate (HDP), but other forms, like amorphous calcium phosphate (ACP), brushite, or octa calcium phosphate, are also possible. The main reaction processes are represented by the following Equations (10)–(12):

$$Mg^{2+} + NH_4^+ + HPO_4^{2-} + 6H_2O \rightarrow MgNH_4PO_4 + 6H_2O + 2H^+, \quad (10)$$

$$5Ca^{2+} + 3HPO_4^{2-} + 4OH^- \rightarrow Ca_5(PO_4)_3(OH) + 3H_2O, \quad (11)$$

$$2Ca^{2+} + HPO_4^{2-} + 2OH^- \rightarrow Ca_2HPO_4(OH)_2, \quad (12)$$

2.4. Research on Influencing Factors of Sludge Treatment by HTO

In order to improve the treatment efficiency of sludge by HTO technology, researchers have conducted a large number of explorations of process optimization. Summarizing the research in recent years, it is found that the factors affecting the treatment of sludge by HTO technology include reaction temperature, time, pressure, pH value, etc. Comparing the performances of various influencing factors in improving the efficiency of HTO technology, it is found that the change in reaction temperature has a more significant improvement on the treatment effect of various sludges. This is because high temperatures can accelerate the dissolution and mineralization of sludge and promote the reaction. In contrast, reaction time and pressure have relatively less influence. The following section will discuss these factors and their impacts on the treatment of various sludges in detail.

(1) Reaction temperature

The reaction temperature of HTO technology is the most important parameter affecting the HTO process. Studies have shown that reaction temperature directly affects the reaction rate and the distribution of products. Generally, a higher temperature can increase the reaction rate of various sludge treatments and promote the degradation of organic matter, thus affecting the removal efficiency of COD and VSS [38]. In addition, it also has a certain influence on the generation of volatile fatty acids. Next, this article will conduct relevant discussions and summaries on the impact of temperature changes on the results when HTO is used to treat two types of sludge, municipal sludge and industrial sludge.

The study on municipal sludge treatment found that the change in the reaction temperature of HTO technology is crucial to sludge treatment, and the increase in reaction temperature is conducive to enhancing the removal ability of organic pollutants and the recovery of materials that can be used as carbon sources. For example, Chen et al. studied the treatment effect of wet oxidation technology on sewage plant sludge after the addition of a catalyst and compared the primary and secondary relationships of three influencing factors: temperature > concentration of $Ca(ClO)_2$ > molar ratio of $FeSO_4{\cdot}7H_2O/Ca(ClO)_2$. The results prove that reaction temperature is an important factor affecting the sludge dewatering performance. The concentration of $Ca(ClO)_2$ and the molar ratio of $FeSO_4{\cdot}7H_2O/Ca(ClO)_2$ have less influence. And they simultaneously used SPSS software to analyze the Pearson correlation coefficients of 18 operating parameters on sludge dewaterability (CST, SRF, and W_c). The results show that temperature is the main influencing factor [39]. Fang et al. studied the effects of temperature and oxygen pressure on HTO reaction and the basic properties of dissolved organic matter (DOM). The results show that, at 260 °C and an oxygen pressure of 0.9 MPa, the removal rates of total chemical oxygen demand (TCOD) and VSS reach the maximum values of 72% and 99%, respectively. And it is concluded that the influence of reaction temperature on the removal of TCOD and VSS is usually more significant in the operation of hydrothermal oxidation technology. An increase in temperature leads to an increase in the removal rates of TCOD and VSS. In addition, formic acid is generally present at lower temperatures, while acetic acid accumulates with the increase in temperature and oxygen pressure [40]. Núnez et al. also found that an increase in the reaction temperature of HTO technology is beneficial to the formation of volatile fatty acids such as acetic acid after municipal sludge treatment. Among them, the concentration at 160 °C for 180 min is 1.28 ± 0.07 g/L (g/L in this paper is equivalent to g/dm^3 in the SI standard), and the concentration at 180 °C for 180 min is 1.65 times. However, the degradation rates of non-volatile fatty acids such as oxalic acid, pyruvic acid, and formic acid accelerate with the increase in temperature, while the degradation rate of formic acid is not so obvious. But the decrease in the maximum concentration of short-chain fatty acids (SCFAs) due to excessive temperature can be attributed to the greater mineralization of organic matter. At higher temperatures, the oxidation pathway leading to the formation of CO_2 is more favorable than the oxidation pathway leading to the formation of acetic acid [41]. Zhang et al. also found that, under optimal conditions, as the reaction temperature increases, the acetic acid concentration increases from 1220 mg/L to 3250 mg/L, and the contribution of protein decreases from 26.7% to 2.3%. The content of carbohydrates also decreases with the increase in temperature due to the oxidation of carbohydrates and the loss of carbonyls related to the "caramel" or Maillard reaction. This study proves that the change in temperature has an important influence on the content changes of some components in sludge [42]. In addition, Wang et al. found that when persulfate and sludge are co-treated hydrothermally, higher temperatures and solid contents will also lead to the melting and deformation of microplastics (MPs), exacerbating the aging of polypropylene MPs, resulting in a rough surface, an increase in the carbonyl index, and a change in crystallinity [43].

For industrial sludge, the reaction temperature of HTO technology also has a significant impact on the removal of pollutants and resource recovery treatment. For example, Qin et al. discussed the influence of reaction conditions on the wet oxidation of excess sludge in the treatment process of caprolactam wastewater. The results show that, as the reaction temperature increases from 180 to 260 °C, the removal of COD and VSS is accelerated because a higher reaction temperature means higher reaction energy, which is beneficial to the increase in COD and VSS removal rates. It is worth noting that when the reaction temperature is too high, the removal rate of COD will not increase significantly because it may induce corrosion and is not conducive to equipment maintenance [35]. Wang et al. designed an improved CWO method for treating excess activated sludge in the process of coal chemical wastewater treatment. The results prove that, as the temperature increases, the solubility of oxygen increases, which increases the removal of organic pollutants in the solution. In addition to enhancing the removal efficiency of organic pollutants, an increase in reaction temperature can also increase the concentration of volatile fatty acids [38]. Chu et al.'s study also found the same conclusion. As the reaction temperature increases, the removal rates of COD and VSS increase significantly, and an increase in reaction temperature can also promote the production of volatile fatty acids [15].

Combining all these results with the results of the impact of temperature changes on sludge treatment by hydrothermal oxidation technology sorted out in Table 3 shows that reaction temperature is a key reaction parameter in the reaction and can significantly affect the relevant treatment effects of sludge, mainly in the following aspects: (1) The higher the reaction temperature, the removal rate of organic pollutants in sludge, and the removal rates of COD and VSS are all significantly enhanced. However, too high a temperature may reduce or stabilize the removal rate. (2) An increase in reaction temperature is beneficial to enhancing the dewatering performance of sludge. This is because the sludge particles become smaller and the pores increase significantly after treatment by HTO technology, which is conducive to the release of intercellular water, so the dewatering performance is improved. (3) As the reaction temperature continues to increase, the concentration of volatile fatty acids in the treated sludge is higher than that of non-volatile fatty acids, and most research results show that the concentration of acetic acid is significantly increased because a higher temperature can promote the solubilization and mineralization of these organic substances. (4) In addition to the above three common conclusions, researchers have also found that as the reaction temperature increases, the generation of NH_3-N will also increase significantly, gaseous carbon w_{CO_2} will rise, and liquid carbon $\omega_{aqueous}$ and ω_{solid} will decrease, proving that an increase in temperature is conducive to the conversion of carbon from liquid to gaseous.

Table 3. Influence of reaction temperature change on sludge treatment by HTO in recent years.

Sludge Type	Treatment Process	Optimal Temperature (°C)	Impact of Temperature (T) Change	Change in Efficiency/Yield (mg/L)	References
Urban Sludge	HTO-Alkaline Hydrolysis	214	T increases, SCOD yield significantly increases then stabilizes, NH_3-N generation significantly increases	(T: 130–210–290 °C) SCOD: 25,000–35,000–30,000; NH_3-N: 400–1200–1800	[11]
Urban Sludge	Thermal hydrolysis -HTO	180	T increases, TOC removal rate significantly improves, carbohydrates and proteins increase then decrease, the concentration of NH_3-N and acetic acid significantly increases	(T: 140–180 °C) TOC: 1400–2500; carbohydrates: 700–900–600; NH_3-N: 50–500	[29]
Municipal Sludge	CWPO	230	T increases, potential toxicity of Pb increases, potential toxicity of Cd increases then decreases (still lower than raw material), the concentration of acetic acid increased, and the concentration of propionic acid, isobutyric acid, and n-butyric acid decreased gradually, the concentration of ammonia nitrogen also increased first and then decreased	(T: 100–250 °C) SCOD: 25,000–35,000–30,000; NH_3-N: 500–2600–2200; Acetic acid: 600–2900–1100	[44]
Urban Sludge	HTO	160	T increases, and it is beneficial to the formation of acetic acid. Volatile fatty acids are produced more than non-volatile fatty acids (the concentration of propionic acid increases with increasing temperature, while oxalic acid, pyruvate, and formic acid are degraded)	SCFA: 6.07 g/L (acetic acid accounted for 28.2%, VFA: 70%); (160–200 °C) oxalic acid: 1300–243; pyruvate: 700–200; formic: 1190–700	[41]
Municipal Sludge	HTO/Thermal hydrolysis	200	T increases, the dissolution of EPS and the release of heavy metals are accelerated; the removal rates of Cu decrease first and then increase, and the removal rate of Zn decrease with the increase in temperature	(160–180–200 °C) Cu: 0.2–56–26.1%; Zn: 11.2–0.8%	[30]
Urban Sludge	HTO	240	T increases, the removal of SCOD decreases, the increase in temperature promotes the conversion of organic nitrogen to NH_3-N and phosphorus from organophosphates and phosphates to non-hydroxyl phosphates, the concentration of humic acid, protein, and polysaccharide increases first and then gradually decreases	(T: 180–260 °C) SCOD: 25,000–10,000; NH_3-N: 1735–2800 mg/L, accounted for 58.3–69.5% of TN, organic nitrogen: 650–778; accounted for 13.3–26.1%, nitrate nitrogen and nitrite nitrogen: 30–40	[45]
Pharmaceutical Sludge	CWO	260	T increases, the removal rates of COD and VSS are significantly increased	(T: 180–260 °C) The removal rate of COD: 20–72.6%; and VSS: 50–87.3%	[22]
Pharmaceutical Sludge	CWO	260	T increases, the removal rates of TCOD and VSS are significantly increased, and the concentration of VFAs is first significantly increased and then stabilizes	(T: 180–260 °C) The removal rate of TCOD: 40–76.5%; and VSS: 60–93.6%; VFA: 1500–5000	[46]

Table 3. *Cont.*

Sludge Type	Treatment Process	Optimal Temperature (°C)	Impact of Temperature (T) Change	Change in Efficiency/Yield (mg/L)	References
Pharmaceutical Sludge	CWO	260	T increases, the removal rate of COD accelerates obviously, the removal rate of VSS increases little, the concentration of acetic acid increases, and the concentration of formic acid and oxalic acid begins to decrease	(T: 180–260 °C) The removal rate of COD: 10–60%; and VSS: 70–97%	[47]
Pharmaceutical Sludge	CWO	260	T increases, COD and VSS removal rates significantly improve	(T: 180–260 °C) The removal rate of COD: 43–81.2%; and VSS: 61–93.8%	[15]
Pharmaceutical Sludge	CWO	260	T increases, the removal rates of TSS and VSS are significantly increased, the removal rates of SCOD and TCOD are significantly increased, the concentration of VFAs rises sharply and then plummets	(T: 200–260 °C) The removal rate of TSS: 72.5–90.1%; and VSS: 80–98.4%; The removal rate of SCOD: 70–81.6%, and TCOD: 43–78.7%; VFA: 1700–4000–3600	[16]
Oilfield Sludge	SCWO	450	T increases, the concentration of TOC and COD decrease first and then became stable, gaseous carbon w_{CO_2} rises while liquid carbon $w_{aqueous}$ and w_{solid} decrease, indicating carbon transitions from liquid to gas, with negligible impact on methane ratio	(T: 390–450 °C) The concentration of TOC: 3000–2600 and COD: 9000–6300; w_{CO_2}: 20–30%, and $w_{aqueous}$: 40–30%; w_{solid}: 20–15%	[48]
Mechanical Processing Wastewater Sludge	CWO	80	T increases, the dewatering performance of sludge is improved	81.41% fell to 42.92%	[39]
Coal Chemical Sludge	CWO	260	T increases, the degradation efficiency of COD, SCOD, and TOC is significantly improved, VFA concentration sharply increases (especially acetic acid)	(T: 180–260 °C) The removal rate of COD: 40–80%; SCOD: 725–83%, and VSS: 61–93%; VFA: 2500–4750 (Acetic acid: 4000)	[38]
Coal Chemical Sludge	HTO	260	T increases, COD removal rate significantly increases then decreases, acetic acid concentration increases, while formic and oxalic acid concentrations decrease	(T: 180–260 °C) The removal rate of COD: 25–78.6%; and VSS: 68–89.3%	[35]

Note: temperature changes, other things being equal. TSS: total suspended solids, VSS: volatile suspended solids, TOC: total organic carbon, COD: chemical oxygen demand, TCOD: total chemical oxygen demand, EPS: extracellular polymers, SCOD: dissolved organic matter, VFA: volatile fatty acids, thermal hydrolysis, HTO: hydrothermal oxidation technology, CWO: catalytic wet oxidation technology, CWPO: catalytic hydrothermal oxidation technology (different oxidants), SCWO: supercritical water oxidation technology.

It is worth noting that if the reaction temperature is too high, excessive acetic acid generated in the reaction is not easily oxidized, which will lead to no significant increase in the COD removal rate. On the other hand, excessive reaction temperature will cause more serious corrosion, which will increase the construction cost of the reaction system [35]. Therefore, considering the impacts of cost and energy saving, it is necessary to find an optimal temperature to make the degradation performance or the performance of recovering substances reach the best. In addition, inevitably, a high temperature will increase the construction and operation costs of the reaction system. Therefore, in the industrial utilization of HTO technology, the reaction temperature should be considered together with the reaction efficiency and the cost of the reaction system.

(2) Reaction time

Aside from reaction temperature, adjusting the reaction time also has a significant impact on the process. Reaction time refers to the duration for which the sludge undergoes oxidation under specific temperature and pressure conditions. This time directly affects the treatment efficiency. Generally, as reaction time increases, the removal rates of COD and VSS gradually improve. The main reason for this is that, during the hydrothermal oxidation process, the sludge undergoes hydrolysis, causing the VSS in the solid phase to quickly dissolve into the liquid phase. Over time, degradable organic matter accumulates in the liquid, and through the action of hydroxyl radicals and other reactive species, it is converted into intermediate products that are harder to oxidize further, such as carboxylic acids like acetic and formic acids [38]. The accumulation of these intermediates may adversely affect COD removal rates. Therefore, determining the optimal reaction time, considering both oxidation efficiency and economic cost, is crucial for enhancing sludge treatment performance. In current research on industrial sludge, notable examples include a study by Shang et al., which optimized conditions at a reaction temperature of 260 °C, with a duration of 60 min, an initial oxygen pressure of 1.0 MPa, and a catalyst concentration of $Cu/\gamma\text{-}Al_2O_3$ at 5.0 g/L. This setup achieved the highest removal rates for VSS at 93.6% and TCOD at 76.5%. The results indicate a linear relationship between the increase in removal rates and reaction time [46]. Similarly, Qin et al. investigated the effects of reaction conditions on the hydrothermal oxidation of excess sludge during the treatment of caprolactam wastewater. They found that at 260 °C, with an initial oxygen pressure of 1.3 MPa and a duration of 60 min, the removal rates of COD and VSS were 78.6% and 89.3%, respectively. Additionally, the accumulation of carboxylic acids resulting from sludge degradation increased with extended reaction time. Both the removal rates of COD and VSS also exhibited gradual increases over time, showing a similar trend. Notably, the VSS removal rate was already high within a short timeframe, likely because the solid-phase organic matter in the sludge transfers to the liquid phase, thereby enhancing VSS removal [35].

For municipal sludge, the effects of varying reaction time in HTO technology are similar, highlighting the need to determine an optimal reaction time for effective sludge treatment. For example, Gong et al. investigated the impact of various HTO conditions in the presence of NaOH on the treatment of urban sludge. They found that a reaction time of 30 min resulted in the highest SCOD yield at 21.67%. As reaction time increased, SCOD yield initially decreased before stabilizing. The concentrations of NH_3-N and NO_3-N gradually declined with extended reaction time. During the wet oxidation process, soluble proteins are degraded, leading to the formation of new organic compounds such as peptides and amino acids. Microbial cells also decompose over time, resulting in a gradual increase in the total phosphorus (TP) concentration. This study demonstrated that changes in reaction time significantly influence the harmlessness and resource recovery of sludge [11]. Additionally, Zhang et al. found that, at temperatures of 200 °C or higher, most organic matter dissolves within 10 min. With prolonged treatment time, the mineralization rate gradually increases due to sustained oxygen supply. The pattern of SCOD in sludge from 0 to 30 min shows an initial increase followed by a decrease, indicating rapid dissolution of organic matter from the solid phase, followed by gradual mineralization [42].

The current research on the effects of varying reaction time on the HTO of sludge reveals several key conclusions, which are summarized in Table 4: (1) the commonly used reaction time is typically around 60 min. (2) Longer reaction times are beneficial for increasing COD removal rates, but there exists a threshold beyond which removal rates plateau. (3) VSS removal rates are generally higher than COD removal rates and can reach their peak in a shorter timeframe. This is because COD removal primarily results from the oxidation of organic matter by soluble oxidants, such as gaseous oxygen, which is more affected by gas–liquid mass transfer, while VSS removal mainly occurs through rapid thermal hydrolysis of sludge. (4) Changes in reaction time also influence ammonia nitrogen concentrations, facilitating the conversion of carbon from liquid to gas. NH_3-N and NO_3-N concentrations initially increase with extended reaction time before gradually declining. These findings indicate that the impact of a changing reaction time is not as significant as that of a changing reaction temperature. Identifying an appropriate reaction time is essential for optimizing sludge treatment efficiency, highlighting the necessity of selecting suitable reaction times to enhance the effectiveness of hydrothermal oxidation technology.

Table 4. Influence of reaction time change on sludge treatment by HTO technology in recent years.

Sludge Type	Treatment Process	Optimal Time	Effect of Time (t) Change	Change in Efficiency/Yield (mg/L)	Reference
Urban Sludge	HTO-Alkaline Hydrolysis	30 min	t increases, the yield of SCOD decreases first and then stabilizes, and the concentrations of NH_3-N and NO_3-N decreases slowly with the extension of reaction time	(t: 10–30–120 min) SCOD: 24,600–26,100–24,900; NH_3-N: 350–360–300	[11]
Urban Sludge	Thermal hydrolysis-HTO	60 min	t increases, the concentration of ammonia nitrogen also increase, the removal rate of carbohydrates, TOC, and protein concentration first increases and then decreases (not significant)	(t: 0–60 min, 180 °C) TOC: 1500–2750–2000; protein: 1000–1500–100; NH_3-N: 200–600	[29]
Coal Chemical Sludge	HTO	60 min	t increases, the removal rate of COD increases gradually, while the removal rate of VSS does not change much	(t: 20–60 min) The removal rate of COD: 43–78.6%; and VSS: 80–89.3%	[35]
Coal Chemical Sludge	CWO	60 min	t increases, the removal rate of COD and VSS increases at the same time, and the increase in removal rate is linear with the reaction time, but there is a node value	(t: 20–60 min) The removal rate of COD: 55–78%; and VSS: 85–93%	[38]
Pharmaceutical Sludge	CWO	60 min	t increases, the removal rate of COD and VSS increases with the increase in temperature, and the removal rate of VSS (which is very high in a short time) is much higher than that of COD	(t: 20–60 min) The removal rate of COD: 43–60%; and VSS: 78–87.3%	[22]
Pharmaceutical Sludge	CWO	60 min	t increases, the removal rate of TCOD and VSS increases with the increase in reaction time; even in a short time, the removal effect of VSS is high, and the increase in the removal rate is linear with the reaction time	(t: 20–60 min) The removal rate of TCOD: 50–75%; and VSS: 85–93.6%	[46]
Pharmaceutical Sludge	CWO	60 min	t increases, the removal rate of VSS and COD increases, and the removal rate of VSS is changed little compared with the removal rate of COD	(t: 20–60 min) The removal rate of COD: 20–60%; and VSS: 80–97%	[47]
Pharmaceutical Sludge	CWO	60 min	t increases, the removal rate of initial VSS is higher, the removal rate of COD increases gradually with the increase in time	(t: 20–60 min) The removal rate of COD: 55–81.2%; and VSS: 85–93.8%	[15]
Oilfield Sludge	SCWO	20 min	t increases, the removal rate of TOC and COD increases, gaseous carbon w_{CO_2} increases, liquid carbon $\omega_{aqueous}$, ω_{solid} decreases, ω_{CO} first increases and then decreases	(t: 5–30 min) The concentration of TOC: 3500–1500 and COD: 6500–4300; w_{CO_2}: 25–53%, and $\omega_{aqueous}$: 40–20%; ω_{solid}: 35–18%	[48]

Note: time changes, other things being the same.

(3) Initial oxygen pressure

The principles of hydrothermal oxidation technology indicate that oxygen supply significantly affects sludge treatment. As oxygen supply increases, the contact opportunities between oxygen and organic matter in the sludge are enhanced, accelerating the reaction rate. Under these conditions, higher oxygen concentrations can provide more reactive oxygen species, facilitating the oxidative decomposition of organic matter and generating various beneficial free radicals, such as hydroxyl radicals, which play a crucial role in removing organic pollutants. The oxygen supply is primarily determined by the initial oxygen pressure.

Research on common industrial sludge treatment has shown that, at elevated temperatures, increasing the initial oxygen pressure is beneficial for the removal of organic pollutants. For instance, Shang et al. demonstrated that the additional oxygen supply plays a vital role in the wet oxidation of pharmaceutical sludge, significantly increasing the TCOD removal rate as oxygen levels rise. This is likely due to the increased production of hydroxyl radicals, which have strong oxidative capabilities. They concluded that the highest TCOD removal rate occurs when the oxygen addition is at 1.0 MPa [46]. Qin et al. explored not only the effects of reaction temperature and time, but also the impact of oxygen supply on hydrothermal oxidation. Their results confirmed that higher oxygen supply effectively accelerates the oxidation reaction rate, thereby facilitating the removal of organic compounds, including COD and VSS [35]. Wang et al. found that the COD removal rate is significantly influenced by the amount of oxygen added; as the initial oxygen pressure increases, the wet oxidation process becomes more favorable. In contrast, the VSS removal rate is only slightly affected by variations in oxygen supply. This may be attributed to the hydrolysis of the sludge, as VSS removal remains above 80% even when the initial oxygen pressure is at 0.2 MPa [38]. Chu et al. also observed that COD removal significantly increases with changes in oxygen pressure, while VSS removal rates show little change, consistent with the findings of previous studies [15].

For municipal sludge, increasing the initial oxygen pressure affects the TOC removal rate and acetic acid generation during HTO. For instance, Fang et al. investigated the effects of reaction temperature and oxygen pressure on dissolved organic matter in the hydrothermal treatment of urban sludge. Their results indicated that the optimal TCOD and VSS removal rates occur at an initial oxygen pressure of 0.9 MPa, with both rates increasing as the reaction temperature and oxygen pressure rise [40]. However, changes in oxygen pressure do not always influence sludge treatment. Malhotra et al. studied the effects of HTO on sludge and found that, while the TOC dissolution rate rapidly increased in the initial reaction stage with rising temperatures, variations in oxygen pressure did not significantly enhance TOC dissolution rates. Notably, at high temperatures, increasing oxygen pressure leads to a significant rise in acetic acid concentration, while changes in other VFAs remain minimal [29]. In summary, based on the discussions from the studies and the results in Table 5, it is evident that changes in initial oxygen pressure have a greater impact on COD removal rates compared to VSS removal rates. Additionally, the effect of oxygen pressure becomes more pronounced with increasing temperature.

Table 5. Influence of initial oxygen pressure (IOP) change on sludge treatment by HTO technology in recent years.

Sludge Type	Treatment Process	Optimal IOP	Effect of Oxygen Pressure (P) Change	Change in Efficiency/Yield (mg/L)	Reference
Urban Sludge	Thermal hydrolysis-HTO	OC: 1	P increases, the change in oxygen pressure does not significantly improve the solubilization rate of TOC but leads to a significant increase in acetic acid concentration	(P: 0.2–1 MPa) TOC: increased by 900	[29]
Coal Chemical Sludge	HTO	1.3 MPa	P increases, the removal rate of COD increases significantly, while the removal rate of VSS increases gradually	(P: 0–1.2 MPa) The removal rate of COD: 10–78.6%; and VSS: 78–89.3%	[35]

Table 5. *Cont.*

Sludge Type	Treatment Process	Optimal IOP	Effect of Oxygen Pressure (P) Change	Change in Efficiency/Yield (mg/L)	Reference
Coal Chemical Sludge	CWO	1 MPa	P increases, the removal rate of COD is greatly affected by the amount of oxygen added, and the removal rate of VSS is only slightly affected by the change in oxygen amount (the possible reason for the removal of VSS is the hydrolysis of sludge)	(P: 0.2–1 MPa) The removal rate of COD: 43–78%; and VSS: 82–93%	[38]
Pharmaceutical Sludge	CWO	1 MPa	P increases, the removal rate of COD increases with increasing oxygen pressure (an increase in the concentration of oxidants usually leads to an increase in the oxidation rate)	(P: 0.2–1 MPa) The removal rate of COD: 30–60%; and VSS: 73–87.3%	[22]
Pharmaceutical Sludge	CWO	1 MPa	P increases, the removal rate of TCOD increases significantly, while the removal rate of VSS increases slightly	(P: 0.2–1 MPa) The removal rate of TCOD: 40–75%; and VSS: 80–93.6%	[46]
Pharmaceutical Sludge	CWO	1 MPa	P increases, the removal rates of VSS and COD increase slowly	(P: 0.2–1 MPa) The removal rate of COD: 30–60%; and VSS: 82–97%	[47]
Pharmaceutical Sludge	CWO	1 MPa	P increases, the removal rate of COD increases significantly, while the removal rate of VSS does not change significantly	(P: 0.2–1 MPa) The removal rate of COD: 43–81.2%; and VSS: 80–93.8%	[15]
Oilfield Sludge	SCWO	OC: 1	P increases, the higher oxidation coefficient OC is conducive to the increase in the TOC removal rate, and the trend is first increased and then decreased; in addition, it is conducive to the existing CH_4 being oxidized into CO_2	(P: 0.2–1 MPa) The concentration of TOC: 4000–1000 and COD: 10,500–4000; w_{CO_2}: 15–60%, and $\omega_{aqueous}$: 40–20%; ω_{solid}: 30–10%	[48]

(4) The effect of pH

In addition to factors such as reaction temperature, time, and initial oxygen pressure that directly affect reaction efficiency, pH also significantly influences the HTO process for various types of sludge or the treatment of industrial sludge, particularly pharmaceutical sludge. HTO under alkaline conditions enhances the potential for resource recovery. Fang et al. found that, when the pH of the process was adjusted to 12.56, the concentration of volatile fatty acids (including acetic acid, propionic acid, isobutyric acid, and isovaleric acid) increased to 4819 mg/L. This suggests that adjusting the pH through the addition of reagents can improve the quality of HTO sludge solution as a potential carbon source [16]. Additionally, the addition of alkaline reagents promotes the removal of organic pollutants and the conversion of ammonia nitrogen during HTO. For example, Hušek et al. discovered that, after adding NaOH, the soluble chemical oxygen demand (SCOD) yield increased initially and then decreased as the NaOH concentration rose. The trends in NH_3-N and NO_3-N levels in the solution were consistent with the SCOD yield. This indicates that the addition of an appropriate amount of alkali promotes ammonia nitrogen conversion and organic pollutant removal from the sludge [14].

For municipal sludge, Guo et al. studied Fe (II)-activated persulfate-assisted hydrothermal conversion of waste-activated sludge (WAS) and found that, without pH adjustment (initial pH is 6.5), Fe^{2+}-activated PMS oxidation led to a 6.14-fold increase in standardized-capillary suction time (SCST) after 2 min of conditioning, a 58.3% increase in centrifuged weight reduction (CWR), and a reduction in moisture content (MC) to 57.5%, which was nearly equal to the value at pH 8. This result demonstrates that an excellent dewatering performance can be achieved under acidic conditions without pH adjustment [21]. Furthermore, Zamisa et al. found that adding NaOH enhanced the removal rates of VSS and COD, indicating that alkaline conditions favor the removal of organic pollutants [47].

In summary, acidic conditions may accelerate reaction rates and facilitate the oxidation of some organic compounds, but they also increase the risk of equipment corrosion. Alkaline conditions reduce oxygen solubility and reactivity, slightly slowing reaction rates, but they promote the removal of organic pollutants from the sludge. A neutral pH allows for moderate reaction rates with less equipment corrosion. Moreover, pH can affect sludge dewatering performance, ammonia nitrogen conversion, and subsequent resource recovery. Therefore, treating sludge under optimal pH conditions can improve both dewatering performance and resource recovery efficiency.

2.5. Issues and Optimization Strategies in HTO of Sludge

2.5.1. Challenges in HTO for Sludge Treatment

Although hydrothermal oxidation can enhance treatment efficiency through process optimization, its effectiveness in sludge treatment remains suboptimal, and several issues need addressing:

(1) Pollution risks: Sludge may contain harmful substances, and improper treatment can pose threats to the environment and public health.
(2) Low resource recovery: Current recovery rates for sludge resources are still low, with much sludge being disposed of through landfilling or incineration.
(3) Outdated equipment: Many reactors involved in HTO are outdated and unable to withstand the high temperatures and pressures required for effective processing over extended periods.
(4) The cost is too high: Hydrothermal oxidation technology requires oxygen assistance and is carried out in a high-temperature and -pressure environment, which inevitably generates some equipment costs and high energy consumption. To improve the efficiency of HTO for sludge treatment, researchers must continue to explore optimization methods to achieve more effective and environmentally friendly sludge management. Currently, various strategies are being implemented to address these three challenges, which will be summarized and discussed in the following sections.

2.5.2. Optimizing the Solution of HTO Technology

(1) Oxidizer assisted HTO technology to treat sludge

As is well known, oxygen is the most commonly used oxidant in sludge HTO. However, controlling oxygen pressure and other parameters can be quite complex. To address this issue, researchers have found that H_2O_2 can serve as an alternative oxidant. Its addition can replace oxygen, eliminating the need for air separation equipment, reducing gas–liquid mass transfer resistance, and accelerating the reaction rate. Moreover, oxidants like H_2O_2 have the advantages of being non-toxic, harmless, and highly oxidative. Additionally, H_2O_2 can react with organic pollutants to generate $\cdot OH$ radicals, which possess strong oxidative power (with a redox potential of 2.8 eV) and high selectivity, making them effective in removing harmful substances. In addition to H_2O_2, persulfates (PMS), such as potassium peroxymonosulfate and potassium peroxydisulfate, have also garnered significant attention due to their higher oxidation potential. These oxidants can react with organic pollutants to generate $\cdot SO_4^{2-}$ radicals, which also positively impact sludge treatment [4]. By adjusting the dosage of the oxidant, the redox potential of the reaction can be controlled, subsequently influencing temperature and pressure. For municipal sludge, the addition of oxidants can improve dewatering performance. For example, Guo et al. studied the effect of PMS on sludge dewatering in hydrothermal oxidation processes. As the amount of PMS increased, both the SCST and CWR initially increased and then decreased with increasing PMS dosage, while the MC first decreased and then increased [21]. For instance, Zhang et al. investigated the effects of HTO temperature and H_2O_2 concentration on the product distribution of municipal sludge, including solid-phase moisture content, heavy metal migration behavior, VFAs, NH_3-N levels, and pH. The results indicated that increasing the amount of H_2O_2 and HTO temperature significantly improved sludge dewaterability. However, the potential toxicity fraction of solid residues containing Pb and Cd increased with higher H_2O_2 concentrations yet remained lower than that of the original material. Additionally, both VFAs and NH_3-N concentrations showed an increasing trend, facilitating further resource recovery from sludge [44]. Moreover, sodium persulfate (SPS) can also serve as an oxidant to assist hydrothermal technology in treating municipal sludge. Xiong et al. found that the active persulfate oxidation method is an effective deep treatment technology for improving sludge dewaterability. Under optimal conditions—a hydrothermal temperature of 145 °C, a reaction time of 2 h, and a persulfate concentration of 150 mg/g DS—the CST reduction efficiency reached a maximum of 90.5%. Furthermore, over 90% of phosphorus and heavy metals remained in the sludge cake after hydrothermal treatment, and the addition of sodium persulfate can convert phosphorus into more stable forms while removing heavy metals, thereby reducing environmental risks [49]. In summary, the incorporation of oxidants into hydrothermal oxidation technology shows tremendous potential and advantages in sludge treatment. Their application not only enhances treatment efficiency, but also improves sludge dewaterability and the degree of harmlessness.

(2) Treatment of sludge with catalyst-assisted HTO

In addition to the fact that the addition of oxidants can improve the treatment of sludge by HTO technology, the addition of catalysts can as well. In recent years, the development of CWO technology has aimed to achieve safe and efficient sludge treatment [46]. By incorporating catalysts into the reaction system, this method effectively lowers the reaction temperature and pressure, enhances the oxidation efficiency, and shortens the reaction time, thus accelerating the oxidative decomposition of organic substances in sludge. This approach allows for harmless treatment while reducing operational costs. The CWO technology is particularly useful for treating sludge contaminated with organic pollutants and heavy metals, offering advantages such as rapid reaction speed, effective treatment, and low energy consumption. In this technology, catalysts play a crucial role, and selecting an appropriate catalyst is essential for improving sludge treatment efficiency. Catalysts can be categorized into two types based on their phase: homogeneous and heterogeneous. Homogeneous catalysts, including soluble transition metal salts, are characterized by high activity and low cost. Among them, Cu-based catalysts are the most widely used; however, they exhibit biotoxicity and are challenging to recover. In contrast, heterogeneous catalysts can be reused and primarily include noble metal catalysts, transition metal catalysts, rare earth element catalysts, and carbon-based catalysts. The following section will introduce the catalysts used in CWO technology in recent years and their catalytic efficiencies, as shown in Table 6.

Table 6. Comparison of catalysts used in CWO technology in recent years.

Catalyst	Original Sludge Type	Reaction Condition	Treatment Results	Reference
Cu-Ce/γ-Al_2O_3	Antibiotic-laden sludge	260 °C, 60 min, 1 MPa, 5 g/L (catalyst dosage)	The removal rate of COD: 81.2% The removal rate of VSS: 93.8%	[15]
Cu-loaded Al_2O_3	Maleic acid (500 mg/L)	70 °C, 60 min, 5 g/L (catalyst dosage), H_2O_2: 0.09 mol/L	The removal rate of TOC: 98%	[50]
Cu/Al_2O_3	Phenol (280 mg/L)	70 °C, 60 min, 5 g/L (catalyst dosage), H_2O_2: 0.09 mol/L	The removal rate of TOC: 96.8%	[51]
Cu-Ce/γ-Al_2O_3	Pharmaceutical sludge	260 °C, 60 min, 1 MPa, 10 g/L (catalyst dosage)	The removal rates of COD and VSS: 62.5%, 86.8%	[22]
Cu-Ce/γ-Al_2O_3	Coal chemical sludge	260 °C, 60 min, 1 MPa, 7 g/L (catalyst dosage)	The removal rate of VSS: 93.2%	[38]
Cu/γ-Al_2O_3	Pharmaceutical sludge	260 °C, 60 min, 1 MPa, 5 g/L (catalyst dosage)	The removal rate of COD and VSS: 76.5%, 93.6%	[46]

Table 6. *Cont.*

Catalyst	Original Sludge Type	Reaction Condition	Treatment Results	Reference
Cu/g-C_3N_4@AC	Phenol	120 °C, 60 min, 0.5 MPa	The removal rate of COD: 92.6%	[52]
FeOCl/Montmorillonite	Fuchsine (200 mg/L)	60 °C, 210 min, 1 g/L (catalyst dosage), H_2O_2: 0.38 mol/L	The removal rate of COD: 70.8%	[53]
CuO doped with Ce	Quinoline (100 mg/L)	75 °C, 85 min, 1 g/L (catalyst dosage), H_2O_2: 196 mmol/L, pH = 7.3	The removal rate of Quinoline and TOC: 98.1%, 86.1%	[54]
Cu/TiO_2	Cresol (1000 mg/L)	140 °C, 120 min, 2 MPa, 10 g/L (catalyst dosage)	The removal rate of COD: 79.1%, the degradation of cresol: 100%	[55]
Ce-modified Cu-based carbon	Phenol (1000 mg/L)	160 °C, 330 min, 3 MPa, 40 mg (catalyst dosage), pH = 9	The removal rate of COD: 89%	[56]
N-La_2CuO_4	Phenol (8000 mg/L)	140 °C, 30 min, 1 MPa, 1 g/L (catalyst dosage), pH = 9	The removal rate of COD: 87.1%	[57]
Fe_3C@NCNT/PSSF	Phenol (1000 mg/L)	80 °C, 7 h, H_2O_2: 5.1 g/L	The removal rate of TOC: 41%, phenol conversion: 90%	[58]
$PrFe_xCo1_{ex}O_3$/Mt	2-Hydroxybenzoic acid	80 °C, 210 min, 0.5 g/L (catalyst dosage), H_2O_2: 20.56 mmol/L, pH = 5.5	Degradation rate 97.6%, the removal rate of COD: 75.23%	[59]
$CoFe_2O_4$@SiO_2	Chlorobenzenes	110 °C, 40 min, 0.3 g/L (catalyst dosage), PS: 15 mM	The removal rate of TOC: 41%	[60]
CuO/γ-Al_2O_3	Reactive Red X-3B (0.3 g/L)	80 °C, 150 min, 5.5 g/L (catalyst dosage), H_2O_2: 0.39 mol/L, pH = 8	Degradation rate 90.72%, the removal rate of TOC: 45.26%	[61]
$FeSO_4$/Ca$(ClO)_2$	Mechanical processing wastewater sludge	180 °C, $FeSO_4 \cdot 7H_2O$/Ca $(ClO)_2$: 1.25, Ca$(ClO)_2$: 0.04, pH: 5.9 ~ 7.4	Moisture content: 51.72% (initial sludge: 76.30%)	[20]

Recent research indicates that non-homogeneous catalysts dominate current studies, reflecting their increasing importance. These catalysts serve diverse functions, enabling the treatment of sludge as well as the removal of organic pollutants from wastewater. Notably, catalysts composed of the same element can yield varying removal efficiencies for different types of sludge. For instance, when comparing the VSS removal rates of Cu-Ce/γ-Al_2O_3 for antibiotic sludge, pharmaceutical sludge, and coal chemical sludge, it was found that the catalyst performed best with antibiotic sludge [15,22,38]. Therefore, selecting an appropriate catalyst is crucial for enhancing the efficiency of catalytic hydrothermal oxidation in sludge treatment.

In addition, the related effects of the addition of catalysts on the treatment of sludge by HTO technology were summarized. The study found that the effects of the addition of catalysts on the treatment of sludge by HTO technology mainly included the following aspects: The first was enhanced removal of pollutants. For instance, Wang et al. utilized a synthesized Cu-Ce/γ-Al_2O_3 catalyst in a CWO process to treat excess activated sludge from coal chemical wastewater. The results showed that, under conditions of 7.0 g/L catalyst, 1.0 MPa initial oxygen pressure, 260 °C temperature, and a reaction time of 60 min, the removal rate of volatile suspended solids reached 93.2%, while the COD removal rate was 78.3% [38]. Xu et al. compared the efficiency of sludge treatment using HTO with and without catalysts, finding that the use of a homogeneous catalyst improved COD removal efficiency by over 10%. This enhancement is attributed to the ability of the catalyst to lower operational temperatures, increase reaction rates, and shorten retention times. Furthermore, when comparing different catalysts, it was observed that using Cu-Ce/γ-Al_2O_3 at 260 °C, 1.0 MPa, and 10 g/L for 60 min yielded the highest removal rates for VSS and COD, reaching 87.3% and 72.6%, respectively [22]. The second was an improved dewatering performance: Li et al. studied the interaction between peroxydisulfate, zero-valent iron (ZVI), and reaction time on sludge dewatering. Under optimal experimental conditions, the moisture content of the sludge was reduced to 54%, enhancing the dewatering performance of the hydrothermal oxidation technology. Additionally, when using the SC-M catalyst, the removal rates for metacresol and TOC were 98.1% and 84.2%, respectively [62]. The third was increased resource utilization of sludge as a carbon source: Shang et al. found that adjusting catalysts and temperature during hydrothermal oxidation promotes the production of VFA from sludge. These VFAs can serve as an organic carbon source, thus reducing the overall costs associated with hydrothermal oxidation technology [46]. It is important to note that the regulation of catalyst dosage also has a significant impact on sludge treatment. Table 7 illustrates the effect of varying additive dosages on the hydrothermal oxidation of sludge in current studies. The research indicates that an appropriate number of additives can enhance COD and VSS removal rates and promote ammonia nitrogen generation. However, excessive additive amounts can lead to reduced dewatering performance or lower SCOD removal rates. Therefore, it is crucial to determine an optimal dosage to avoid secondary pollution caused by excessive additions.

Table 7. Influence of additive content (AC) change on sludge treatment by HTO technology in recent years.

Original Sludge Type	Treatment Process	Optimal AC	Effect of Additive Concentration (C) Change	References
Pharmaceutical Sludge	CWO	5.0 g/L Cu-Ce/γ-Al_2O_3	C increases, the removal rate of COD and VSS remain basically unchanged after increasing gradually	[15]
Machining Wastewater Sludge	CWO	$FeSO_4 \cdot 7H_2O/Ca(ClO)_2$: 1.25	C increases, the dewatering property increased first and then decreased	[20]
Coal Chemical Sludge	CWO	7.0 g/L Cu-Ce/γ-Al_2O_3	C increases, the removal rate of COD is significantly increased, while the effect of VSS is slight	[38]
Municipal Sludge	CWPO	H_2O_2: 15%	C increases (H_2O_2 from 0% to 15%); with the increase in H_2O_2 mass fraction, lower solid yield and higher liquid gas yield can be achieved, which can improve the dehydration performance; the concentration of ammonia nitrogen increases first, then stabilizes and then increases, and the concentration of acetic acid increases	[44]
Urban Sludge	CWPO	0.2M $K_2S_2O_8$	C increases, the concentration of TP in liquid products increases rapidly from 103.85 to 422.87 mg/L, an increase of 307.19%	[63]
Pharmaceutical Sludge	CWO	10 g/L Cu-Ce/γ-Al_2O_3	C increases, the removal rate of COD is significantly increased, while the removal efficiency of COD is only slightly increased compared with that of VSS	[22]
Pharmaceutical Sludge	CWO	5 g/L Cu/γ-Al_2O_3	C increases, the removal rate of TCOD increases significantly, but the removal rate of VSS does not change much	[46]
Oilfield Sludge	SCWO	20 wt% Ni/Al_2O_3	C increases, the removal rate of TOC increases, the CO_2 production rate increases, and the production of CO and CH_4 decreases	[48]

Overall, the addition of catalysts significantly enhances the effectiveness of HTO for sludge treatment, leading to improved pollutant removal, better dewatering, and greater resource recovery. In summary, CWO demonstrates unique advantages in sludge treatment. Compared to traditional hydrothermal oxidation, it achieves higher reaction efficiency, enabling deep oxidative decomposition of organic matter in sludge in a shorter time frame, significantly reducing processing time and energy consumption. Moreover, this technology is environmentally friendly, converting organic substances in sludge into harmless materials and facilitating resource recovery, thereby minimizing environmental pollution and treatment costs. As environmental protection standards continue to rise,

CWO has substantial development potential. It can be integrated with other methods to further enhance sludge treatment efficacy. In the future, ongoing optimization of catalysts and processes is expected to establish this technology as one of the mainstream approaches for sludge treatment, providing robust support for the harmless, reduced, and resource-efficient management of sludge.

Moreover, oxidants and catalysts can work synergistically in HTO technology to achieve more significant treatment effects on sludge containing high levels of antibiotics. For instance, Yang et al. studied the use of acidified red mud (ARM) in a catalytic wet peroxide oxidation (CWPO) system to degrade antibiotic-laden sludge and produce the hydrogen carrier formic acid. In their experiments, under conditions of 90 °C, a reaction time of 30 min, H_2O_2 concentration of 20 mL/L, ARM dosage of 0.8 g/L, pH = 7, and stirring at 500 rpm, the yield of formic acid reached 792.38 mg/L [32]. Similarly, Ning et al. investigated the use of Fe (II)-activated persulfate-assisted hydrothermal treatment (Fe (II)-PS-HT) to enhance nitrogen removal efficiency from sewage sludge (SS) under relatively mild conditions (150 °C, 20 min). Their findings revealed that the nitrogen removal rate from SS treated under these optimal conditions was 35.0% higher than that of SS subjected to conventional high-temperature treatment [64]. These studies demonstrate that the incorporation of oxidants and catalysts enables HTO technology to achieve excellent sludge treatment outcomes without the need for high temperatures. This approach offers new insights into the degradation and resource recovery of sludge.

(3) Supercritical water oxidation sludge treatment

Subcritical water (100 °C < T ≤ 373.946 °C), also known as superheated water, thermal pressurized water, or pressurized hot water (PHW), is a liquid with pressure above its saturated vapor pressure (SVP). Chen et al. concluded that, in the presence of air, subcritical water at relatively high temperatures (200–325 °C) and pressures (5–15 MPa) can generate large amounts of highly reactive hydroxyl radicals with high oxidation potential. As a result, subcritical water oxidation has demonstrated excellent performance as an advanced oxidation process (AOP) for degrading refractory wastewater and sludge [65]. Based on this example and the above-mentioned content, it was found that the previous optimization efforts were all subcritical; although the treatment effect of sludge was enhanced, the reaction times were not fast, all of them being around 1 h or even longer. In order to speed up the reaction rate, researchers also discovered the supercritical water oxidation technology for sludge treatment. It has been reported that, at pressures and temperatures above the critical point (P = 22.064 MPa, T = 373.946 °C), water behaves as a typical fluid known as supercritical water. SCWO offers advantages over conventional hydrothermal oxidation, including faster reaction rates, thorough treatment, and no secondary pollution. It can convert heteroatoms such as chlorine, sulfur, and phosphorus into their respective inorganic acids or salts, and it can also mineralize heavy metals, depositing them as stable solid residues [36]. Li et al. reviewed the treatment effects and kinetic parameters of various nitrogen-containing organic compounds and NH_3-N in the SCWO environment, analyzing the transformation pathways of organic nitrogen in different nitrogen groups to understand its degradation mechanisms. Their results indicated that nitro-containing organic compounds primarily convert nitrogen into nitrogen oxides (NO_x^-) in the form of nitrogenous intermediates, while the main intermediate for amino-containing or nitrogen–heteroatom organic compounds is NH_3-N. For organic compounds containing both amino and nitro groups, the generated NO_x^- can react with NH_3-N through redox processes to form N_2. Furthermore, supercritical water oxidation, when coupled with conventional oxidants like H_2O_2 or O_2 under relatively mild SCWO conditions, enhances the synergistic removal efficiency of COD or TOC and NH_3-N while avoiding the generation of NO_x, NO_3^-, and NO_2^-, converting organic nitrogen into environmentally friendly products such as N_2 and N_2O [36]. Yan et al. investigated the effect of adding a Ni/Al_2O_3 catalyst in the SCWO process, with reaction conditions set at 450 °C for 20 min, achieving TOC removal and carbon conversion to gas phase efficiency of 95.2% and 68.2%, respectively [48]. The study revealed that the advantage of SCWO technology lies in its ability to achieve efficient conversion of organic matter in a short time, far surpassing traditional hydrothermal oxidation in treatment efficiency. Additionally, SCWO does not produce harmful by-products such as dioxins, and inorganic salts can be recovered in solid form, reducing the complexity and cost of subsequent treatment processes.

2.5.3. Optimization Measures for the Reactor

In addition to the optimization of the hydrothermal oxidation process itself, improvements in reactor design and process flow are also crucial for enhancing sludge treatment efficiency. Traditional HTO technologies operate at high temperatures and pressures, leading to expensive equipment costs and safety concerns. Therefore, optimizing reactors and refining process designs can significantly increase treatment efficiency, reduce operational costs, and promote the widespread application of HTO in sludge treatment. WO technology is an early form of HTO. Initially, reactors were designed as high-pressure vessels where the organic matter in sludge reacts with oxygen under controlled temperatures (150–374 °C) and pressures (2–15 MPa). The Zimpro wet oxidation reactor is a classic example of this technology; it efficiently removed organic pollutants from sludge. However, its high energy consumption and equipment costs limited its widespread use [66]. As demand grew, reactor designs evolved from simple high-pressure vessels to catalytic wet oxidation reactors, where metal catalysts were introduced to lower reaction temperatures and pressures, improving efficiency. Later, SCWO technology marked a significant breakthrough. The unique properties of supercritical water enabled the effective decomposition of organic matter at higher temperatures and pressures, although issues such as equipment cost and material corrosion remained [36]. For instance, Fidel A. Mato and colleagues proposed a novel energy recovery method based on injecting reactor effluents into gas turbines, utilizing ultrafiltration and evaporation units for front-end concentration while employing SCWO reactors as the main treatment unit. This method was found to be more efficient compared to simpler direct-use methods [67].

In recent years, further advancements in HTO technology have included the development of tubular reactors, fixed-bed reactors, and continuous-flow reactor designs, all aimed at making the process more efficient. For example, the Vertech wet oxidation reactor significantly reduced energy consumption through heat recovery [68]. The introduction of multiphase reactors, which incorporate solid catalysts in liquid-phase reactions, has notably increased the reaction rate between oxygen and organic matter. The latest wet peroxide oxidation reactors (WPO) utilize hydrogen peroxide as a strong oxidizing agent, achieving high removal efficiencies for organic matter at lower temperatures and pressures [69]. Consequently, researchers have increasingly focused on reactor optimization. For instance, Yamasaki et al. proposed a hydrothermal catalytic plasma reactor to expand the treatment capacity for water-soluble volatile organic compounds (VOCs), such as toluene, acetaldehyde, acetic acid, and ammonia. They also installed a manganese dioxide catalytic unit downstream to remove ozone, further improving efficiency. Mato et al. developed an energy recovery system utilizing gas turbines to optimize the entire process flow by recovering energy from reactor effluents [67,70]. Additionally, Qiu et al. designed a dual-shell reactor using hydrothermal flames as an internal heat source, effectively avoiding corrosion, salt blockages, and overheating while demonstrating advantages in both economic and energy efficiency [27].

Overall, reactors used in wet oxidation and HTO technologies have evolved from simple high-pressure vessels to more advanced, environmentally friendly designs that incorporate catalysts, supercritical conditions, and energy recovery innovations. These developments have greatly enhanced sludge treatment efficiency while reducing costs. As these optimizations continue, the potential for HTO technology in sludge treatment will expand significantly.

2.6. Comparison of Sludge Before and After HTO Treatment

In order to study the changes before and after the reaction of sludge treated by hydrothermal oxidation technology, including its treatment efficiency of organic pollutants and dewatering performance of sludge and product composition, the following summary was made, as shown in Table 8:

Table 8. Comparison of sludge changes before and after hydrothermal oxidation treatment of sludge.

Original Sludge Type	Treatment Process	Before Treatment	After Treatment	Reference
Pharmaceutical Sludge	CWO	COD: 15,000–16,000	COD and VSS removal rates: 81.2%, 93.8%	[15]
Municipal Sludge	HTO-Alkaline Hydrolysis	TCOD: 12,000, COD: 285–320, TP: 25.52, pH: 6.52	COD: 69,500, TP: 1145, pH: 11.8, TN: 2150 (NH_3-N: 1080, NO_3-N: 693)	[11]
Municipal Sludge	HT-HTO	Soluble TOC and COD: 65, 164; total TOC and COD: 3880, 13,000 ± 260; TSS, VSS: 14,448 ± 28.3, 9310 ± 12.5, ammonium nitrogen and protein (liquid phase): 14 and 0.5	Total COD and VSS removal rate 55%, 90%, TOC: 2750, soluble COD, protein, carbohydrate concentration: 5800, 2250, 900, NH_4^+-N concentration is 600, acetic acid, propionic acid, butyric acid and isobutyric acid concentration: 400, 100, 10–15.	[29]
Coal Chemical Sludge	HTO	Total COD: 20,000–25,000, pH: 7.42–8.56	COD and VSS removal rates: 78.6%, 89.3%, acetic acid, formic acid, and oxalic acid concentrations: 7000, about 1500	[35]
Mechanical Processing Wastewater Sludge	CWO	Initial sludge with particle size 100–200 μm, moisture content 76.30%	Sludge particle size significantly reduced, moisture content 51.72%	[20]
Coal Chemical Sludge	CWO	COD: 16,500–17,500, VSS/SS 81.3%, pH: 8.47	COD, VSS, and SCOD removal rates: 78.3%, 93.2%, 83%; acetic and VFAs concentrations: 4000, 4750	[38]
Municipal Sludge	CWPO	Moisture content: 95.5%, TN: 12.5, NH_4^+: 10.3, CST: 163.5	Moisture content: 38.5%, SCST: 8.74, CWR: 89.2%, TOC: 285.6, TN: 153.4, NH_4^+: 52.3	[21]
Municipal Sludge	CWPO	N: 4.24%, volatile: 45%, fixed carbon: 9%	Denitrification rate: 76.2%, protein: 4184	[64]
Municipal Sludge	CWPO	Ammonia nitrogen: 1230, moisture content: 46.41%	Acetic acid: 2923.41, moisture content: 45.70%, ammonia Nitrogen: 2100	[44]
Urban Sludge	HTO	TCOD: 22,030 ± 40, SCOD: 350 ± 150, TOC: 6100 ± 800, TSS: 33,000 ± 1000, VSS: 26,800 ± 400	SCFA: 6070, acetic acid accounts for 28.2%, 2830 ± 80, non-volatile lactic acid: 750 ± 20, lactic acid bacteria, malic acid, oxalic acid, pyruvate and formic acid concentration: 160 ± 50, 110 ± 10, 1300 ± 100, 700 ± 6, 1190 ± 80	[41]
Municipal Sludge	HTO/HT	Ni, Cu, Hg, Cr, Zn: 28, 150, 0.6, 19, 340 mg/g	Ni, Cu, Hg, Cr, Zn removal rates: 86.7%, 56.1%, 35.7%, 14%, 11.2%	[30]
Urban Sludge	CWPO	Organic matter content: 44.22%, TN: 18.89 mg/g, TP: 18.36 mg/g	98% phosphorus recovery in liquid phase (solid recycling water: 9.47% ± 0.21%), liquid TOC removal rate: 376.45–744.80 mg/L, solid TP: 0.96 ~ 5.02 mg/g	[63]
Urban Sludge	HTO	TCOD: 53,540–60,677, TN: 2976–3255, TP: 1733–1955, Soluble Ammonia Nitrogen: 45–50	COD and TOC removal rates: 60–70%, 68.1%; amino acids, proteins, humic acid derivatives, VFAs, acetic acid concentrations: 3000, 5151, 2400, 4199, 3900; SCOD: 10,000, TN: 5000	[45]

Table 8. *Cont.*

Original Sludge Type	Treatment Process	Before Treatment	After Treatment	Reference
Pharmaceutical Sludge	CWO	COD: 15,000–16,000, VSS: 13,500–13,800, VSS/SS ratio: 39–40%, pH: 7.5–8.0	COD and VSS removal rates: 72.6%, 87.3%	[22]
Pharmaceutical Sludge	CWO	TCOD: 16,500–18,000, TSS: 16,300–17,800, VSS: 13,200–14,100, pH: 7.3–7.8	TCOD and VSS removal rates: 76.5%, 93.6%	[46]
Pharmaceutical Sludge	CWO	TCOD: 19,000–20,000 g/L; VSS: 15,500–15,800 g/L, pH: 7.5–8.5	COD and VSS removal rates: 57.3%, 95.2%; acetic, formic, and oxalic acid concentrations: 8000, 2000, 2250	[47]
Oilfield Sludge	SCWO	Volatile: 7.3 wt%, fixed carbon: 73.84 wt%, C: 48.18 wt%, N: 0.36 wt%, S: 0.28 wt%	TOC removal rate: 95.2%, TOC concentration: 1368.5, COD: 3832.5; carbon conversion rate (CE): 68.2%, w_{CO_2}: 53.6%, $\omega_{aqueous}$: 29.7%, w_{CH_4} : 6.7%, ω_{CO}: 10.8%	[48]
Pharmaceutical Sludge	CWO	VSS: 160.6 g/L; TSS: 192.3 g/L; VSS/TSS: 83.5%; pH: 8.17; TCOD: 206.3 g/L	TSS: 90.1%, VSS: 98.4%, TCOD: 78.7%, COD: 81.6%, volatile fatty acids: 4819	[16]
Urban Sludge	HTO	TS: 1.7 ± 0.1%; VS/TS: 58.5 ± 0.5%; pH: 7.8 ± 0.1; TCOD: 12.0 ± 0.3 g/L; SCOD: 7.1 ± 0.3 g/L; TN: 3.3 ± 0.2 g/L	Ammonia nitrogen: 3609, amino acids, humic acid derivatives, acetic acid concentrations: ~3000, 2400, 3900	[42]

Note: The above ununited data use a unit of uniform concentration—mg/L.

From Table 8, it is evident that sludge undergoes significant changes after treatment with hydrothermal oxidation technology, which is reflected in several key aspects: first, the dewatering performance of the sludge improves, leading to a reduction in moisture content. Second, the removal rates of TOC and VSS increase, enhancing the removal of organic pollutants from the sludge. Finally, the products after treatment are mainly volatile fatty acids (acetic acid content is the most), nitrogen (ammonia nitrogen content is the most), phosphorus, carbon dioxide, etc.

The development of the above optimization measures can be summarized as follows: first, HTO technology uses oxidants to replace oxygen, avoiding the safety risks brought by oxygen injection. Second, the addition of catalysts reduces the operating temperature of the equipment while enhancing the degradation of organic matter in sludge. Third, supercritical oxidation technology further improves the harmless disposal of sludge. Finally, advancements in reactor design and upgrades play a crucial role. In addition to these optimization methods, pretreatment techniques such as ultrasonic processing or acid/alkali pretreatment can also enhance the efficiency of HTO in sludge treatment. However, due to the complex composition of sludge, some byproducts are inevitably generated after treatment. To reduce processing costs, resource recovery from these byproducts can be employed to improve economic viability.

2.7. Resource Recycling

From the literature reviewed, it is evident that the sludge treated by hydrothermal oxidation contains rich nutritional components such as proteins, nitrogen (N), and phosphorus (P). Below, we summarize several potential and common applications of the products obtained from treated sludge, categorized into three main aspects: solid products, liquid products, and gaseous products, as shown in Figure 4.

Figure 4. Application of sludge-treated products.

2.7.1. Solid Products

The solid products obtained from the hydrothermal oxidation of sludge typically possess the following characteristics: first, the pollutants, mostly inorganic substances, are effectively removed; second, they contain essential nutrients such as nitrogen, phosphorus, potassium, calcium, and magnesium. Based on these characteristics, the solid products can be utilized in several ways.

(1) Construction Materials

After hydrothermal oxidation treatment, the sludge demonstrates not only an improved dewatering performance, but also significantly reduced organic pollutants. The inorganic content of the solid products is relatively high and stable. Therefore, these solid products can be further processed for use as construction materials, such as brick-making raw materials. Their particle size and composition characteristics can meet the strength requirements for bricks, facilitating resource recovery from sludge and reducing the need for natural raw materials. For example, Zeng et al. demonstrated that, after catalyst treatment, the COD and VSS of sludge were reduced to a certain extent, and the resulting inorganic residues could be used for the production of construction materials [22]. Wang et al. found that, following hydrothermal oxidation, the sludge particles decreased in size and showed increased porosity, which aided in the release of intercellular moisture, thereby enhancing dewatering performance. The solids can be used to produce ceramic aggregates or permeable bricks. Additionally, these materials can serve as fillers for wastewater treatment, effectively utilizing waste [38].

(2) Fertilizers and Soil Conditioners

The treated solid products contain certain amounts of nutrients and minerals, such as nitrogen, phosphorus, potassium, calcium, and magnesium, allowing for their use as fertilizers. For instance, Cañas et al. studied phosphorus recovery from wastewater sludge treated by hydrothermal oxidation. Under high-temperature (up to 300 °C) and high-pressure (up to 200 bar) conditions, the process achieved high removal rates of organic matter (up to 85% COD) and total solids (up to 75%). Liquid and solid fractions recovered from the effluent contained significant amounts of recoverable phosphorus. In the liquid effluent, phosphorus was recovered as struvite ($MgNH_4PO_4{\cdot}6H_2O$), a slow-release fertilizer, reaching concentrations of up to 90 mg P/L. The recovered struvite can be utilized as fertilizer [37]. Similarly, Liu et al. found that high reaction temperatures converted organic phosphorus and phosphate into non-hydroxyl phosphorus. The sludge cake enriched with non-hydroxyl phosphorus post-hydrothermal oxidation can serve as a soil conditioner. Notably, researchers also validated this hypothesis using seed germination tests, finding that the effluent from treated sludge promoted plant germination and growth, although different seeds required varying concentrations. The results indicated that the effluent had a beneficial effect on seed growth [45].

2.7.2. Liquid Phase Products

The liquid products consist of partially oxidized organic substances and dissolved inorganic salts, mainly volatile fatty acids (mainly acetic acid, formic acid, propionic acid, etc.), ammonia nitrogen, etc. These liquids can be further processed to recover energy and nutrients, with several main applications as follows:

(1) Soil Amendments

The liquid products obtained from hydrothermal oxidation contain essential nutrients. After appropriate treatment and purification, they can be used as liquid fertilizers for agricultural irrigation, providing nutrients to crops and promoting growth while reducing reliance on freshwater resources. For instance, Liu et al. proposed a resource recovery method, revealing that, after treatment at 240 °C, phosphorus, potassium, and heavy metals remained in the solid phase. The effluent contained ammonium nitrogen at 2800 mg/L, amino acids near 3000 mg/L, humic acid derivatives at 2400 mg/L, and acetic acid at 3900 mg/L. Acetic acid aids in seed germination, while the other components are crucial for root growth. Thus, the solid products from sludge treatment have potential as soil amendments. Applying these products to impoverished soils can enhance organic matter content and promote plant growth. Their study demonstrated that diluted effluent significantly improved germination and development of three plant species, especially scallions, with a germination index exceeding 200% [42]. It is important to strictly control pollutant levels during application, employing methods such as filtration, sedimentation, and biological treatment to remove harmful substances and ensure compliance with standards.

(2) Carbon Source

Under hydrothermal conditions, sludge undergoes hydrolysis and oxidation, generating small organic acids such as acetic acid, formic acid, and oxalic acid as primary products, which can serve as carbon sources in wastewater treatment. Xu et al. noted that using a Cu-Ce/γ-Al_2O_3 catalyst at an initial oxygen pressure of 1.0 MPa and 10 g/L of catalyst allowed the exothermic wet oxidation reaction to sustain itself. The produced low molecular weight carboxylic acids have potential commercial applications as organic carbon sources in biological wastewater treatment [22]. Similarly, Qin et al. found that hydrothermally treated coal chemical sludge yielded primarily acetic acid, formic acid, and oxalic acid, which can also serve as carbon sources in wastewater treatment. Therefore, wet oxidation can facilitate the resource recovery of organic carbon [35]. Additionally, research by Chu et al. indicates that sludge treated via catalytic hydrothermal oxidation has potential as a carbon source, as the degradation products mainly include acetic acid and propionic acid. Biological treatment experiments showed that the degradation liquid after catalytic wet oxidation has the potential to meet the carbon source requirements for biological treatment, aiding in the removal of COD and TN [15]. This work confirms the effectiveness of the catalyst in treating antibiotic-laden sludge, providing new insights for the rational disposal of such sludge.

2.7.3. Gaseous Phase Products

The gaseous products primarily consist of carbon dioxide (CO_2), nitrogen (N_2), and other small molecular gases, possibly including trace amounts of combustible gases such as methane and hydrogen. These gases can be collected and purified for various applications. For instance, the produced carbon dioxide can be utilized for greenhouse gas management, as a resource in the production of carbonated beverages, or dry ice, or to enhance plant photosynthesis. Nitrogen, being an inert gas, finds use in packaging, metal processing, and the electronics industry. Additionally, the gases generated during hydrothermal oxidation can be harnessed for thermal energy recovery, providing energy for electricity generation and heating within the plant, thereby reducing overall energy consumption while generating economic benefits. For example, Zhang et al. found that, during wet air oxidation (WAO) of digested sludge, insoluble organic matter is converted into volatile fatty acids (VFAs), ammonia, nitrates, nitrites, organic nitrogen, and carbon dioxide. The study also indicated that the WAO method can decompose digested sludge, with subsequent anaerobic digestion (AD) further producing methane [42].

Overall, the solid-phase products of sludge treated with HTO technology are primarily inorganic compounds. The liquid-phase products mainly include non-volatile fatty acids (with acetic acid as the main component), aldehydes, ketones, and aromatic compounds, while the gas-phase products consist mostly of carbon dioxide and nitrogen. Under ideal reaction conditions, these intermediate products can eventually be converted into acetic acid and carbon dioxide. However, in engineering practice, excessively pursuing organic conversion can significantly increase energy consumption and operational costs, making it impractical for real-world production needs. As a result, achieving the ideal outcomes may not always be feasible.

This highlights the necessity of predicting the products of sludge treatment under different conditions, designing well-planned treatment schemes, and preventing the formation of unexpected byproducts. For example, Arrian Prince-Pike developed a detailed kinetic model for the wet oxidation of municipal sludge. This model can predict the concentrations of intermediate products that are critical to the biological nutrient removal (BNR) process under various reaction conditions, providing insights into the concentration of key intermediates during wet oxidation [71].

Therefore, to successfully apply HTO technology in industrial settings, continuous experimental research is essential to further refine and optimize the technology.

3. Conclusions and Prospect

Hydrothermal oxidation technology shows remarkable promise in the field of sludge treatment. Current research indicates that this technology offers a highly potential solution to various sludge disposal challenges. On one hand, hydrothermal oxidation effectively achieves sludge reduction and harmless treatment. By conducting reactions under optimized conditions, it rapidly decomposes organic pollutants in the sludge and enhances dewatering performance, thereby addressing land occupation issues associated with sludge. The removal of harmful substances from the sludge also mitigates risks for subsequent disposal and utilization, transforming sludge from waste into useful materials—an essential consideration in today's increasingly strained land resources. On the other hand, resource recovery stands out as a significant advantage of hydrothermal oxidation in sludge treatment. The products recovered after treatment can be repurposed as building materials, carbon sources, or fertilizers, contributing to successful waste utilization and promoting resource recycling.

However, hydrothermal oxidation technology is not without its challenges. Several issues remain: (1) Cost: The implementation costs of these emerging technologies are currently high, particularly for industrial applications, which require substantial investment in equipment and operational expenses, potentially limiting widespread adoption. (2) Technological maturity: Compared to traditional wastewater treatment technologies, hydrothermal oxidation is still in a developmental stage, lacking sufficient long-term industrial application data to support its broader use. (3) Byproduct management: During the HTO process, the addition of certain additives may lead to the formation of toxic intermediate products. Additionally, some catalysts contain heavy metals that can leach into the products due to losses during use. These byproducts can hinder the subsequent resource recovery and utilization of the treated sludge.

In light of the future development of this research, several prospects and key areas of focus are proposed:

(1) Achieving sludge self-catalysis: Given the excellent performance of CWO, future research should focus on developing new catalysts that promote the hydrothermal oxidation

of sludge. Beyond enhancing the catalyst's function, it is also crucial to explore whether the catalyst can enable the sludge to self-catalyze during the CWO process. This would allow the sludge itself to promote the removal of harmful substances, such as organic pollutants and heavy metals.

(2) Prediction, planning, and control of products: Developing precise techniques for detecting product composition, such as advancements in liquid and gas chromatography, is essential. Alternatively, new kinetic models can be designed to accurately analyze the treated products. Furthermore, when utilizing these products as resources, it is necessary to test relevant parameters to ensure they meet usage standards. For example, the concentration of heavy metals must remain below the acceptable thresholds.

(3) Elimination of ammonia nitrogen: At present, there is a large amount of ammonia nitrogen in the liquid phase of sludge treated by hydrothermal oxidation process, so it is very important to solve the issue regarding the potential harm of ammonia nitrogen. Future efforts should focus on further eliminating ammonia nitrogen and potentially converting it into nitrogen gas. This would improve the safety of sludge treatment and enhance the decontamination process.

(4) Improving sludge dewatering performance: While hydrothermal oxidation technology can improve sludge dewatering, the results are not yet optimal. Future research should investigate how to enhance the dewatering efficiency to achieve more effective sludge reduction.

(5) Producing a single carboxylic acid: A key finding of this study is that hydrothermal treatment of sludge can produce various volatile fatty acids such as acetic acid and propionic acid, along with non-volatile acids like lactic acid. This raises the question of whether it is possible to regulate the sludge treatment process to produce a single type of carboxylic acid. For example, if only acetic acid were produced, it could be entirely used as a carbon source. Moreover, the detection technology for different acids in the sludge effluent requires further optimization. Advanced chromatographic techniques could be employed to distinguish and separate these acids more effectively.

(6) Cost reduction: Successfully treated sludge products can be utilized for resource recovery, generating economic benefits and reducing overall costs. Additionally, using inexpensive oxidants and catalysts, as well as developing new technologies to enhance treatment efficiency, can further lower costs.

(7) Diversifying applications: Currently, the applications of treated sludge are somewhat limited, mainly focused on construction materials or soil conditioners. Further exploration is needed in areas such as electricity generation, hydrogen production, and methane production. Expanding the applications of sludge products and forming a comprehensive industrial chain would maximize the utility of treated sludge.

Additionally, equipment innovation and improvement will be a key focus of future research to enhance stability, reliability, and cost-efficiency. Looking ahead, hydrothermal oxidation technology has the potential to play an increasingly important role in sludge treatment. As the technology continues to evolve, costs will decrease, and efficiency will improve. Furthermore, combining this technology with other advanced methods could open new possibilities for sludge treatment. In the near future, hydrothermal oxidation is expected to become a mainstream technology in sludge management, contributing significantly to environmental protection and sustainable development.

Author Contributions: Conceptualization, H.Y. and Y.L.; methodology, H.Y., Y.L. and N.G.; software, Z.P. and W.P.; validation, J.W., L.W. and B.Z.; formal analysis, B.Z. and Y.Z.; investigation, J.W. and Y.Z.; resources, L.W.; writing—original draft preparation, H.Y., Y.L. and N.G.; writing—review and editing, W.P., H.Y. and H.W.; visualization, Z.P.; supervision, H.Y.; project administration, H.W.; funding acquisition, H.Y. and H.W. All authors have read and agreed to the published version of the manuscript.

Funding: This research was funded by the National Natural Science Foundation of China (No. 52370127), Young Elite Scientists Sponsorship Program by China Association for Science and Technology (No. 2022QNRC001), Project LJ222410151015, supported by Fundamental Research Funds of higher education Department of Liaoning Province (Program for Liaoning Innovative Research Team in University), and Fundamental Research Funds for the Central Universities (No. 3132023518). This work was financially supported by the National Key R&D Program of China (2023YFE0101000) and Innovation Research Fund of Dalian Institute of Chemical Physics (DICP I202320).

Data Availability Statement: The authors confirm that the data supporting the findings of this study are available from the authors upon reasonable request.

Conflicts of Interest: Author Libo Wu and Jinling Wan were employed by the company Hangzhou Sunrise Environment Co., Ltd. The remaining authors declare that the research was conducted in the absence of any commercial or financial relationships that could be construed as a potential conflict of interest.

References

1. Khalidi-Idrissi, A.; Madinzi, A.; Anouzla, A.; Pala, A.; Mouhir, L.; Kadmi, Y.; Souabi, S. Recent advances in the biologica treatment of wastewater rich in emerging pollutants produced by pharmaceutical industrial discharges. *Int. J. Environ. Sci Technol.* **2023**, *20*, 11719–11740. [CrossRef] [PubMed]
2. Filipovic, K.; Petrovic, M.; Najdanovic, S.; Velinov, N.; Hurt, A.; Bojic, A.; Kostic, M. Highly efficient nano sorbent as a superio material for the purification of wastewater contaminated with anthraquinone dye RB19. *J. Water Process Eng.* **2024**, *67*, 106118 [CrossRef]
3. Radovic, M.D.; Mitrovic, J.Z.; Kostic, M.M.; Bojic, D.V.; Petrovic, M.M.; Najdanovic, S.M.; Bojic, A.L. Comparison of ultraviolet radiation/hydrogen peroxide, Fenton and photo-Fenton processes for the decolorization of reactive dyes. *Hem. Ind.* **2015**, *69* 657–665. [CrossRef]
4. Mitrovic, J.; Vučić, R.M.; Kostic, M.; Velinov, N.; Najdanovic, S.; Bojic, D.; Bojic, A. Sulfate radicals-based degradation of antraquinone textile dye in a plug flow photoreactor. *J. Serbian Chem. Soc.* **2019**, *84*, 1041–1054. [CrossRef]
5. Petrović, M.; Rančev, S.; Đorđević, M.P.; Najdanović, S.; Velinov, N.; Vučić, M.R.; Bojić, A. Electrochemically synthesized Molybdenum oxides for enhancement of atmospheric pressure non-thermal pulsating corona plasma induced degradation of ar organic compound. *Chem. Eng. Sci.* **2021**, *230*, 116209. [CrossRef]
6. Petrović, M.; Jovanović, T.; Rančev, S.; Kovač, J.; Velinov, N.; Najdanović, S.; Kostić, M.; Bojić, A. Plasma modified electrosynthe sized cerium oxide catalyst for plasma and photocatalytic degradation of RB 19 dye. *J. Environ. Chem. Eng.* **2022**, *10*, 107931 [CrossRef]
7. Najdanović, S.M.; Petrović, M.M.; Slipper, I.J.; Kostić, M.M.; Prekajski, M.D.; Mitrović, J.Z.; Bojić, A.L. A New Photocatalyst Bismuth Oxo Citrate: Synthesis, Characterization, and Photocatalytic Performance. *Water Environ. Res.* **2018**, *90*, 719–728 [CrossRef] [PubMed]
8. Liang, C.-Z.; Sun, S.-P.; Li, F.-Y.; Ong, Y.-K.; Chung, T.-S. Treatment of highly concentrated wastewater containing multiple synthetic dyes by a combined process of coagulation/flocculation and nanofiltration. *J. Membr. Sci.* **2014**, *469*, 306–315. [CrossRef]
9. Haque, M.M.; Smith, W.T.; Wong, D.K.Y. Conducting polypyrrole films as a potential tool for electrochemical treatment of azo dyes in textile wastewaters. *J. Hazard. Mater.* **2015**, *283*, 164–170. [CrossRef] [PubMed]
10. Liu, P.; Zhang, Y.; Tang, H.; Wang, Y.M.; Sun, X.D. Prevalence of Streptococcus suis in pigs in China during 2000–2021: A systematic review and meta-analysis. *One Health* **2023**, *16*, 100513. [CrossRef]
11. Gong, W.J.; Huang, L.; Ji, Y.; Lv, J.J.; Guo, L.A.; Li, Y.; Li, J.X.; Ren, Y.L.; Zhang, J.J. Study on the reuse of municipal sludge treated with wet oxidation coupled with alkali hydrolysis. *Desalination Water Treat.* **2023**, *294*, 40–50. [CrossRef]
12. Hou, J.C.; Hong, C.; Ling, W.; Hu, J.S.; Feng, W.B.; Xing, Y.; Wang, Y.J.; Zhao, C.W.; Feng, L.H. Research progress in improving sludge dewaterability: Sludge characteristics, chemical conditioning and influencing factors. *J. Environ. Manag.* **2024**, *351*, 119863 [CrossRef]
13. Ding, A.; Zhang, R.R.; Ngo, H.H.; He, X.; Ma, J.; Nan, J.; Li, G.B. Life cycle assessment of sewage sludge treatment and disposal based on nutrient and energy recovery: A review. *Sci. Total Environ.* **2021**, *769*, 144451. [CrossRef] [PubMed]
14. Hušek, M.; Moško, J.; Pohořelý, M. Sewage sludge treatment methods and P-recovery possibilities: Current state-of-the-art. *J. Environ. Manag.* **2022**, *315*, 115090. [CrossRef] [PubMed]
15. Chu, S.Y.; Lin, H.; Zeng, X. The Treatment of Antibiotic Excess Sludge via Catalytic Wet Oxidation with Cu-Ce/γ-Al_2O_3 and the Production of a Carbon Source. *Water* **2024**, *16*, 1249. [CrossRef]
16. Fang, K.Y.; Zeng, X.; Yao, G.D.; Xia, S.Q.; Zhao, J.F.; Zhou, Y.Y.; Zhu, Y.T.; Li, X.J.; Qu, C.J. Wet Oxidation of Pharmaceutical Sludge for the Treatment and Production of Value-Added Materials. *Processes* **2023**, *11*, 2747. [CrossRef]
17. Bertanza, G.; Galessi, R.; Menoni, L.; Salvetti, R.; Slavik, E.; Zanaboni, S. Wet oxidation of sewage sludge: Full-scale experience and process modeling. *Environ. Sci. Pollut. Res.* **2015**, *22*, 7306–7316. [CrossRef] [PubMed]
18. Liu, Z.M.; Yin, D.N.; Ge, H.J.; Zhou, G.; Song, T. Techno-economic analysis of municipal sludge co-combustion power generation with different fuels. *J. Environ. Chem. Eng.* **2023**, *11*, 111593. [CrossRef]
19. Guo, Y.Q.; Gong, H.; Shi, W.J.; Fang, N.; Tan, Y.Q.; Zhou, W.Q.; Huang, J.L.; Dai, L.L.; Dai, X.H.; Guo, Y.L. Insights into multisource sludge distributed in the Yangtze River basin, China: Characteristics, correlation, treatment and disposal. *J. Environ. Sci.* **2023**, *126*, 321–332. [CrossRef] [PubMed]

0. Chen, D.D.; Dou, Y.H.; Tang, Q.; Huang, Y.J.; Song, M.; Peng, F.R. New Insight on the Combined Effects of Hydrothermal Treatment and $FeSO_4/Ca(ClO)_2$ Oxidation for Sludge Dewaterability Improvement: Moisture Distribution and Noncovalent Interaction Calculation. *ACS OMEGA* **2020**, *5*, 15891–15900. [CrossRef]
1. Guo, J.Y.; Zhou, Y.L. Transformation of heavy metals and dewaterability of waste activated sludge during the conditioning by Fe^{2+}-activated peroxymonosulfate oxidation combined with rice straw biochar as skeleton builder. *Chemosphere* **2020**, *238*, 124628. [CrossRef]
2. Zeng, X.; Liu, J.; Zhao, J. Wet oxidation and catalytic wet oxidation of pharmaceutical sludge. *Sci. Rep.* **2023**, *13*, 2544. [CrossRef] [PubMed]
3. Duan, Y.H.; Gao, N.B.; Sipra, A.T.; Tong, K.; Quan, C. Characterization of heavy metals and oil components in the products of oily sludge after hydrothermal treatment. *J. Hazard. Mater.* **2022**, *424*, 127293. [CrossRef] [PubMed]
4. Chen, R.J.; Dai, X.H.; Dong, B. Mechanism insights into hydrothermal-activated tannic acid (TA) for simultaneously sewage sludge deep dewatering and antibiotics removal. *Water Res.* **2024**, *256*, 121619. [CrossRef] [PubMed]
5. Kossińska, N.; Grosser, A.; Kwapińska, M.; Kwapiński, W.; Ghazal, H.; Jouhara, H.; Krzyżyńska, R. Co-hydrothermal carbonization as a potential method of utilising digested sludge and screenings from wastewater treatment plants towards energy application. *Energy* **2024**, *299*, 131456. [CrossRef]
6. Thomsen, L.B.S.; Anastasakis, K.; Biller, P. Wet oxidation of aqueous phase from hydrothermal liquefaction of sewage sludge. *Water Res.* **2022**, *209*, 117863. [CrossRef] [PubMed]
7. Qiu, Y.; Zhang, F.; Yuan, Y.; Zhao, Y.; Liu, Y.; Rong, W. Thermodynamic and economic comparisons of supercritical water oxidation and gasification of oily sludge under hydrothermal flames. *Int. J. Hydrogen Energy* **2024**, *85*, 571–585. [CrossRef]
8. Jiang, G.Y.; Xu, D.H.; Hao, B.T.; Liu, L.; Wang, S.Z.; Wu, Z.Q. Thermochemical methods for the treatment of municipal sludge. *J. Clean. Prod.* **2021**, *311*, 127811. [CrossRef]
9. Malhotra, M.; Garg, A. Characterization of value-added chemicals derived from the thermal hydrolysis and wet oxidation of sewage sludge. *Front. Environ. Sci. Eng.* **2021**, *15*, 13. [CrossRef]
30. Pola, L.; Fernández-García, L.; Collado, S.; Oulego, P.; Díaz, M. Heavy metal solubilisation during the hydrothermal treatment of sludge. *J. Environ. Manag.* **2021**, *286*, 112120. [CrossRef]
31. Zhang, X.; Li, X.X.; Li, R.; Wu, Y.L. Hydrothermal Carbonization and Liquefaction of Sludge for Harmless and Resource Purposes: A Review. *Energy Fuels* **2020**, *34*, 13268–13290. [CrossRef]
32. Leng, L.J.; Zhang, W.J.; Leng, S.Q.; Chen, J.; Yang, L.H.; Li, H.L.; Jiang, S.J.; Huang, H.J. Bioenergy recovery from wastewater produced by hydrothermal processing biomass: Progress, challenges, and opportunities. *Sci. Total Environ.* **2020**, *748*, 142383. [CrossRef] [PubMed]
33. Wang, L.P.; Chang, Y.Z.; Li, A.M. Hydrothermal carbonization for energy-efficient processing of sewage sludge: A review. *Renew. Sustain. Energy Rev.* **2019**, *108*, 423–440. [CrossRef]
34. Huber-Benito, D.; Martin-Martinez, M.; Larriba, M.; Agueda, I.; García, J. Enhanced prednisone removal by catalytic wet air oxidation using sewage sludge derived catalyst. *Process. Saf. Environ. Prot.* **2024**, *190*, 475–485. [CrossRef]
35. Qin, S.L.; Wang, Z.Q.; Zheng, W.C.; Zeng, X. Effects of reaction conditions on the wet oxidation of excess sludge from the caprolactam wastewater treatment process. *Water Sci. Technol.* **2023**, *88*, 2491–2498.
36. Li, J.N.; Wang, S.Z.; Li, Y.H.; Jiang, Z.H.; Xu, T.T.; Zhang, Y.S. Supercritical water oxidation and process enhancement of nitrogen-containing organics and ammonia. *Water Res.* **2020**, *185*, 116222. [CrossRef] [PubMed]
37. Xu, H.; He, P.; Gu, W.; Wang, G.; Shao, L. Recovery of phosphorus as struvite from sewage sludge ash. *J. Environ. Sci.* **2012**, *24*, 1533–1538. [CrossRef]
38. Wang, Z.Q.; Qin, S.L.; Zheng, W.C.; Lou, X.D.; Zeng, X.; Wu, T.H. The Catalytic Wet Oxidation of Excess Activated Sludge from a Coal Chemical Wastewater Treatment Process. *Catalysts* **2023**, *13*, 1352. [CrossRef]
39. Chen, D.D.; Dou, Y.H.; Tang, Q.; Huang, Y.J.; Song, M.; Zhou, J.; Fu, L.C. New insight on the combined effects of hydrothermal treatment and $FeSO_4/Ca(ClO)_2$ oxidation for sludge dewaterability improvement: From experimental to theoretical investigation. *Fuel Process. Technol.* **2020**, *197*, 106196. [CrossRef]
40. Fang, K.Y.; Yao, G.D.; Zhou, Y.Y.; Zhao, J.F.; Xia, S.Q. Effects of reaction temperature and oxygen pressure on dissolved organic matter in hydrothermal reactions of municipal sludge: A comprehensive analysis. *Chem. Eng. J.* **2024**, *495*, 153279. [CrossRef]
41. Núñez, D.; Zabatta, M.; Oulego, P.; Collado, S.; Riera, F.A.; Díaz, M. Production and purification of organic acids from partially oxidised waste activated sludge. *Sep. Purif. Technol.* **2024**, *343*, 126815. [CrossRef]
42. Zhang, Y.; Li, H.; Li, D. Maximize methane recovery from sludge anaerobic digestion by combining an optimal wet air oxidation process. *Renew. Energy* **2021**, *179*, 359–369. [CrossRef]
43. Wang, Y.; Shi, Y.F.; Fang, L.Y.; Wang, Z.P.; Wu, P.; Yang, X.; Shi, X.; Pi, K.W. Characteristics and aging of microplastics in waste activated sludge under persulfate and hydrothermal co-treatment: Impact of solid content and temperature. *Environ. Pollut.* **2024**, *358*, 124498. [CrossRef] [PubMed]
44. Zhang, Y.-F.; Zhang, S.-Y.; Li, H.; Wang, C.-W.; Jiang, F.-H.; Lyu, J.-F. Treatment of municipal sludge by hydrothermal oxidation process with H_2O_2. *Chemosphere* **2020**, *257*, 127140. [CrossRef]
45. Liu, J.Y.; Wang, L.; Li, D.B.; Li, H. Nutritional ingredients and fertilizer potential of the dewatering effluent after sludge wet air oxidation. *Environ. Technol. Innov.* **2023**, *32*, 103443. [CrossRef]

46. Chu, S.Y.; Zeng, X.; Lin, H.; Zhu, Y.T. High-Performance Catalytic Wet Oxidation of Excess Activated Sludge Derived from Pharmaceutical Wastewater Treatment Process over a Cu/γ-Al_2O_3 Catalyst. *Water* **2023**, *15*, 3494. [CrossRef]
47. Zhu, Y.T.; Zeng, X.; Fang, K.Y. Enhanced Wet Oxidation of Excess Sludge from Pharmaceutical Wastewater Treatment by NaOH. *Catalysts* **2023**, *13*, 1070. [CrossRef]
48. Yan, M.; Chen, C.; Zhong, L.; Hantoko, D.; Kanchanatip, E. Experimental study on the catalytic supercritical water oxidation of oilfield sludge. *Process. Saf. Environ. Prot.* **2022**, *168*, 582–590. [CrossRef]
49. Xiong, Q.; Xia, J.; Wu, X.; Wu, X.; Hou, H.; Lv, H. Influence of persulfate on transformation of phosphorus and heavy metals for improving sewage sludge dewaterability by hydrothermal treatment. *Environ. Sci. Pollut. Res.* **2022**, *29*, 33252–33262. [CrossRef]
50. Zhang, W.M.; Guo, Y.; Ding, Q.Y.; Jin, J.X.; Xi, Y.Y.; Lin, X.F. The degradation of maleic acid with wet peroxide oxidation catalyzed by Al_2O_3-supported Cu catalyst: Effect of inorganic ions. *React. Kinet. Catal. Lett.* **2024**, *137*, 3175–3189. [CrossRef]
51. Zhang, W.; Zhuang, H.; Guo, Y.; Chi, H.; Ding, Q.; Wang, L.; Xi, Y.; Lin, X. Wet peroxide oxidation process catalyzed by Cu/Al_2O_3: Phenol degradation and Cu^{2+} dissolution behavior. *Environ. Sci. Pollut. Res.* **2024**, *31*, 26916–26927. [CrossRef] [PubMed]
52. Wei, W.X.; Cui, H.Y.; Zhao, J.Y.; Song, F.; Wang, J.H. g-C3N4 Coating Enhanced Cu@AC for Catalytic Wet Air Oxidation of Phenol under Mild Conditions. *Chemistryselect* **2024**, *9*, e202304800. [CrossRef]
53. Hun, T.C.; Zhao, B.X.; Zhu, T.T.; Liu, L.X.; Li, Z.L.; Sun, L. Catalytic wet peroxide oxidation degradation of magenta wastewater and preparation of FeOCl/montmorillonite. *Water Sci. Technol.* **2021**, *84*, 596–608. [CrossRef] [PubMed]
54. Jiao, Z.J.; Zhang, X.M.; Gong, H.F.; He, D.L.; Yin, H.; Liu, Y.Q.; Gao, X. CuO-doped Ce for catalytic wet peroxide oxidation degradation of quinoline wastewater under wide pH conditions. *J. Ind. Eng. Chem.* **2022**, *105*, 49–57. [CrossRef]
55. Feng, B.X.; Hao, L.N.; Deng, C.T.; Wang, J.Q.; Song, H.B.; Xiao, M.; Huang, T.T.; Zhu, Q.H.; Gai, H.J. A highly hydrothermal stable copper-based catalyst for catalytic wet air oxidation of m-cresol in coal chemical wastewater. *Chin. J. Chem. Eng.* **2023**, *57*, 338–348. [CrossRef]
56. Li, G.Q.; Chai, S.Q.; Zhang, G.J.; Liu, J.; Zhang, Y.F.; Lv, Y.K.; Wang, Y.; Zhao, Y.Q. Deactivation characteristics of Ce-modified Cu-based carbon materials for catalytic wet air oxidation of phenol wastewater. *J. Environ. Chem. Eng.* **2022**, *10*, 108228. [CrossRef]
57. Ding, Z.T.; Liu, C.; Yang, B.G.; Ding, C.; Mao, S.; Shi, M.X.; Hong, X.Y.; Wang, F.Y.; Xia, M.Z. The efficient degradation of high concentration phenol by Nitrogen-doped perovskite La_2CuO_4 via catalytic wet air oxidation: Experimental study and DFT calculation. *Sep. Purif. Technol.* **2023**, *322*, 124310. [CrossRef]
58. Huang, H.X.; Zhang, H.P.; Yan, Y. Preparation of novel catalyst-free Fe_3C nanocrystals encapsulated NCNT structured catalyst for continuous catalytic wet peroxide oxidation of phenol. *J. Hazard. Mater.* **2021**, *407*, 124371. [CrossRef]
59. Zhao, B.; Gao, Y.; Hun, T.; Fan, X.; Shao, N.; Chen, X. Preparation of $PrFe_xCo_{1-x}O_3$/Mt catalyst and study on degradation of 2-hydroxybenzoic acid wastewater by catalytic wet peroxide oxidation. *Chin. J. Chem. Eng.* **2024**, *65*, 286–297. [CrossRef]
60. Görmez, O. Mineralization of clofibric acid by persulfate-promoted catalytic subcritical water oxidation process using $CoFe_2O_4$@SiO_2 catalyst. *J. Mater. Sci.* **2024**, *59*, 12793–12811. [CrossRef]
61. Zhao, P.; Zhao, Y.W.; Guo, Y.P.; Guo, R.; Tian, Y.M.; Zhao, W.G. Preparation of CuO/γAl_2O_3 catalyst for degradation of azo dyes (reactive brilliant red X–3B): An optimization study. *J. Clean. Prod.* **2021**, *328*, 129624. [CrossRef]
62. Yu, L.; Chang, H.; Wu, S.; Zhu, J.; Zhao, Y.; Wang, L.; Wei, H. Efficient degradation of m-cresol during catalytic wet peroxide oxidation with biochar derived from the pyrolysis of persulfate-ZVI treated sludge. *J. Environ. Manag.* **2022**, *324*, 116388. [CrossRef] [PubMed]
63. Wang, R.C.; Zhu, W.; Zhao, S.; Cao, J. Hydrothermal oxidation-precipitation method for recovering phosphorus from dewatered sludge and the mechanisms involved. *Sep. Purif. Technol.* **2022**, *298*, 121580. [CrossRef]
64. Ning, H.; Zhai, Y.B.; Li, S.H.; Liu, X.M.; Wang, T.F.; Wang, B.; Liu, Y.L.; Qiu, Z.Z.; Li, C.T.; Zhu, Y. Fe(II) activated persulfate assisted hydrothermal conversion of sewage sludge: Focusing on nitrogen transformation mechanism and removal effectiveness. *Chemosphere* **2019**, *244*, 125473. [CrossRef] [PubMed]
65. Chen, Z.; Zheng, Z.; He, C.; Liu, J.; Zhang, R.; Chen, Q. Oily sludge treatment in subcritical and supercritical water: A review. *J. Hazard. Mater.* **2022**, *433*, 128761. [CrossRef]
66. Haering, K.C.; Daniels, W.L.; Evanylo, G.K. Soybean phytotoxicity from land-applied biosolids. *J. Residuals Sci. Technol.* **2008**, *5*, 1–12.
67. Mato, F.A.; Peña, M.; García-Rodríguez, Y.; Bermejo, M.-D.; Martín, A. Analysis of the Energy Flow in a Municipal Wastewater Treatment Plant Based on a Supercritical Water Oxidation Reactor Coupled to a Gas Turbine. *Processes* **2021**, *9*, 1237. [CrossRef]
68. Kolaczkowski, S.T.; Plucinski, P.; Beltran, F.J.; Rivas, F.J.; McLurgh, D.B. Wet air oxidation: A review of process technologies and aspects in reactor design. *Chem. Eng. J.* **1999**, *73*, 143–160. [CrossRef]
69. Márquez, J.J.R.; Levchuk, I.; Sillanpää, M. Application of Catalytic Wet Peroxide Oxidation for Industrial and Urban Wastewater Treatment: A Review. *Catalysts* **2018**, *8*, 673. [CrossRef]
70. Yamasaki, H.; Kishimoto, K.; Shimada, T.; Kuroki, T.; Kang, J.Y.; Kim, D.; Yagi, T.; Okubo, M. Toward Ideal VOCs and Nanoparticle Emission Control Technology Using a Wet-Type Catalysis Nonthermal Plasma Reactor. *IEEE Trans. Ind. Appl.* **2022**, *58*, 6591–6598. [CrossRef]
71. Prince-Pike, A.; Wilson, D.I.; Baroutian, S.; Andrews, J.; Gapes, D.J. A kinetic model of municipal sludge degradation during non-catalytic wet oxidation. *Water Res.* **2015**, *87*, 225–236. [CrossRef] [PubMed]

applied sciences

MDPI

Article

Clinker-Free Cement: A New Reality

Marcel Demarco de Souza Oliveira, Sara de Carvalho Zago and Fernando Vernilli Junior *

Lorena School of Engineering, University of São Paulo, EEL-USP, Lorena 12602-810, Brazil; marcel.demarco@usp.br (M.D.d.S.O.)
* Correspondence: fernando.vernilli@usp.br

Abstract: The search for binders with a lower environmental impact has grown, especially for those with reduced clinker content or even for formulations not used due to the high CO_2 emissions associated with their production. In this context, the steel industry, which generates a significant amount of waste and co-products, especially steel aggregates, such as blast furnace slag and basic oxygen slag, can be a source of raw materials for cement plants. For this reason, in search of an innovative alternative with low environmental impact, in this work, a clinker-free cement was developed by mixing waste from steel and civil construction. The results of this study demonstrated the possibility of producing a Portland clinker-free cement, using only solid waste, through a mixture of steel slag from blast furnaces, basic oxygen slag and gypsum residue from civil construction, presenting hydraulic properties compatible with national cement CPIII-32 and CPII E-32, and the European cement CSS 32.5 N. The cement mixtures were also characterized as resistant to sulfate attack. It was found that the basic oxygen slag acted as an alkaline activator, thus enhancing the hydration process of the mixture of blast furnace slag and gypsum.

Keywords: granulated blast furnace slag; BOF slag; clinker-free cement; solid waste

1. Introduction

Committed to mitigating environmental impacts, there is a growing search for alternative cement mixtures to Portland cement (PC), which offer compatible or superior performance to the binder. This is justified because the cement industry is responsible for both the large consumption of natural raw materials and the emission of gases, mainly CO_2, that contribute to the greenhouse effect and the emission of particulate matter [1,2].

Important studies seek supplementary cementitious materials (SCM) based on solid waste to partially replace Portland cement and to produce geopolymer or alkali-activated cement that uses solid waste as a substitute for natural raw materials [2–4].

Granulated blast furnace slag (GBFS), a byproduct of pig iron production, is not only an important SCM widely used in the production of Portland cement due to its latent hydraulicity but is also studied in the search for alternative cement as an alkali-activated material. GBFS consists of a material rich in amorphous silica content and aluminum oxide, making them a potential cement alternative material [2].

In terms of cement, supersulfated cement (SSC) stands out for its high-content use of blast furnace slag, which can vary between 70 and 90%, in addition to 10–20% calcium sulfate and up to 5% alkaline activator, for which alkaline hydroxides, clinker or PC can be used [5,6].

SSC is considered a low-carbon cement and environmentally friendly material with superior mechanical and durability performance compared to PC. The formation of ettringite and C-S-H crystals during the hydration process confers SSC a higher late mechanical strength with a lower heat of hydration release, as well as superior chemical resistance, especially against sulfate attacks [2,5,6].

However, it has qualities negatively associated with the use of Portland clinker as an alkaline activator for GBFS. The decomposition of blast furnace slag is achieved in a high alkalinity environment, which commonly occurs through the use of clinker or Portland cement, creating an economic and environmental disadvantage for the binder [5,6].

Basic Oxygen Furnace (BOF) slag, a steelmaking waste generated during steel production, is rich in calcium oxide and hydroxide. It also has similar phases in its composition to that of Portland clinker, such as dicalcium and tricalcium silicates (Ca_2SiO_4 and Ca_3SiO_5), in addition to tetracalcium ferroaluminate ($Ca_4Al_2Fe_2O_{10}$), providing hydraulicity to the slag, although less significantly compared to blast furnace slag and Portland clinker. These characteristics of BOF slag allow the residue to be a possible replacement for the alkaline activator for blast furnace slag in formulations such as SSC [7–10].

In this sense, this research proposed not only not using clinker in the production of cement, but also producing cement mixtures exclusively from industrial waste, composed of blast furnace slag and hydrated gypsum residue, as well as BOF slag as an alkaline activator of the mixtures, to evaluate the mechanical performance and durability of cement mixtures without clinker in comparison to those of PC and SSC.

2. Experimental Program

Cementitious mixture formulations were prepared with different slags and gypsum contents, ranging from 90 to 100% blast furnace slag and gypsum mixture, and 0 to 10% basic oxygen furnace slag. All tests were conducted at Companhia Siderúrgica Nacional (CSN), in Volta Redonda, Brazil.

2.1. Raw Materials

The materials used in this process comprise co-products and waste from various sources. The steel aggregates, granulated blast furnace slag (GBFS), and basic oxygen furnace slag (BOF) were produced at Companhia Siderúrgica Nacional (CSN) (Volta Redonda, Brazil). The samples were gathered following the Brazilian waste sampling standard NBR 10007 [11], also used for calcium sulfate, and were obtained from piles in the CSN Cimentos raw materials receiving yard. The samples were dried and fragmented to obtain particles measuring 2.36 mm in diameter.

The dried BOF slag underwent a process to remove the remaining metallic elements by using stainless steel 3480 Gauss neodymium magnets. Hydrated gypsum waste (GW) from civil construction was utilized to obtain calcium sulfate, including waste from pre-cast elements and drywall sheets.

The conditions for the chemical and mineralogical analyses of the GBFS and BOF slags are presented in Table 1, along with the gypsum waste, determined using X-ray fluorescence (XRF) and X-ray diffractometry (XRD) techniques, and the degree of vitrification.

Table 1. Techniques for characterizing raw materials.

Raw Materials	Preparation	XRF	XRD	Vitrification Degree
GBFS	Grinding up to 2.36 mm	Lithium tetraborate and lithium metaborate fused pellets/Panalytical Spectrometer, model Axios	Rigaku diffractometer, Windmax 1000 model, Cu Kα radiation with 40 KV—20 mA and scan speed of 2°/min	Optical microscopy of transmitted light, in the Zeiss petrographic microscope, model AxioCam ERc 5S
BOF	Metal fraction removal and grinding up to 2.36 mm			
GW				-

The Rietveld refinement method was used to obtain quantitative results from XRD, using an internal standard of 10% rutile to quantify the amorphous phase.

To evaluate the reactivity of slag, slag mortars with a specific surface area of 350 m^2/Kg were molded, using a 20% NaOH solution as mixing water, following NBR 7215 [12] under accelerated curing conditions and normal curing conditions. For accelerated curing, the slag mortars were kept in an oven at 55 °C for 24 h. For normal curing, the slag mortars were kept in climate-controlled room at 23 °C and relative humidity $\geq$ 50% for 48 h. After curing, the compressive strength of the mortars was obtained using a compression testing machine model F-25EX-F-CPILOT, Forney. The specific surface area of the slag was obtained using an Automatic Blaine Permeabilitymeter, model BSA 1 ACMEL.

2.2. Cement Mixtures

Cementitious mixtures with BOF slag were prepared, replacing the GBFS/gypsum mixture, according to the proportions presented in Table 2. Due to the difference in the grindability of ironmaking slags, the cement grinding was carried out by fine grinding the GBFS and GW mixture, with increasing replacements of (GBFS + GW) by ground BOF slag to a Blaine fineness of 550 m^2/Kg.

Table 2. Formulations of cement mixtures.

Cement Mixtures	GBFS + GW [wt%]	BOFS [% p]
M0	100	0
M3	97	3
M6	94	6
M10	90	10

All mixtures were characterized using physical–chemical and mechanical analysis as shown in Table 3.

The thermal behavior of the cement mixtures was evaluated during the hydration period (3, 7, 28, and 91 days) by thermogravimetric analysis. The Netzsch—STA 449 F3 Jupiter equipment was used for simultaneous thermal analyses (TG-DTA/DSC).

To evaluate cement chemical resistance, the linear expansion of mortar bars in sulfate solution was carried out according to standard NBR 13583 [13]. The unmolded and previously measured bars were immersed in a saturated sodium sulfate solution and kept at 40 °C for measurements at 7, 14, 21, 28, 35, and 42 days, in a Solotest extensometer.

Table 3. Characterization techniques of cement mixtures.

Techniques	Reference Standards	Analysis Conditions and Equipment
Air Permeability Method	NBR NM 76 [14]	Blaine air permeability apparatus, manufactured by Acmel, model BSA 1
Particle Size Distribution	n/a	Laser granulometer, CILAS, model 1064L
Paste Alkalinity	n/a	Pastes with a water/cement ratio of 0.31, pH meter Dgimed, model DM2
Normal Consistency of Cement Paste	NBR 16606 [15]	Vicat apparatus, manufactured by Soloteste
Setting Times	NBR NM 65 [16]	
Le Chatelier Expansion	NBR 11582 [17]	Le Chatelier needle moulds, immersed in boiling water and water
Pozzolanic Activity	NBR 5733 [18]	Frattini test
Compressive Strength	NBR 7215 [12]	Specimens were cured for 3, 7, 21, and 28 days, Forney compression testing machine, model F-25EX-F-CPILOT
Cement Paste Preparation [1]	n/a	Water/cement ratio of 0.31, saturated lime curing (3 g/L) for 3, 7, 28, and 91 days
Hydration stoppage [1]	n/a	Immersion in acetone for 2 h and oven drying at 50 °C/30 min

n/a: not applicable as it is a non-standardized test; [1]: performed for thermogravimetric analysis.

3. Results and Discussions

3.1. Raw Materials Results

According to Table 4, the primary chemical components of GBFS are SiO_2, CaO, Al_2O_3, and MgO, which make up more than 90% of the composition. BOF slag, on the other hand, has high levels of CaO and ferric phases, as well as high quaternary basicity indices—IB4 (NBR 16697) [19], and ternary—IB3 (EN 15743) [20]. These high basicity levels of BOF slag help in the chemical activation of GBFS and the formation of hydration products. Gypsum waste's chemical composition consists mainly of CaO and SO_3, which means that its purity is greater than 95%, stoichiometrically calculated.

Table 4. Chemical and mineralogical composition of BOFS and BFS and GW, in mass percentage.

Chemical Constituent	BOF	GBFS	GW	Mineralogical Phases	BOF	GBFS	GW	BOF	GBFS
LOI [A]	0.44	0.38	-	Belite—βC_2S	33.3	-	-	Vitrification degree by microscopy	
Fe_{total}	21.2	0.89	-	Brownmillerite—C_4AF	24.2	-	-		
Fe_2O_3 [B]	10.4	<0.01	-	Wustite—FeO	23.7	-	-	98.4	1.2
FeO [C]	17.5	1.02	-	Calcium oxide	5.6	-	-		
Fe^0 [D]	0.22	0.21	-	Calcite	5.5	0.1	-	Reactivity—Accelerated curing (MPa)	
CaO	42.9	40.4	29.6	Quartz	3.4	0.8	-		
K_2O	0.14	0.4	-	Guellenite C_2AS	1.9	-	-		
SiO_2	11.0	39.0	0.9	Dolomite	1.3	-	-	13.8	4.2
Al_2O_3	4.07	9.89	-	Portlandite—CH	0.8	-	-		
MgO	8.03	6.81	-	Metallic iron	0.3	-	-	Reactivity—Normal curing (MPa)	
SO_3	0.1	0.6	43.6	Gypsum	-	-	91.3		
				Bassanite	-	-	0.4		
B4 > 1.0	4.3	1.6	-	Anhydrite	-	-	1.5	8.1	2.2
B3 > 1.0	4.1	1.2	-	Amorphous phase	22.6	99.1	-		

Note: [A] Test carried out by gravimetry at a temperature of 750 °C; [B] Fe_2O_3 calculated by stoichiometry; [C] Determination performed by volumetry; [D] Determination performed by atomic absorption spectrometry.

The mineralogical gypsum waste composition indicates that it is a byproduct of the hydration process of anhydrous or hemihydrate phases, consisting mainly of gypsum. Additionally, the cement grinding did not result in the removal of water.

BOF slag is composed of crystalline materials such as belite (Ca_2SiO_4—C_2S) and brownmillerite ($Ca_4Al_2Fe_2O_{10}$—C_4AF). It also has a high amount of the wustite phase (FeO) which is typical of BOF slag. These mineral phases can significantly affect the reactivity of the reaction environment, especially when it comes to its hydraulic behavior and the formation of hydrates resulting from the hydration of BOF slag phases. Additionally, BOF slag also functions as an alkaline activator due to the presence of phases such as lime and portlandite.

On the other hand, due to the granulation process of blast furnace slag, the material is mostly amorphous, meaning that it lacks a definite shape and is not crystalline in nature. According to the Rietveld method, the material has a vitreous degree of 99%, indicating that it is predominantly glassy. Hence, it was not possible to identify significant crystalline peaks in the material.

The degree of vitrification of the slags, based on the counting technique using transmitted light optical microscopy, corroborates the XRD results since blast furnace slag presented a predominantly vitreous crystallinity degree, greater than 98%. The BOF slag presented a predominantly crystalline structure attributed to its chemical composition, which had a high content of alkaline oxides, and the function of its slow cooling after leaving the converter.

In terms of reactivity, GBFS displayed better mechanical behavior than BOF slag for both normal and accelerated curing conditions. However, despite having inferior performance, the mortar composed of BOF slag exhibited hydraulic activity, mainly due to the presence of C_4AF and C_2S phases [21].

The utilization of an accelerated method reduced the difference in mechanical resistance between the slags, which led to a faster hydration process. However, although the accelerated curing conditions favored the reactivity of BOF slag, the stability of its crystalline atomic structure plays a crucial role in its reactivity. In the context of cement hydration, the dissolution and crystallization reactions will be slower for this material.

3.2. Cementitious Mixtures Results

The results of the specific surface, alkalinity and setting times of the cement mixtures are shown in Figure 1.

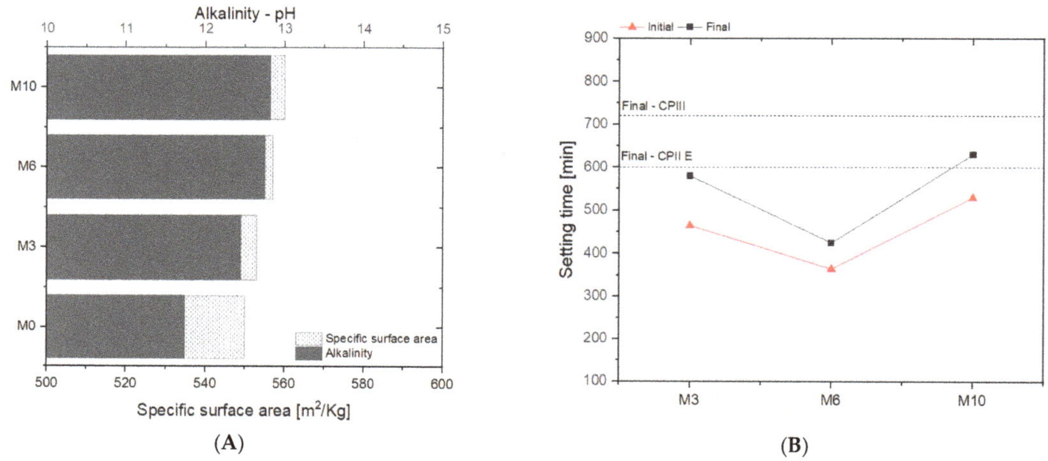

Figure 1. (**A**) Alkalinity and (**B**) setting times of cement mixtures. Note: The cement mixtures were compared to the standardized Brazilian cement CPIII and CPIIE, which have the highest GBFS content, despite being composed of Portland clinker.

The cement mixtures produced in the study had particles with a specific surface area of approximately 550 m^2/kg. This ensured that the cement had a similar water demand, as observed in the results of the normal consistency paste, as they all remained within the range of 29%.

Although the cement mixtures in this study had particles with a similar specific surface area, the hydration speed resulted, in addition to the cement's fineness, from the solution's alkalinity and the chemical composition of the cement mixtures [5].

The hydration process of these cements relies on the decomposition conditions of the GBFS. It is essential to have an alkaline reaction environment as the slag dissolution occurs through a hydroxyl attack. This environment allows for the dissolution of aluminum, calcium, and silicon ions, to react with the sulfate calcium, forming ettringite, hydrated calcium aluminates, and silicates (C-A-H and C-S-H).

In that sense, it has been confirmed (Figure 1A) that the cement blends with BOF slag have higher alkalinity levels when compared to M0. This is due to the high basicity of BOF slag caused by its alkaline oxides. As a result, when these oxides dissolve, the environment becomes more alkaline.

The M0 mixture had a low hydration rate, which made it impossible to characterize setting times. The other mixtures, however, performed well within normative limits for CPIIE and CPIII cements, except for M10, which had a longer final setting time than the normative limit for CPIIE, a behavior attributed to the highest content of BOF slag and the slower slag dissolution and hydrated product formation reaction.

The alkalinity influence effect is also noticeable when evaluating the pozzolanic activity of cement blends. The result of the pozzolanic test, exhibited in Figure 2, determined that all cement exhibit pozzolanic behavior, as $Ca(OH)_2$ is fixed by the formation of C-S-H, reducing the total alkalinity and CaO content in the solution, especially for cements M3, M6 and M10, which demonstrates that the mixtures with BOF slag had greater availability of $Ca(OH)_2$ to react with the silica from GBFS.

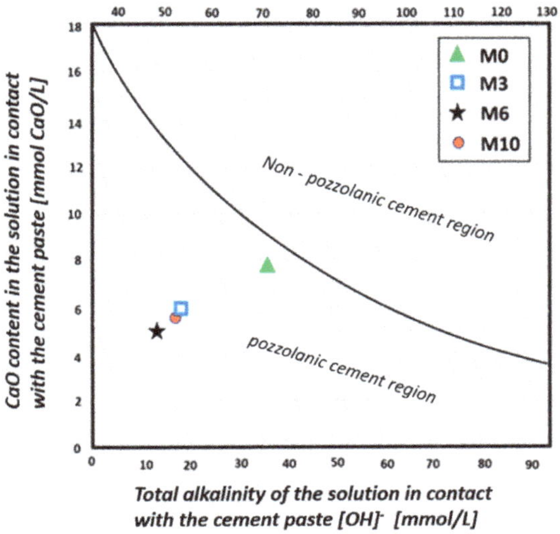

Figure 2. Pozzolanicity of the pastes of the cement mixtures.

The impact of the alkaline activation of BOF slag was also verified through mechanical tests. As shown in Figure 3, the setting time results of M0 are supported by the fact that compressive strength results could not be obtained until after 3 days of curing for the M0 mixture. This is because the specimens did not possess sufficient mechanical resistance to be unmolded.

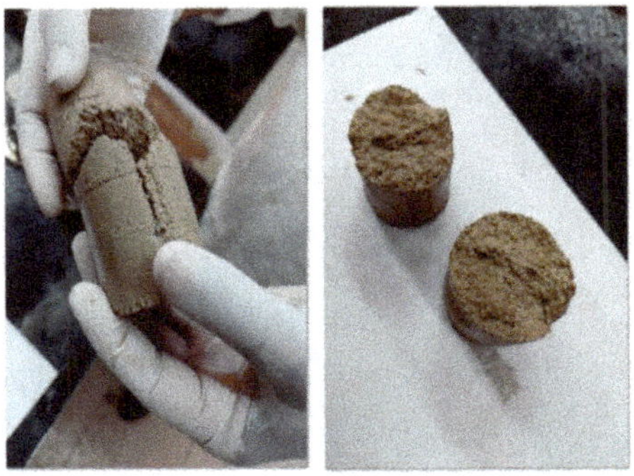

Figure 3. Fragmentation of the specimens during demolding of M0 after 24 h of curing.

The mechanical performance of cement mixtures, evaluated by compressive strength, is shown in Figure 4 and compares them to the normative limits of cement CP III-32 and CPII E-32 [19], and CSS 32.5 N [20], as these may be considered the normalized types of cement most similar to the mixtures produced.

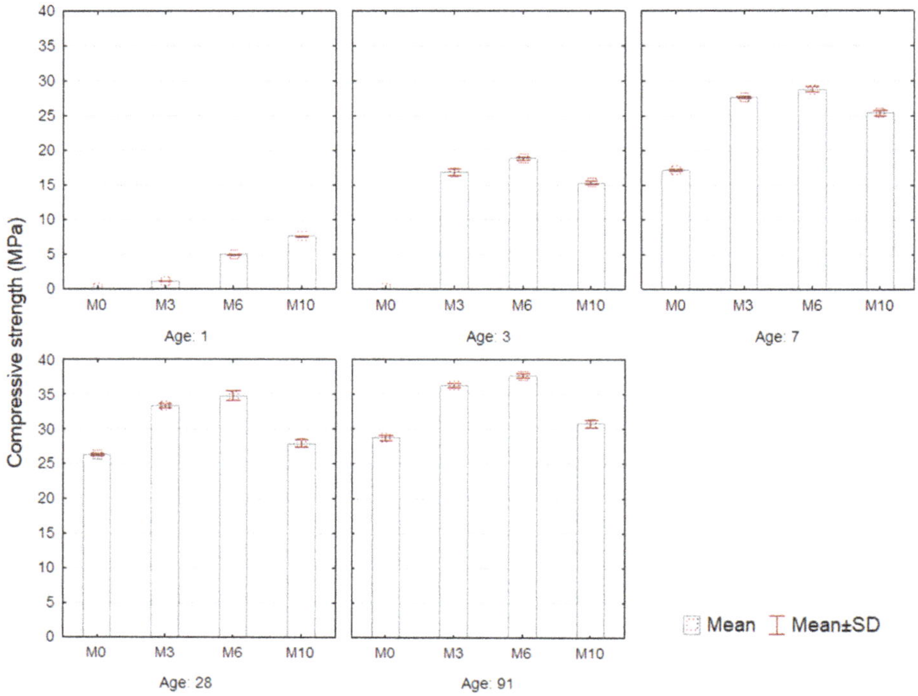

Figure 4. Compressive strength of the cement mixtures.

Unlike M0, all mixtures containing BOF slag showed the development of mechanical strength at the initial curing ages (Figure 4). Additionally, at all ages evaluated, the cementitious mixtures exhibited a mechanical behavior superior to the M0 cement mixture.

The results demonstrate the chemical activating action that the BOF slag provided to the environment through the effect of alkalinity and pozzolanicity. The dissolution of the BOF slag and, consequently, the release of OH^- ions, which act to disrupt the film formed by the C-S-H and C-A-H phases on the surface of the GBFS particle, provided an alkaline environment that boosted the dissolution of GBFS and the formation of products of hydration at early ages, mainly from ettringite [22,23].

In general, mixtures of cement presented a mechanical performance superior to the normative limits required to produce Brazilian—CPIII and European—CSS cement, except the M10 mixture, at 28 days (32 MPa and 32.5 MPa, respectively). This behavior corroborates what was observed for the setting time of the same cement mixture, as shown in Figure 1B.

It is understood that the Basic Oxygen Furnace (BOF) slag content ensures the alkalinity of the reaction medium, enabling the development of an initial mechanical resistance beyond normative limits [24]. However, for this replacement content of 10% BOF slag, the development of mechanical resistance was delayed.

The reduction in the main component of hydraulicity in the mixture, the Ground Granulated Blast Furnace Slag (GBFS), may reduce the mechanical strength at the final ages (R28 and R91). This results from the depletion of $A_{12}O_3$ and SiO_2, which makes it challenging to develop resistances at later ages since the continuous decomposition of the GBFS and the formation of hydration products would lead to the densification of the structure of the hydrated cement matrix.

Figures 5 and 6 show the thermogravimetric curves of the cement mixture pastes after 3, 7, 28, and 91 days of curing to evaluate the hydration process through the amount of chemically combined water referring to the hydrated cement products.

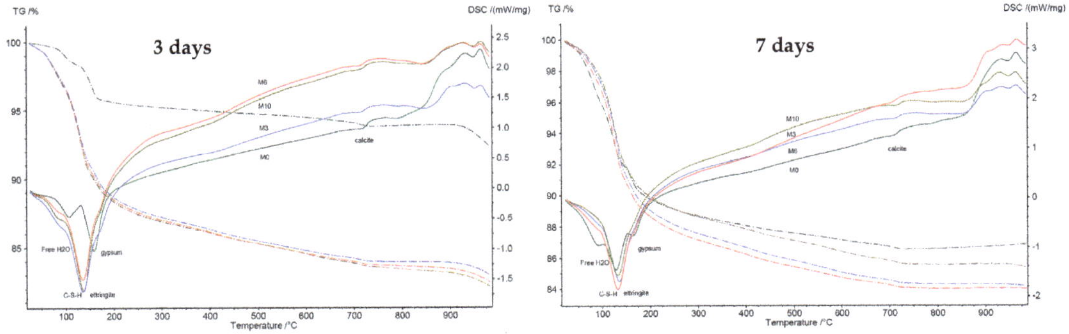

Figure 5. Thermogravimetric curve of cement mixtures after 3 and 7 days of curing.

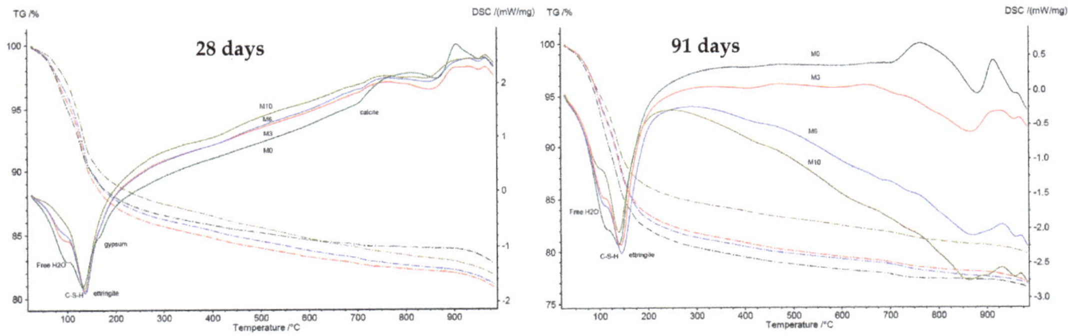

Figure 6. Thermogravimetric curve of cement mixtures after 28 and 91 days of curing.

Three main thermal events stand out in the curves after 3 days of curing, as well as the fact that the mass loss curve of the M0 mixture differs significantly from the others. Initially, up to a temperature close to 110 °C, for mixtures containing BOF slag, there is a loss of mass attributed to the release of free water, and, up to 200 °C, a sequence of events with endothermic peaks that include the decomposition of hydrated compounds.

Close to a temperature of 135 °C, the thermal events are attributed to the decomposition of ettringite, with mass losses increasing due to the additions of BOF slag. However, forM0, this phenomenon was not observed. This result corroborates the lack of results in the setting time and compressive strength tests at ages up to 3 days of curing since the hydration process had not yet generated hydration products and significant resistance to the mortar tested.

Between approximately 160 and 170 °C, gypsum decomposition was observed for all mixtures, including M0, demonstrating the phase's availability to conduct the hydration process and the formation of new hydrated products.

After 7 days of curing (Figure 5), a thermal behavior similar to the first age evaluated was identified for the samples, except for the presence of an endothermic peak that demonstrates the beginning of ettringite decomposition for sample M0, at approximately 130 °C.

After 28 days of curing, characteristic peaks of gypsum decomposition were observed only in samples M0 and M3, at around 170 °C. The ettringite decomposition was observed in all samples, with endothermic peaks of higher intensity for the mixtures with BOF slag. That is, for a higher energy absorption, greater decomposition of the hydrates is attributed to this temperature range, and, consequently, the release of larger amounts of combined water results in gains in the strength of cement mixtures.

The results at 91 days demonstrate that M0 still showed a loss of mass related to the decomposition of gypsum, due to the slow hydration of the cement mixture which, consequently, leads to the slow development of mechanical resistance, proven by the worst mechanical performance among all mixtures cement (28.8 MPa). Compared to other ages, for all cementitious mixtures, the mass losses related to the decomposition of ettringite are more significant, as are the results of compressive strength.

Although the entire discussion was carried out due to the decomposition of ettringite, it is known that in addition to ettringite, hydrated calcium silicates and aluminates are the hydrated products present in the cement matrix that lose chemically combined water in characteristic peaks in the ranges of temperature between 100 and 300 °C [25].

Therefore, the discussion around ettringite was due to this phase being one of the mainly produced hydrates in this cement. However, through TG/DSC techniques, it was possible to verify the evolution of the formation of different cement hydration products over the ages, mainly evidenced in the increase in heat absorption and the broadening of DSC peaks due to the increase in decomposition temperature range.

Figure 7A,B demonstrate the behavior of mixtures containing BOF slag when cured in water and sulfate solution, respectively. It is important to note that the use of BOF slag in construction raises concerns about its tendency to expand. However, it is worth noting that in both cases, the results are within the established normative limits.

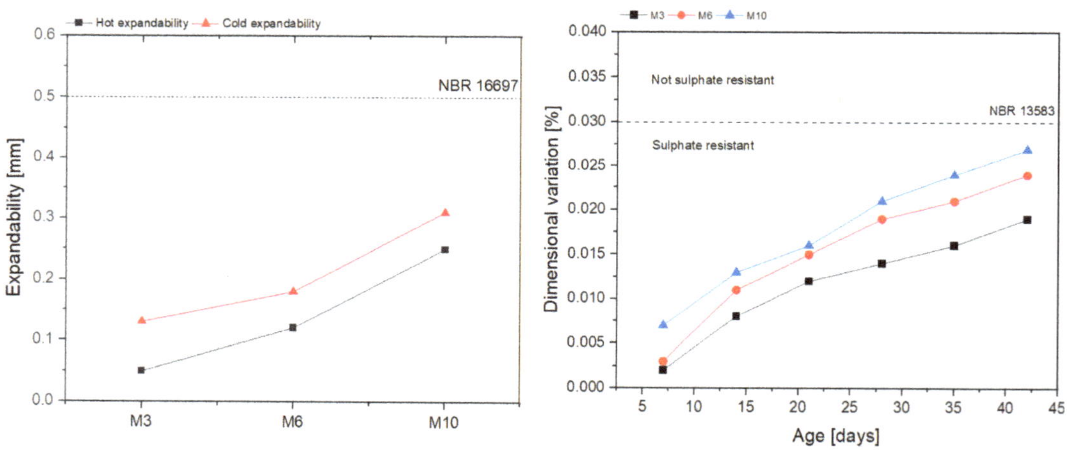

(A) (B)

Figure 7. (**A**) Le Chatelier expandability, and (**B**) dimensional variations of cement bars.

It was found that for both cold and hot curing in water, the fineness of BOF slag and its contribution to the development of pozzolanic reactions provided the mixtures with control over the expandability of free CaO, resulting in expansibility results below the normative limits for slag cement.

In specific sulfate ion attack conditions (Figure 7B), the cementitious mixtures were classified as resistant to sulfate attack with a dimensional variation of lower than 0.03%. The results are attributed to the fact that their formulations do not contain the tricalcium aluminate (C_3A) phase. The absence of this phase prevents the growth of secondary ettringite crystals in the hardened matrix, which could lead to the development of cracks due to generated tensions. Furthermore, the cement mixtures studied do not contain CH as one of their main hydration products, which could undergo leaching due to sulfate attack since GBFS phases tend to react with the alkaline component to form C-S-H (calcium silicate hydrate) [26].

4. Conclusions

This study proposed the production of clinker-free cementitious mixtures using steel slag and hydrated gypsum residue. The results indicated that these cementitious mixtures exhibited hydraulic properties compatible with CPIII-32, CPII E-32, and CSS 32.5 N standards. The increase in strength was attributed to the hydration of blast furnace slag, which was chemically activated by BOF slag and reacted with calcium sulfate. This process facilitated the formation of hydration products such as ettringite, C-S-H and C-A-H, as verified by thermogravimetric analyses. The addition of BOF slag was crucial to the hydration mechanism due to its alkalinity, which accelerated the dissolution and crystallization rates of the hydrated compounds. While gains in compressive strength were observed at both the early and later stages, a 10% replacement of BOF slag resulted in a delay in the development of mechanical strength. No detrimental effects concerning expansion were observed in the cement matrix of any mixture. Furthermore, the cementitious mixtures were classified as sulfate-resistant cement.

Author Contributions: Conceptualization, F.V.J. and M.D.d.S.O.; Formal analysis, F.V.J., M.D.d.S.O. and S.d.C.Z.; Investigation, M.D.d.S.O.; Writing—original draft, S.d.C.Z. All authors have read and agreed to the published version of the manuscript.

Funding: This research received no external funding.

Institutional Review Board Statement: Not applicable.

Informed Consent Statement: Not applicable.

Data Availability Statement: The original contributions presented in the study are included in the article, further inquiries can be directed to the corresponding author.

Conflicts of Interest: The authors declare no conflict of interest.

References

1. Terán-Cuadrado, G.; Tahir, F.; Nurdiawati, A.; Almarshoud, M.A.; Al-Ghamdi, S.G. Current and potential materials for the low-carbon cement production: Life cycle assessment perspective. *J. Build. Eng.* **2024**, *96*, 110528. [CrossRef]
2. Pacios, A.; Tejado, E.; Martín, A.; Fernández, A.; Pastor, J.Y. High-temperature performance of clay-enhanced alkali-activated binders: A sustainable alternative to Portland cement. *Constr. Build. Mater.* **2024**, *453*, 138860. [CrossRef]
3. Li, X.; Wang, D.; Chen, Q.; Qi, C. Alkali activation of blast furnace slag using Bayer red mud as an alternative activator to prepare cemented paste backfill. *Constr. Build. Mater.* **2024**, *456*, 139061. [CrossRef]
4. Maldonado-Alameda, A.; Mañosa, J.; Giro-Paloma, J.; Formosa, J.; Chimenos, J.M. Alkali-Activated Binders Using Bottom Ash from Waste-to-Energy Plants and Aluminium Recycling Waste. *Appl. Sci.* **2021**, *11*, 3840. [CrossRef]
5. Juenger, M.C.G.; Winnefeld, F.; Provis, J.L.; Ideker, J.H. Advances in alternative cementitious binders. *Cem. Concr. Res.* **2011**, *41*, 1232–1243. [CrossRef]
6. Qi, G.; Zhang, Q.; Sun, Z. Effect of Calcium Aluminate and Carbide Slag on Mechanical Property and Hydration Mechanism of Supersulfated Cement. *Buildings* **2024**, *14*, 930. [CrossRef]
7. Carvalho, S.Z.; Vernilli, F.; Almeida, B.; Demarco, M.; Silva, S.N. The recycling effect of BOF slag in the Portland cement properties. *Resour. Conserv. Recycl.* **2017**, *127*, 216–220. [CrossRef]
8. Zepper, J.C.O.; Van Der Laank, S.R.; Schollbach, K.; Brouwers, H.J.H. Reactivity of BOF slag under autoclaving conditions. *Constr. Build. Mater.* **2023**, *364*, 129957. [CrossRef]
9. Costa, H.N.; Noberto, C.C.; Almeida, L.A.; Nogueira, R.E.F.Q.; Cabral, A.E.B. Alkaline activation of cement pastes with desulfurization slag. *Cerâmica* **2021**, *67*, 399–405. [CrossRef]
10. Ahmed, M.J.; Santos, W.F.; Brouwers, H.J.H. Air granulated basic Oxygen furnace (BOF) slag application as a binder: Effect on strength, volumetric stability, hydration study, and environmental risk. *Constr. Build. Mater.* **2023**, *367*, 130342. [CrossRef]
11. *NBR 10007*; Amostragem de Resíduos Sólidos. Associação Brasileira de Normas Técnicas: Rio de Janeiro, Brazil, 2004.
12. *NBR 7215*; Cimento Portland—Determinação da Resistência à Compressão de Corpos de Prova Cilíndricos. Associação Brasileira de Normas Técnicas: Rio de Janeiro, Brazil, 2019.
13. *NBR 13583*; Cimento Portland—Determinação da Variação Dimensional de Barras de Argamassa de Cimento Portland Expostas à Solução de Sulfato de Sódio. NORMATÉCNICA: Rio de Janeiro, Brazil, 2014.
14. *NBR NM 76*; Cimento Portland—Determinação da Finura Pelo Método de Permeabilidade Ao Ar (Método de Blaine). Associação Brasileira de Normas Técnicas: Rio de Janeiro, Brazil, 2015.
15. *NBR 16606*; Cimento Portland—Determinação da Pasta de Consistência Normal. Associação Brasileira de Normas Técnicas: Rio de Janeiro, Brazil, 2018.
16. *NBR NM 65*; Cimento Portland—Determinação do Tempo de Pega. Associação Brasileira de Normas Técnicas: Rio de Janeiro, Brazil, 2017.

17. *NBR 11582;* Cimento Portland—Determinação da Expansibilidade Le Chatelier. Associação Brasileira de Normas Técnicas: Rio de Janeiro, Brazil, 2016.
18. *NBR 5733;* Cimento Portland—Cimento Portland de alta Resistência Inicial. Associação Brasileira de Normas Técnicas: Rio de Janeiro, Brazil, 2018.
19. *NBR 16697;* Cimento Portland—Requisitos. Associação Brasileira de Normas Técnicas: Rio de Janeiro, Brazil, 2018.
20. *EN 15743:2010;* Supersulfated Cement—Composition, Specifications and Conformity Criteria. European Committee for Stardardization—CEN: Brussels, Belgium, 2010.
21. Kaja, A.M.; Melzer, S.; Brouwers, H.J.H.; Yu, Q. On the optimization of BOF slag hydration kinetics. *Cem. Concr. Compos.* **2021**, *124*, 104262. [CrossRef]
22. Wu, Q.; Xue, Q.; Yu, Z. Research status of super sulfate cement. *J. Clean. Prod.* **2021**, *294*, 126228. [CrossRef]
23. Liu, Z.; Ni, W.; Li, Y.; Ba, H.; Li, N.; Ju, Y.; Zhao, B.; Jia, G.; Hu, W. The mechanism of hydration reaction of granulated blast furnace slag-steel slag-refining slag-desulfurization gypsum-based clinker-free cementitious materials. *J. Build. Eng.* **2021**, *44*, 103289. [CrossRef]
24. Demarco, M.; Vernilli, F.; Zago, S.C. Clinker-Free Cement Manufactured with Metallurgical Slags. *Buildings* **2024**, *14*, 1739. [CrossRef]
25. Scrivener, K. A Practical Guide to Microstructural Analysis of Cementitious Materials. In *A Practical Guide to Microstructural Analysis of Cementitious Materials;* Routledge: London, UK, 2018. [CrossRef]
26. Escadeillas, G.; Hornain, H. *A Durabilidade do Concreto Frente a Meios Agressivos;* Quarcioni, V.A., Translator; Ibracon: São Paulo, Brazil, 2014.

applied sciences

MDPI

Article

Synergistic Chemical Modification and Physical Adsorption for the Efficient Curing of Soluble Phosphorus/Fluorine in Phosphogypsum

Junsheng Zhou [1,2], Yue Yang [1,3], Huiquan Li [1,2], Ganyu Zhu [1,*] and Haoqi Yang [1,4,*]

1 CAS Key Laboratory of Green Process and Engineering, National Engineering Research Center of Green Recycling for Strategic Metal Resources, Institute of Process Engineering, Chinese Academy of Sciences, Beijing 100190, China; zhoujunsheng22@ipe.ac.cn (J.Z.)
2 School of Chemical Engineering, University of Chinese Academy of Sciences, Beijing 100049, China
3 College of Resources and Environmental Engineering, Wuhan University of Science and Technology, Wuhan 430081, China
4 Institute of Technology for Carbon Neutralization, College of Electrical, Energy and Power Engineering, Yangzhou University, Yangzhou 225127, China
* Correspondence: gyzhu@ipe.ac.cn (G.Z.); yanghq@yzu.edu.cn (H.Y.)

Abstract: Phosphogypsum (PG), a by-product of phosphoric acid production, contains high levels of fluorine and phosphorus impurities, which negatively impact the strength and setting time of PG-based cement materials and pose environmental risks. This study explores a dual approach combining physical adsorption using zeolite powder and chemical modification with quicklime (CaO) to immobilize these impurities. The composition of 90 wt.% PG, 5 wt.% zeolite powder, and 5 wt.% quicklime reduces the soluble phosphorus to below the detection limits and significantly lowers the free water content in the PG. Through SEM, XRD, and FT-IR analyses, it was found that zeolite powder adsorbs fluorine and phosphorus through encapsulation, while quicklime chemically reacts to form insoluble calcium phosphate and calcium fluoride. This transformation decreases the solubility, mitigating potential environmental contamination. The combination of physical adsorption and chemical conversion provides a sustainable strategy to reduce environmental hazards and enhance PG's suitability for cement-based materials. The findings from this research offer a promising pathway for the sustainable utilization of PG, providing a mechanism for its safe incorporation into building materials, while addressing both environmental and material performance concerns.

Keywords: phosphogypsum; physical adsorption; chemical modification; fluorine/phosphorus curing; curing mechanism

1. Introduction

The wet phosphoric acid (WPA) process is the predominant industrial method for producing phosphoric acid (H_3PO_4) from phosphate ore using sulfuric acid [1–3]. The main chemical reaction can be expressed as:

$$Ca_5F(PO_4)_3 + 5H_2SO_4 + nH_2O \rightarrow 3H_3PO_4 + 5CaSO_4{\cdot}nH_2O + HF$$

Phosphogypsum (PG) is the primary by-product of phosphoric acid production [2,4], with approximately 4.55 tons of PG generated for every ton of P_2O_5 produced (calculated as P_2O_5) [5,6]. Globally, the annual production of PG reaches around 300 million metric tons [7]. However, the comprehensive utilization rate of PG remains alarmingly low at only 15%, with the remaining 85% being landfilled, stockpiled, or discarded [8]. This practice not only occupies vast land areas but also poses significant environmental risks [9]. The environmental threat is primarily driven by the presence of soluble phosphates and fluorides, which can leach into soil and water, causing severe ecological damage [10,11]. Consequently, the low utilization rate and environmental hazards highlight the urgent need for effective strategies to stabilize these impurities and enable the sustainable use of PG [12,13].

Various efforts have explored PG's application in soil conditioning [14], new materials [15,16], and construction [17,18]. While these approaches offer some mitigation, the actual demand for PG in these areas is insufficient to achieve large-scale utilization. Moreover, impurities in PG, such as heavy gold ions [19], fluoride [11,20], phosphoric acid [21], and radioactive elements [7,16,22], are the main reasons limiting its application. Particularly, phosphorus and fluorine impurities directly affect the compressive strength and setting time of PG-based cement materials [23–25]. Therefore, the effective stabilization of these harmful substances is key to solving the environmental pollution problems associated with PG and improving its utilization rate. The current PG pretreatment methods, such as water washing, are widely used [25–27]. However, these techniques are resource-intensive, requiring large amounts of water and energy, further exacerbating water scarcity issues. Additionally, water washing generates significant wastewater, posing risks of secondary pollution if not adequately treated. Moreover, these methods involve complex processes, high equipment costs, and significant labor requirements, reducing their economic viability [28]. In contrast, dry treatment techniques present several advantages, including resource conservation, cost reductions, and the avoidance of wastewater generation [29]. These methods also simplify the operational process, lowering the equipment and labor costs. Given their efficiency, environmental compatibility, and economic feasibility, dry reaction methods show significant potential for large-scale PG treatment, offering a sustainable and cost-effective solution for the utilization of this industrial by-product.

Phosphorus and fluorine impurities in PG can be effectively stabilized through advanced curing technologies [29–31]. These techniques immobilize impurities within specific crystalline structures or chemical compounds [32,33], significantly reducing their mobility and solubility. As a result, curing treatments enhance the environmental safety and utilization potential of PG. Curing technology has been recognized as an efficient, rapid, and cost-effective method for managing phosphorus and fluorine impurities [34]. By encapsulating these impurities in dense matrices or converting them into stable, insoluble compounds, the risk of leaching is substantially mitigated. Recent advancements have shown that incorporating specific materials can significantly improve the curing performance [35–37]. For instance, Al_2O_3 demonstrates exceptional adsorption capacity for phosphorus and fluorine due to its large specific surface area and efficient mass transfer properties [38]. Similarly, integrating polymeric aluminum chloride (PAC) into PG binder formulations has shown promising results in enhancing the fluoride retention through high adsorption capacity and encapsulation efficiency [39]. While these studies demonstrate that the leaching rate of fluorine and phosphorus can be effectively reduced by adding binders to PG, it is important to note that the current methods require significant quantities of other materials, such as fly ash [40], red clay [41], and lime [42], to prepare PG-based cement materials. This substantially reduces the PG consumption, increases process costs, and limits the environmental compatibility. Therefore, identifying suitable curing materials for fluorine and phosphorus is a critical scientific challenge in unlocking the full resource potential of PG and advancing its large-scale utilization.

Zeolite has emerged as a widely utilized material for pollutant treatment, owing to its unique nanoscale pore structure and excellent adsorption properties [43–45]. Its large specific surface area and well-developed pore network provide a promising platform for the solidification of phosphorus and fluorine impurities, effectively reducing their mobility and environmental risks. In addition to zeolite, inexpensive and readily available curing agents such as quicklime (CaO) and slaked lime ($Ca(OH)_2$) have been identified as key materials for stabilizing phosphorus and fluorine impurities. These agents react with phosphorus and fluorine impurities to form insoluble composite precipitates, thereby achieving effective immobilization [46–48]. The synergistic use of zeolite and lime-based additives offers an efficient approach to addressing the challenges associated with impurity leaching in PG. By leveraging their complementary properties, these materials hold significant potential for scalable, cost-effective, and environmentally friendly applications in PG treatment and resource recovery.

In this study, quicklime and zeolite powder were selected as curing agents to address the challenges of stabilizing soluble phosphorus and fluorine impurities in PG. This selection was guided by their complementary properties; zeolite powder offers exceptional physical adsorption capabilities due to its nanoscale pore structure and large specific surface area, while quicklime is highly reactive and facilitates chemical modification through the formation of insoluble compounds. By examining the individual and combined roles of quicklime and zeolite powder, this study provides critical insights into the optimization of curing conditions. To deepen our understanding, the curing mechanisms of quicklime and zeolite powder were thoroughly analyzed using advanced characterization techniques. These analyses revealed the specific pathways through which physical adsorption and chemical reactions contribute to impurity stabilization. The findings of this study provide a theoretical foundation and technical guidance for optimizing PG treatment processes. They highlight the potential of integrating physical and chemical methods to develop economical, efficient, and environmentally friendly strategies for large-scale impurity stabilization and PG resource utilization.

2. Materials and Methods

2.1. Materials

The PG used in this study was sourced from Hubei Xingfa Chemical Group Co., Ltd., Yichang City, Hubei Province, China. The material appeared off-white in color (Figure 1), indicative of its typical composition and purity. The primary chemical composition of PG, as analyzed and detailed in Table 1, includes calcium sulfate dihydrate ($CaSO_4{\cdot}2H_2O$) as the dominant phase, along with impurities such as soluble phosphorus and fluorine compounds. These impurities were key targets for stabilization in this study due to their environmental mobility and associated risks.

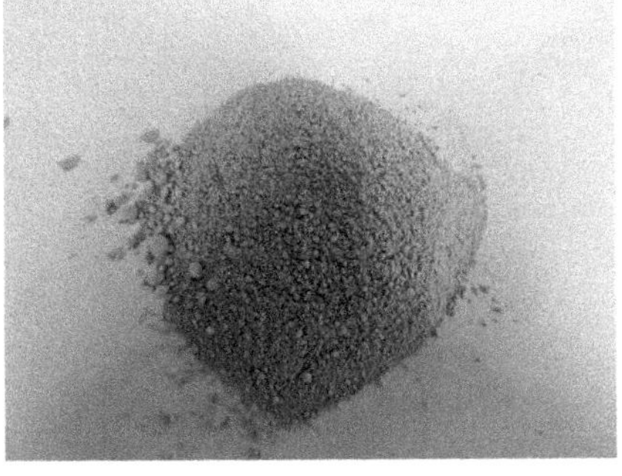

Figure 1. Picture of Hubei Xingfa Chemical Group Co., Ltd.'s PG.

Table 1. Analysis of major chemical constituents in PG raw materials (wt.%).

SO_3	CaO	SiO_2	P_2O_5	F	Al_2O_3	K_2O	Fe_2O_3
43.3207	36.2531	1.7975	0.5932	0.4028	0.2285	0.1315	0.1244

The zeolite powder, utilized as a key curing agent, was procured from Chaoyang Xinhe Zeolite Science and Technology Co., Ltd., Chaoyang, China. For experimental consistency and enhanced adsorption performance, 100–120 mesh zeolite powder was selected. This particle size was chosen to maximize the material's surface area and facilitate efficient interaction with impurities. The chemical composition of the zeolite powder, listed in Table 2, highlights its significant content of aluminosilicate minerals, which are critical for adsorption and encapsulation mechanisms. The well-developed porous structure and high cation exchange capacity of the zeolite contribute to its effectiveness in stabilizing impurities. All other chemicals used in the experiments, including the quicklime and analytical-grade reagents, were sourced from certified suppliers to ensure the reliability and reproducibility of results.

Table 2. Analysis of the main chemical composition of zeolite powder raw material (wt.%).

SiO_2	Al_2O_3	CaO	K_2O	Fe_2O_3	MgO	Na_2O	TiO_2
57.2450	10.8209	2.7942	2.4337	1.8661	0.8454	0.4164	0.1378

2.2. Experimental Methods

In this study, several experimental and control groups were established to investigate the effects of different material combinations on the stabilization of phosphorus and fluorine impurities in PG (Scheme 1). Samples of PG, zeolite powder, and quicklime were mixed in a mass ratio of 90:5:5 and subjected to thorough grinding using a mortar and pestle to ensure uniformity in the particle size distribution. After grinding, the samples were placed in an oven at 45 °C for 24 h to facilitate the curing process, forming what was designated as experimental group-1. This experimental setup aimed to evaluate the combined effect of zeolite powder and quicklime on the stabilization of impurities in PG. As part of the study, additional sample combinations were prepared for comparison purposes. In control group-1, the PG and zeolite powder were mixed in a 95:5 mass ratio, ground thoroughly, and then placed in the oven under the same conditions as experimental group-1. This group was designed to isolate the effects of zeolite on impurity stabilization, without the influence of quicklime. In control group-2, a mixture of PG and quicklime in a 95:5 mass ratio was prepared and ground, and then subjected to the same curing conditions. This group aimed to assess the impact of quicklime alone on the stabilization of the phosphorus and fluorine impurities. Finally, control group-3 consisted solely of PG, which was ground and placed in the oven without the addition of any curing agents. This group served as the baseline for evaluating the natural behavior of PG under the same temperature conditions.

2.3. Methods of Analysis

2.3.1. Chemical Composition and Mineral Phase Analysis of Materials

The elemental composition of the PG was analyzed using an X-ray fluorescence spectrometer (XRF) with an AXIOS-MAX instrument (PANalytical B.V., Tokyo, Japan). This non-destructive technique allows for the quantitative analysis of the elemental composition of PG, providing detailed information on the concentrations of key elements such as calcium (Ca), sulfur (S), phosphorus (P), fluorine (F), and other trace elements. The mineralogical composition of the PG was determined using an X-ray diffractometer (XRD), employing an Empyrean diffractometer from PAN-alytical, Tokyo, Japan. The scanning range of 2θ was set between 5° and 90°, with a scanning speed of 5°/min. This mineralogical analysis provides essential insights into the dominant crystalline phases present in PG, such as calcium sulfate dihydrate ($CaSO_4 \cdot 2H_2O$), and helps identify other phases that may influence the material's stability, reactivity, and potential for reuse in various applications.

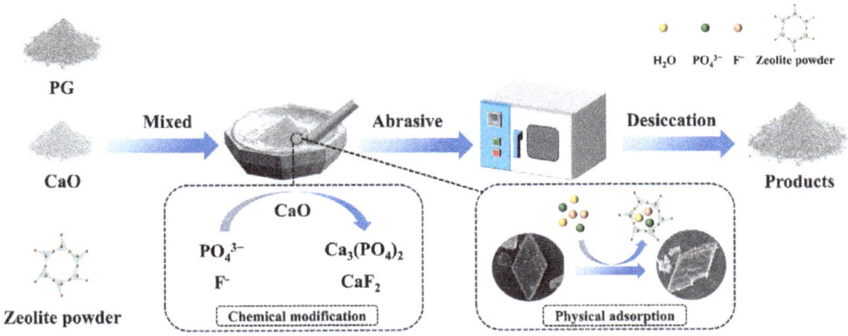

Scheme 1. Schematic illustration of physical adsorption and chemical modification method to immobilize soluble phosphorus and fluorine in PG.

2.3.2. Micro-Morphological Analysis

The microstructural characteristics of the PG were investigated using a scanning electron microscope (SEM) with a thermal field emission electron microscope (JSM-7610F, JEOL, Tokyo, Japan). A secondary electron image (SEI) detector was used to capture high-quality images of the PG's surfaces. The SEM analysis was conducted at an accelerating voltage of 15 kV and a working distance of 10.0 mm to optimize the image resolution and depth of focus. The images obtained through the SEM provide valuable information about the texture, porosity, and surface features of the PG particles, which are crucial for understanding the material's reactivity and its interactions with curing agents such as zeolite powder and quicklime.

2.3.3. Infrared Spectroscopy Analysis

Changes in the reactive functional groups in PG, particularly those involved in the stabilization of phosphorus and fluorine impurities, were analyzed using Fourier transform infrared (FTIR) spectroscopy, performed with a Nicolet iS50 spectrometer (Thermo Fisher Scientific, Waltham, MA, USA). The FTIR spectra were collected in the range of 4000 to 500 cm^{-1}, and changes in peak intensities or shifts in absorption bands were carefully analyzed to assess the impact of curing treatments on the chemical structure of the PG.

2.3.4. Particle Size and Surface Area Analysis

The specific surface area of the PG was determined using the Brunner–Emmett–Taylor (BET) method, employing a Micromeritics ASAP 2020 PLUS HD88 analyzer (ASAP2460, Micromeritics, Norcross, GA, USA). The BET method is based on nitrogen adsorption, and the surface area is calculated from the amount of nitrogen gas adsorbed at different relative pressures (P/P_0 = 0.01 to 0.95). The BET analysis provides critical data on the surface area available for adsorption, which is a key factor in evaluating the efficiency of curing agents such as zeolite powder in immobilizing phosphorus and fluorine impurities. Additionally, the pore size distribution of the PG was obtained using the Barret–Joyner–Halenda (BJH) method, which is based on nitrogen desorption data from the isotherm. These data are critical for understanding how the porosity of PG influences the adsorption of curing agents and the immobilization of impurities. By analyzing both the surface area and pore size distribution, a more comprehensive understanding of the physical properties can be gained, informing the design of more efficient treatment strategies.

2.3.5. Specimen Preparation

We weighed 5 g of test material, accurate to 0.0001 g, placed this into the mortar, added 50 mL of water, and ground this in the mortar for 15 min. After, grinding all specimens were transferred to a 250 mL volumetric flask to be fixed, and after 30 min of resting were filtered with a medium-speed qualitative filter paper for the determination of the water-soluble phosphorus pentoxide (P_2O_5) and water-soluble fluorine (F^-)

2.3.6. Analysis of Water-Soluble Phosphorus Content

We accurately withdrew 50 mL of the solution prepared in step 2.3.5, placed it into a 300 mL beaker, added 10 mL of nitric acid solution ($V(H_2O)$: $V(HNO_3)$ = 1:1), and diluted it with water to 100 mL. At the same time, we performed a blank test (the blank test was the same as the above except that no reagent solution was added), covered the surface dish, and heated it up to boiling on the electric stove, then took it off and used a small amount of water to rinse the surface dish and the wall of the glass. Under constant stirring, we added 30 mL of quinhydrone solution, then continued to heat it gently to a slight boil for 1 min. We removed the beaker, stirring 3~4 times during the cooling process, and left it to precipitate.

A filter was drawn using a glass crucible pre-dried to a constant weight in a constant temperature oven at 180 °C. The upper layer of clear liquid was first filtered, and then the precipitate was washed using the pouring method 1~2 times (each with about 25 mL of water). The precipitate was then transferred to the crucible and washed with water 5~6 times.

The water at the bottom of the crucible was sucked up with filter paper and then placed in a constant-temperature drying oven at 180 °C, dried to a constant weight, placed in a desiccator and cooled for 30 min, and weighed accurately to 0.0001 g.

Calculation of water-soluble phosphorus content:

$$\omega_1 = \frac{[(m_{14} - m_{13}) - (m_{16} - m_{15})] \times 0.03207}{m_{12}(1 - \omega)\frac{50}{250}}$$

Here, ω indicates that the mass fraction of attached water was measured for the same sample; m_{12} = the mass of the specimen in grams (g); m_{13} = the mass of the crucible in grams (g); m_{14} the mass of the quinone phosphomolybdate precipitate and crucible in grams (g); m_{15} the mass of the crucible used for the blank test in grams (g); m_{16} the mass of the blank test precipitate and crucible in grams (g).

The test results were taken as the arithmetic mean of two parallel tests, accurate to 0.01%.

2.3.7. Analysis of Water-Soluble Fluoride Content

We accurately withdrew 10 mL of the solution prepared in Section 2.3.5, added 5 drops of citric acid–sodium citrate buffer (pH: 5.5~6) and two drops of bromocresol green indicator, neutralized the solution with sodium hydroxide solution (200 g/L) until the solution turned blue, and then adjusted the solution with nitric acid solution ($V(H_2O)$:$V(HNO_3)$ = 1:5) until it was precisely yellow. We then added 20 mL of citric acid–sodium citrate buffer solution, diluted it with water to scale, shook it well, and poured it into a dry 50 mL beaker. Next, we inserted the fluoride-ion-selective electrode and the saturated calomel electrode, turned on the magnetic stirrer and stirred it at a constant speed for 2 min, stopped for 30 s, measured the potential value at equilibrium, and identified the mass of the fluorine (F^-) on the working curve c_4.

Calculation of water-soluble fluoride content:

$$\omega_2 = 2.5 \times \frac{c_4}{m_{12}(1-\omega)}$$

Here, ω indicates that the mass fraction of the attached water was measured for the same sample; m_{12} = the mass of the specimen in grams (g).

The test results were taken as the arithmetic mean of two parallel tests, accurate to 0.01%.

3. Results

3.1. Curing Performance of Zeolite Powder Synergized with CaO on Pollutants in PG

As shown in Figure 2, the raw PG material used in this study contained approximately 0.3894 wt.% soluble phosphorus impurities. This high concentration of soluble phosphorus is problematic, as it can negatively impact the properties of PG-based products, such as the strength and durability, and poses a potential environmental risk due to leaching. The use of zeolite powder as a physical adsorbent for soluble phosphorus impurities in PG has proven to be highly effective. Zeolite, with its well-developed nanoscale pore structure and large specific surface area, adsorbs soluble phosphorus, significantly reducing its concentration. Following the treatment with zeolite powder, the soluble phosphorus content in the PG decreased to 0.3432 wt.%. A further improvement in the removal of soluble phosphorus impurities was achieved through the chemical modification of PG with CaO. When the PG was treated with both zeolite powder and CaO, the soluble phosphorus content was reduced to below the detection limit, demonstrating the combined effectiveness of physical adsorption and chemical modification.

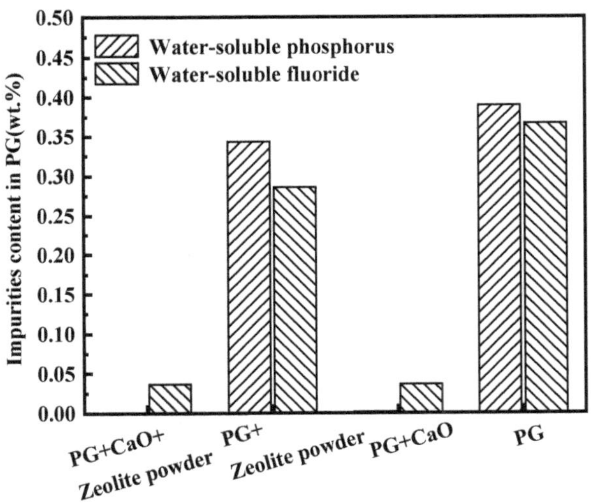

Figure 2. Comparison of soluble phosphorus and fluorine impurity contents in PG before and after treatment with zeolite powder and CaO.

In addition to phosphorus, PG also contains a considerable amount of soluble fluorine impurities, with an initial concentration of approximately 0.3659 wt.%. Fluorine compounds, particularly soluble fluoride, are known to be environmentally hazardous and can lead to soil and water contamination if not properly treated. Zeolite powder was again employed to adsorb the soluble fluorine impurities present in PG, achieving a reduction in fluorine content to 0.2857 wt.%. Although this represents a reduction in fluorine concentration, the removal efficiency was not as high as that achieved for phosphorus. To further reduce the fluorine content, the PG was subjected to chemical modification with CaO. The application of CaO alone reduced the soluble fluorine content to 0.0365 wt.%. When zeolite powder and CaO were simultaneously used in a synergistic treatment process, both physical adsorption and chemical modification contributed to a more substantial reduction in soluble fluorine content. This combined treatment reduced the fluorine concentration to 0.0361 wt.%, a significant improvement over either treatment applied individually. These results demonstrate that the physical adsorption of pollutants, when coupled with chemical modification, offers a highly effective method for reducing both soluble phosphorus and fluorine impurities in PG. The combination of zeolite powder and CaO treatment provides a robust, synergistic approach to improving the environmental safety and resource potential of PG.

3.2. Mechanism of Zeolite Powder Synergized with CaO for Immobilization of Pollutants in PG

The original PG sample exhibited a characteristic tetragonal rhombic structure (Figure 3a), which is typical of PG. This structure is formed during the production process when PG crystals stack in horizontal layers. While this arrangement imparts some structural integrity to the material, it significantly restricts the flow of interstitial liquids. This limitation hinders the efficient removal of soluble impurities, such as phosphorus and fluorine, through traditional methods such as water washing. Furthermore, PG contains a large number of free hydroxyl groups on its surface, which enhance its reactivity but also promote the agglomeration of particles, especially in the presence of high impurity concentrations. This agglomeration results in a more heterogeneous material, where particles tend to cluster together, leading to an uneven distribution of the impurities throughout the sample. Our microscopic observation of the PG structure revealed that it predominantly consisted of plate-like crystals, which form a spatial network through weak physical interactions, such as van der Waals forces, at their contact points. The weak intercrystal bonding and low compaction of PG contribute to its high porosity, which allows for fluid circulation and pollutant retention. However, this also makes PG susceptible to stress concentration when subjected to external forces, leading to material fracture and reduced mechanical strength. The low compaction and large pore size further contribute to poor water resistance and make it prone to rapid erosion and degradation, especially under wet conditions.

CaO has been demonstrated to effectively react with soluble phosphorus and fluorine impurities in PG, converting them into insoluble phosphorus and fluorine precipitates. This chemical reaction occurs during the solid-phase milling process, where it interacts with the impurities, facilitating the formation of stable compounds such as calcium phosphate ($Ca_3(PO_4)_2$) and calcium fluoride (CaF_2). However, while this chemical modification is effective in reducing the solubility of these pollutants, it does not significantly alter the overall microscopic morphology of PG (Figure 3b). After treatment, the PG particles largely retain their characteristic rhombohedral plate-like shape. Some fine particles, likely consisting of the newly formed insoluble precipitates, can be observed to be dispersed around the edges of the PG crystals but the core structure remains largely unchanged. In contrast, when PG is treated with zeolite powder, a noticeable change in the microstructure can be observed. Zeolite particles appear to surround the PG crystals, as the zeolite physically adsorbs soluble impurities, such as phosphorus and fluorine, from the surface of the PG particles (Figure 3c). When both zeolite powder and CaO are used in combination to treat PG, a more complex morphological transformation occurs (Figure 3d). The crystals of PG exhibit fine cracks, suggesting that both the physical adsorption by the zeolite and the chemical reaction with CaO contribute synergistically to the immobilization of soluble impurities. This dual approach of physical adsorption coupled with chemical modification not only reduces the mobility of pollutants but also alters the microstructure of the PG in a way that enhances its environmental stability and suitability for further applications.

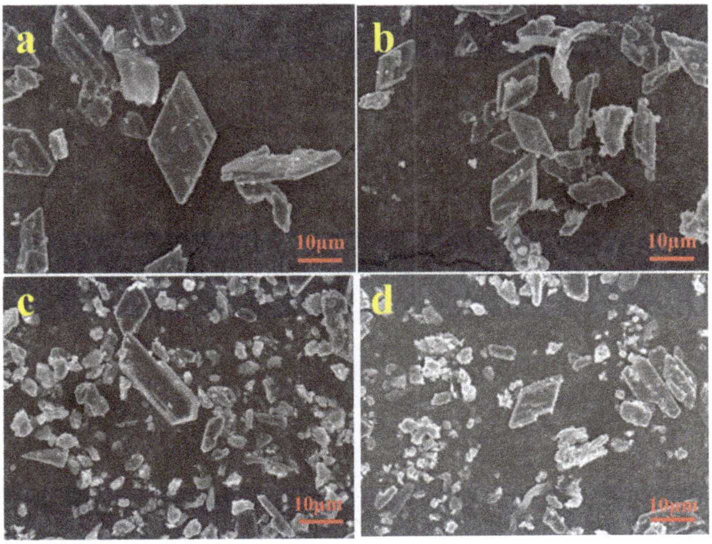

Figure 3. Morphological changes in PG under different treatment conditions: (**a**) untreated PG; (**b**) PG treated with CaO; (**c**) PG treated with zeolite powder; (**d**) PG treated with a combination of zeolite powder and CaO.

The treatment of PG using zeolite powder and CaO does not significantly alter the overall crystal structure of the PG, as confirmed by our XRD analysis (Figure 4). The XRD patterns of the treated PG samples closely resemble those of the untreated PG, indicating that the primary crystalline phases remain intact after treatment. This observation is consistent with our SEM imaging, which showed no drastic changes in the morphology of the PG particles upon treatment. In the XRD spectra, several characteristic peaks were identified. The peaks at 11.52°, 20.64°, 23.43°, 31.15°, 47.86°, and 56.81° correspond to the calcium sulfate dihydrate ($CaSO_4{\cdot}2H_2O$) phase, which is the main crystalline component of PG. The presence of additional peaks at 29.34°, 31.14°, 40.23°, and 50.55° corresponds to the anhydrite phase of calcium sulfate ($CaSO_4{\cdot}0.5H_2O$), which is also a significant constituent of PG. Furthermore, the peaks at 26.65° and in the range of 32°~37° are attributed to the presence of silicon dioxide (SiO_2) crystals, Additionally, the XRD pattern revealed some minor peaks between 60° and 90°, indicating the presence of trace impurities such as calcium fluorophosphate ($Ca_5F(PO_4)_3$) and sodium hexafluorosilicate (Na_2SiF_6). These impurities, although present in small amounts, may contribute to the environmental concerns associated with PG. Despite the treatment with zeolite powder and CaO, these impurities remained detectable, suggesting that the treatment process primarily focuses on reducing the solubility and mobility of soluble pollutants, while the overall crystalline framework of the PG remains largely unaltered.

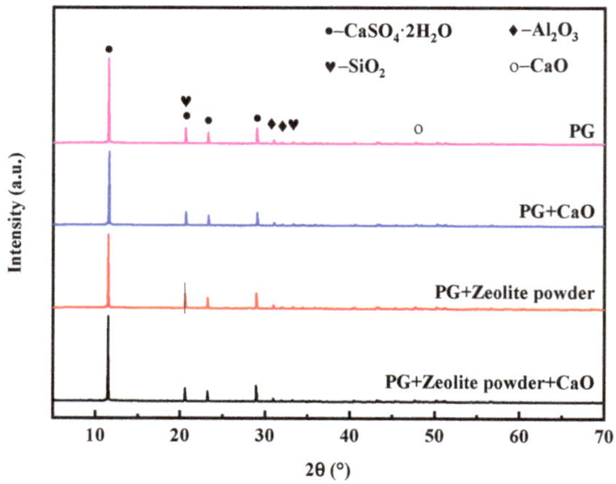

Figure 4. XRD spectra of PG under different treatment conditions.

As illustrated in Figure 5, FT-IR spectroscopy was employed to analyze the functional groups and molecular interactions within the PG samples before and after treatment. In the FT-IR spectra of the untreated PG, the absorption peaks observed at 597 cm^{-1} and 665 cm^{-1} are attributed to the asymmetric bending vibrations of sulfate ions (SO_4^{2-}), which are a prominent feature of PG due to its calcium sulfate content. Additionally, the absorption peak at 1096 cm^{-1} corresponds to the asymmetric stretching vibration of SO_4^{2-}, further confirming the presence of calcium sulfate and its characteristic functional groups [49,50]. The two absorption bands at 1620 and 1682 cm^{-1} are indicative of the symmetric stretching vibrations of hydroxyl groups (OH^-) in free water molecules, which are present in PG as part of its hydrated structure [51]. These peaks suggest the presence of adsorbed water in the crystalline structure of PG, which is often linked to the crystallization water in $CaSO_4{\cdot}2H_2O$. However, when zeolite powder and CaO were used in combination to treat PG, noticeable changes in the FT-IR spectra were observed. Specifically, the intensity of the broad absorption peak at 3392 cm^{-1}, which corresponds to the stretching vibration of hydroxyl groups in free water, was significantly weakened. This suggests that the treatment effectively reduced the amount of free water in the PG structure. The decrease in the intensity of this absorption band is attributed to the removal or immobilization of free water molecules during the physical adsorption by zeolite and the chemical reaction with CaO. The reduction in free water content highlights the successful interaction between the treatment agents and the PG, leading to a more stable material with less water retention, which could enhance the material's mechanical properties and reduce its environmental impact.

Following the treatment of PG with zeolite powder and CaO, the nitrogen adsorption–desorption isotherms of all treated samples showed typical IV-type isotherms, characteristic of mesoporous materials (Figure 6a). The corresponding pore size distributions were primarily in the range of 8–12 nm (Figure 6b), indicating the presence of mesopores in the PG samples. This behavior suggests that the treatment methods, including physical adsorption and chemical modification, did not significantly alter the mesoporous nature of the PG but rather enhanced its adsorption properties without substantially changing its overall pore structure. The specific surface area of the untreated PG was measured at 21.63 $m^2\ g^{-1}$, with an average pore size of 11.54 nm, reflecting the inherent characteristics of the PG. After treatment with zeolite powder, the specific surface area slightly increased to 22.02 $m^2\ g^{-1}$, accompanied by a small reduction in the average pore size to 10.81 nm. In contrast, the treatment with CaO resulted in a minor decrease in the specific surface area to 20.94 $m^2\ g^{-1}$, with an increase in the average pore size to 12.25 nm. This may suggest that the CaO treatment had a slight effect on the structure of the PG, possibly due to the formation of insoluble precipitates that did not significantly block or alter the mesopores. When both zeolite powder and CaO were applied synergistically, the specific surface area was 20.84 $m^2\ g^{-1}$, with an average pore size of 8.71 nm. Although there was a slight decrease in specific surface area compared to the untreated PG, the combined effect of zeolite and CaO did not dramatically alter the pore size distribution or surface area, indicating that the physical adsorption and chemical modification strategies employed had a minimal impact on the fundamental textural properties of the PG. Overall, these results suggest that while the treatment methods effectively reduce the soluble impurities in PG, they do not significantly affect the material's mesoporosity, preserving its potential for further applications.

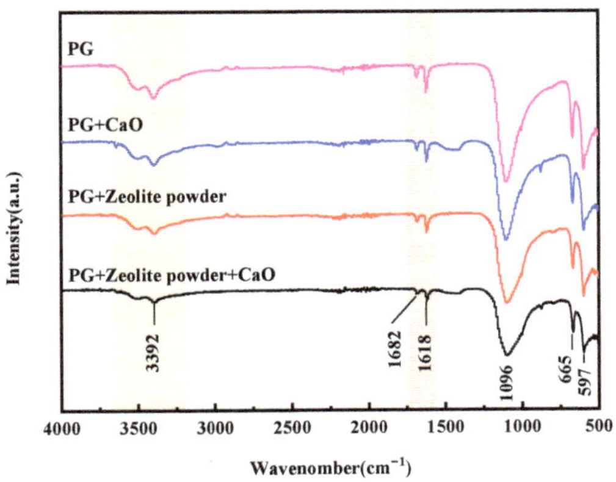

Figure 5. FT-IR spectra of PG under different treatment conditions.

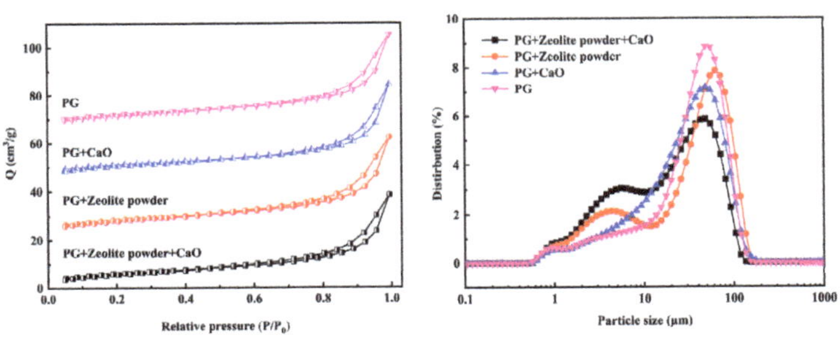

(a) (b)

Figure 6. (**a**) Specific surface area and (**b**) pore size analyses of PG samples under different treatment conditions.

4. Discussion

The results of this study are consistent with previous research on phosphogypsum PG treatments, demonstrating the effectiveness of zeolite powder and CaO in reducing soluble phosphorus and fluorine impurities. The observed reductions in impurity concentrations support the hypothesis that physical adsorption and chemical modification can work synergistically to mitigate the environmental risks associated with PG. However, the findings also highlight areas requiring further investigation. For instance, while the study successfully reduced the impurity concentrations, it did not assess the long-term stability of these reductions under variable environmental conditions. Additionally, the minimal impact on the PG's structural properties suggests that although the treatment preserves its porosity, further studies are needed to evaluate how these changes might affect the mechanical strength of PG, particularly in construction or soil applications. PG-based cementitious materials are becoming increasingly important for large-scale PG utilization. This technology typically involves using PG as an aggregate, mixed with cementitious materials and water to prepare a paste, which is then used for backfilling in underground mines or open pits. However, phosphorus and fluoride can retard the hardening of PG and weaken its structure. The soluble phosphorus also significantly reduces the hydration rate. Fluoride, in particular, impacts the strength development by interfering with the pore structure and formation of hydration products, which results in the need for increased amounts of cementitious materials to compensate for the strength loss, thereby raising costs. Future research should focus on optimizing the treatment process for large-scale applications, evaluating the long-term leaching behavior of treated PG, and assessing the economic viability and environmental impact of using zeolite and CaO on an industrial scale. Furthermore, exploring alternative, more sustainable treatment agents could further enhance the effectiveness and cost-efficiency of PG resource recovery efforts.

5. Conclusions

In this paper, we have presented a novel approach combining physical adsorption with chemical modification, demonstrating its effectiveness in reducing soluble impurities in phosphogypsum (PG), particularly soluble phosphorus and fluorine. By adding only 5 wt.% zeolite powder and 5 wt.% CaO, the soluble phosphorus content in the PG was reduced to below the detection limit. This treatment not only immobilized the fluorine and phosphorus but also enhanced the compressive strength and setting time of the PG-based cementitious materials, making it one of the most viable options for large-scale PG remediation. Importantly, the treatment did not significantly alter the crystal structure or micromorphology of the PG, preserving its potential for direct use in the production of α-hemihydrate gypsum and maintaining its suitability for applications such as building materials. Additionally, the inclusion of CaO reduced the free water content in the PG, improving its handling and storage properties, which is beneficial for subsequent processing and utilization. Our solid-phase treatment method avoids complications such as wastewater disposal, further simplifying the process. Moreover, the proposed curing technology is cost-effective, easy to operate, and well-suited for large-scale industrial applications. These findings underscore the environmental benefits of this approach and open new pathways for the sustainable utilization of PG across various industries. Our future research will focus on optimizing the process for large-scale operations and assessing the long-term performance of treated PG under real-world conditions.

Author Contributions: Conceptualization, G.Z. and H.Y.; methodology, J.Z. and Y.Y.; validation, J.Z., Y.Y. and H.Y.; investigation, J.Z.; data curation, J.Z. and Y.Y.; writing—original draft preparation, J.Z.; writing—review and editing, H.L. and H.Y.; supervision, G.Z.; project administration, H.Y.; funding acquisition, G.Z. All authors have read and agreed to the published version of the manuscript.

Funding: This study was funded by the 14th Five-Year National Key Research and Development Plan Project (Grant No. 2022YFC3902704) and National Natural Science Foundation of China (22078344).

Institutional Review Board Statement: Not applicable.

Informed Consent Statement: Not applicable.

Data Availability Statement: Data are contained within the article.

Acknowledgments: We would like to express our heartfelt gratitude to the National Engineering Research Center of Green Recycling for Strategic Metal Resources for providing the experimental instruments and workstations, as well as to Fade Wu and Ziheng Meng for their contributions to the experimental research and discussion.

Conflicts of Interest: The authors declare no conflicts of interest. The funders had no role in the design of the study; in the collection, analyses, or interpretation of the data; in the writing of the manuscript; or in the decision to publish the result.

References

1. Zhang, P. Comprehensive Recovery and Sustainable Development of Phosphate Resources. *Procedia Eng.* **2014**, *83*, 37–51. [CrossRef]
2. Fang, K.; Xu, L.; Yang, M.; Chen, Q. One-step wet-process phosphoric acid by-product $CaSO_4$ and its purification. *Sep. Purif.* **2023**, *309*, 123048. [CrossRef]
3. Ryszko, U.; Rusek, P.; Kolodynska, D. Quality of Phosphate Rocks from Various Deposits Used in Wet Phosphoric Acid and P-Fertilizer Production. *Materials* **2023**, *16*, 793. [CrossRef] [PubMed]
4. Yang, J.; Han, C.; Shao, L.; Nie, R.; Dong, S.; Liu, H.; Ma, L. Chemical looping gasification of lignite to syngas using phosphogypsum: Overview and prospects. *J. Clean. Prod.* **2024**, *445*, 141329. [CrossRef]
5. Rashad, A.M. PG as a construction material. *J. Clean. Prod.* **2017**, *166*, 732–743. [CrossRef]
6. Wu, F.; Ren, Y.; Qu, G.; Liu, S.; Chen, B.; Liu, X.; Zhao, C.; Li, J. Utilization path of bulk industrial solid waste: A review on the multi-directional resource utilization path of phosphogypsum. *J. Environ. Manag.* **2022**, *313*, 114957. [CrossRef] [PubMed]
7. Bilal, E.; Bellefqih, H.; Bourgier, V.; Mazouz, H.; Dumitraş, D.-G.; Bard, F.; Laborde, M.; Caspar, J.P.; Guilhot, B.; Iatan, E.-L.; et al. Phosphogypsum circular economy considerations: A critical review from more than 65 storage sites worldwide. *J. Clean. Prod.* **2023**, *414*, 137561. [CrossRef]
8. Neto, J.S.A.; Bersch, J.D.; Silva, T.S.M.; Rodríguez, E.D.; Suzuki, S.; Kirchheim, A.P. Influence of phosphogypsum purification with lime on the properties of cementitious matrices with and without plasticizer. *Constr. Build. Mater.* **2021**, *299*, 123935. [CrossRef]
9. Shu, J.; Chen, M.; Wu, H.; Li, B.; Wang, B.; Li, B.; Liu, R.; Liu, Z. An innovative method for synergistic stabilization/solidification of Mn^{2+}, NH_4^+-N, PO_4^{3-} and F^- in electrolytic manganese residue and phosphogypsum. *J. Hazard. Mater.* **2019**, *376*, 212–222. [CrossRef] [PubMed]
10. Cao, W.; Yi, W.; Peng, J.; Li, G.; Yin, S. Preparation of anhydrite from phosphogypsum: Influence of phosphorus and fluorine impurities on the performances. *Constr. Build. Mater.* **2022**, *318*, 126021. [CrossRef]
11. Valančius, Z.; Vaickelionienė, R.; Vaickelionis, G.; Makčinskas, P. Use of an industrial by-product phosphogypsum in the production of white textured paints. *J. Clean. Prod.* **2022**, *380*, 134888. [CrossRef]
12. Xie, Y.; Huang, J.; Wang, H.; Lv, S.; Jiang, F.; Pan, Z.; Liu, J. Simultaneous and efficient removal of fluoride and phosphate in phosphogypsum leachate by acid-modified sulfoaluminate cement. *Chemosphere* **2022**, *305*, 135422. [CrossRef]
13. Cui, S.F.; Fu, Y.Z.; Zhou, B.Q.; Li, J.X.; He, W.Y.; Yu, Y.Q.; Yang, J.Y. Transfer characteristic of fluorine from atmospheric dry deposition, fertilizers, pesticides, and phosphogypsum into soil. *Chemosphere* **2021**, *278*, 130432. [CrossRef] [PubMed]
14. Qi, J.; Zhu, H.; Zhou, P.; Wang, X.; Wang, Z.; Yang, S.; Yang, D.; Li, B. Application of phosphogypsum in soilization: A review. *Int. J. Environ. Sci. Technol.* **2023**, *20*, 10449–10464. [CrossRef]
15. Du, M.; Wang, J.; Dong, F.; Wang, Z.; Yang, F.; Tan, H.; Fu, K.; Wang, W. The study on the effect of flotation purification on the performance of alpha-hemihydrate gypsum prepared from phosphogypsum. *Sci. Rep.* **2022**, *12*, 95.
16. Contreras, M.; Teixeira, S.R.; Santos, G.T.A.; Gázquez, M.J.; Romero, M.; Bolívar, J.P. Influence of the addition of phosphogypsum on some properties of ceramic tiles. *Constr. Build. Mater.* **2018**, *175*, 588–600. [CrossRef]
17. Liu, S.; Wu, F.; Qu, G.; Kuang, L.; Liu, Y.; Ren, Y.; Chen, B.; Li, J. Preparation of high-performance lightweight materials based on the phosphogypsum -inorganic material system. *Sustain. Chem. Pharm.* **2022**, *30*, 100901. [CrossRef]
18. Xiang, J.; Qiu, J.; Song, Y.; Miao, Y.; Gu, X. Synergistic removal of phosphorus and fluorine impurities in phosphogypsum by enzyme-induced modified microbially induced carbonate precipitation method. *J. Environ. Manag.* **2022**, *324*, 116300. [CrossRef] [PubMed]
19. Pliaka, M.; Gaidajis, G. Potential uses of PG: A review. *J. Environ. Sci. Health A Tox Hazard. Subst. Environ. Eng.* **2022**, *57*, 746–763. [CrossRef]
20. Rao, S.M.; Asha, K. Role of Fly Ash Pozzolanic Reactions in Controlling Fluoride Release from phosphogypsum. *J. Mater.* **2013**, *25*, 999–1005. [CrossRef]
21. Tiwari, S.B.; Veksha, A.; Chan, W.P.; Fei, X.; Liu, W.; Lisak, G.; Lim, T.-T. Acidic hydrothermal carbonization of sewage sludge for enhanced alkaline extraction of phosphorus and reduced co-extraction of trace elements. *Resour. Conserv. Recycl.* **2025**, *212*, 107936. [CrossRef]
22. Attallah, M.F.; Metwally, S.S.; Moussa, S.I.; Soliman, M.A. Environmental impact assessment of phosphate fertilizers and phosphogypsum waste: Elemental and radiological effects. *Microchem. J.* **2019**, *146*, 789–797. [CrossRef]

23. Gaidajis, G.; Anagnostopoulos, A.; Garidi, A.; Mylona, E.; Zevgolis, I.E. Laboratory evaluation of phosphogypsum for alternative uses. *Environ. Geotech* **2018**, *5*, 310–323. [CrossRef]
24. Singh, M. Role of PG impurities on strength and microstructure of selenite plaster. *Constr. Build. Mater.* **2005**, *19*, 480–486 [CrossRef]
25. Zhou, Y.; Li, X.; Shi, Y.; Zhu, Q.; Du, J. Reuse of phosphogypsum pretreated with water washing as aggregate for cemented backfill. *Sci. Rep.* **2022**, *12*, 16091. [CrossRef]
26. Pérez-Moreno, S.M.; Romero, C.; Guerrero, J.L.; Barba-Lobo, A.; Gázquez, M.J.; Bolívar, J.P. Evolution of the waste generated along the cleaning process of PG leachates. *J. Environ. Chem. Eng.* **2023**, *11*, 111485. [CrossRef]
27. Akfas, F.; Elghali, A.; Toubri, Y.; Samrane, K.; Munoz, M.; Bodinier, J.L.; Benzaazoua, M. Environmental assessment of phosphogypsum: A comprehensive geochemical modeling and leaching behavior study. *J. Environ. Manag.* **2024**, *359*, 120929 [CrossRef]
28. Qin, X.; Cao, Y.; Guan, H.; Hu, Q.; Liu, Z.; Xu, J.; Hu, B.; Zhang, Z.; Luo, R. Resource utilization and development of PG-based materials in civil engineering. *J. Clean. Prod.* **2023**, *387*, 135858. [CrossRef]
29. Chen, X.; Gao, J.; Liu, C.; Zhao, Y. Effect of neutralization on the setting and hardening characters of hemihydrate phosphogypsum plaster. *Constr. Build. Mater.* **2018**, *190*, 53–64. [CrossRef]
30. Xue, S.; Li, M.; Jiang, J.; Millar, G.J.; Li, C.; Kong, X. Phosphogypsum stabilization of bauxite residue: Conversion of its alkaline characteristics. *J. Environ. Sci.* **2019**, *77*, 1–10. [CrossRef]
31. Wu, F.; Chen, B.; Qu, G.; Liu, S.; Zhao, C.; Ren, Y.; Liu, X. Harmless treatment technology of phosphogypsum: Directional stabilization of toxic and harmful substances. *J. Environ. Manag.* **2022**, *311*, 114827. [CrossRef]
32. Liu, Y.; Zhang, L.; Chen, L.; Xue, B.; Wang, G.; Zhu, G.; Gou, W.; Yang, D. Potential of artificial soil preparation for vegetation restoration using red mud and phosphogypsum. *Sci. Total Environ.* **2024**, *941*, 173553. [CrossRef] [PubMed]
33. Huang, Y.; Qian, J.; Lu, L.; Zhang, W.; Wang, S.; Wang, W.; Cheng, X. Phosphogypsum as a component of calcium sulfoaluminate cement: Hazardous elements immobilization, radioactivity and performances. *J. Clean. Prod.* **2020**, *248*, 119287. [CrossRef]
34. Wu, Y.; Xu, F.; Wu, X.; Jiao, Y.; Sun, T.; Li, Z.; Yang, F.; Li, H.; Li, B.; Xu, J.; et al. Retardation mechanism of phosphogypsum in PG-based excess-sulfate cement. *Constr. Build. Mater.* **2024**, *428*, 136293. [CrossRef]
35. Chen, L.; Liu, T.; Cheng, M.; Zhang, L.; Liu, Y.; Fu, Z.; Zhou, T.; Yang, J.; Xiao, P.; Huang, Q.; et al. Enhanced strength and fluoride ion solidification/stabilization mechanism of modified phosphogypsum backfill material. *Constr. Build. Mater.* **2024**, *449*, 138572 [CrossRef]
36. Lin, S.; Zheng, Y.; Liu, W.; Ma, H.; Rao, F.; Yang, L.; Zhong, S. Consolidation of phosphorus tailings and soluble fluorine & phosphorus with calcium carbide residue-mirabilite waste as a green alkali activator. *Case Stud. Constr. Mater.* **2023**, *18*, e01779.
37. Huang, S.; Liu, W.; Li, W.; Xi, Z.A. Study on water resistance of phosphogypsum solidified by magnesium oxychloride cement (MOC) and $NaHCO_3$. *Constr. Build. Mater.* **2024**, *451*, 138877. [CrossRef]
38. Liu, Y.; Chen, Q.; Dalconi, M.C.; Molinari, S.; Valentini, L.; Wang, Y.; Sun, S.; Wang, P.; Artioli, G. Enhancing the sustainable immobilization of phosphogypsum by cemented paste backfill with the activation of γ-Al_2O_3. *Constr. Build. Mater.* **2022**, *347*, 128624. [CrossRef]
39. Chen, Q.; Zhang, Q.; Wang, Y.; Zhang, Q.; Liu, Y. Highly-efficient fluoride retention in on-site solidification/stabilization of phosphogypsum: Cemented paste backfill synergizes with poly-aluminum chloride activation. *Chemosphere* **2022**, *309*, 136652 [CrossRef] [PubMed]
40. Novytskyi, Y.; Topylko, N.; Marushchak, U.; Turba, Y. Composite Materials Based on phosphogypsum for Constructive Layers of Road Pavement. *Chem. Chem. Technol.* **2024**, *18*, 7–15. [CrossRef]
41. Liu, Z.-Y.; Chen, K.-S. Research on Road Characteristics and the Microscopic Mechanism of Lime–phosphogypsum -Stabilized Red Clay. *Appl. Sci.* **2023**, *13*, 8057. [CrossRef]
42. James, J.; Arthi, C.; Balaji, G.; Chandraleka, N.; Kumar, R.H.M.N. Lime activated flyash- phosphogypsum blend as a low-cost alternative binder. *Int. J. Environ. Sci. Technol.* **2022**, *19*, 8969–8978. [CrossRef] [PubMed]
43. Nizevičienė, D.; Vaičiukynienė, D.; Vaitkevičius, V.; Rudžionis, Ž. Effects of waste fluid catalytic cracking on the properties of semi-hydrate phosphogypsum. *J. Clean. Prod.* **2016**, *137*, 150–156. [CrossRef]
44. Dehghan, R.; Anbia, M. Zeolites for adsorptive desulfurization from fuels: A review. *Fuel Process. Technol.* **2017**, *167*, 99–116. [CrossRef]
45. Chen, W.; Zhao, H.; Xue, Y.; Chang, X. Adsorption Effect and Adsorption Mechanism of High Content Zeolite Ceramsite on Asphalt VOCs. *Materials* **2022**, *15*, 6100. [CrossRef] [PubMed]
46. Chen, B.; Wu, F.; Qu, G.; Ning, P.; Ren, Y.; Liu, S.; Jin, C.; Li, H.; Zhao, C.; Liu, X.; et al. Waste control by waste: A comparative study on the application of carbide slag and quicklime in preparation of phosphogypsum -based ecological restoration materials. *Chem. Eng. Process.* **2022**, *178*, 109051. [CrossRef]
47. Zhou, Z.; Lu, Y.; Zhan, W.; Guo, L.; Du, Y.; Zhang, T.C.; Du, D. Four stage precipitation for efficient recovery of N, P, and F elements from leachate of waste phosphogypsum. *Miner. Eng.* **2022**, *178*, 107420. [CrossRef]

48. Wu, M.; Wang, C.; Zuo, Y.; Yang, S.; Zhang, J.; Luo, Y. Study on strength prediction and strength change of phosphogypsum-based composite cementitious backfill based on BP neural network. *Mater. Today Commun.* **2024**, *41*, 110331. [CrossRef]
49. Wang, C.-Q.; Chen, S.; Huang, D.-M.; Huang, Q.-C.; Li, X.-Q.; Shui, Z.-H. Safe environmentally friendly reuse of red mud modified phosphogypsum composite cementitious material. *Constr. Build. Mater.* **2023**, *368*, 130348. [CrossRef]
50. Zhang, X.; Huang, R.; Cao, Y.; Wang, C. Rapid conversion of red mud into soil matrix by co-hydrothermal carbonization with biomass wastes. *J. Environ. Chem. Eng.* **2021**, *9*, 106039. [CrossRef]
51. Lan, X.; Gao, J.; Qu, X.; Guo, Z. An environmental-friendly method for recovery of soluble sodium and harmless utilization of red mud: Solidification, separation, and mechanism. *Resour. Conserv. Recycl.* **2022**, *186*, 106543. [CrossRef]

applied sciences

MDPI

Article

Co-Hydrothermal Carbonization of Sawdust and Sewage Sludge: Assessing the Potential of the Hydrochar as an Adsorbent and the Ecotoxicity of the Process Water

Matheus Cavali [1,*], Thuanne Braúlio Hennig [1], Nelson Libardi Junior [1], Boram Kim [2], Vincent Garnier [3], Hassen Benbelkacem [2], Rémy Bayard [2], Adenise Lorenci Woiciechowski [4], William Gerson Matias [1] and Armando Borges de Castilhos Junior [1]

1 Department of Sanitary and Environmental Engineering, Federal University of Santa Catarina, Florianópolis 88040-970, SC, Brazil; thuannebrauliohennig@gmail.com (T.B.H.); nelson.libardi@ufsc.br (N.L.J.); william.g.matias@ufsc.br (W.G.M.); armando.borges@ufsc.br (A.B.d.C.J.)
2 DEEP Laboratory, EA 7429, National Institute of Applied Sciences of Lyon, 69621 Villeurbanne, France; boram.kim@insa-lyon.fr (B.K.); hassen.benbelkacem@insa-lyon.fr (H.B.); remy.bayard@insa-lyon.fr (R.B.)
3 MATEIS Laboratory, UMR 5510, National Institute of Applied Sciences of Lyon, 69621 Villeurbanne, France; vincent.garnier@insa-lyon.fr
4 Department of Bioprocess Engineering and Biotechnology, Federal University of Paraná, Curitiba 81531-908, PR, Brazil; adenise@ufpr.br
* Correspondence: cavali.matheus@gmail.com

Abstract: Hydrothermal carbonization (HTC) is a promising thermochemical process to convert residues into hydrochar. While conventional HTC utilizes one type of residue as raw material only, Co-HTC generally combines two. By mixing dry and wet wastes, Co-HTC can advantageously avoid water addition. Therefore, this work investigated the potential of hydrochar derived from the Co-HTC of sawdust and non-dewatered sewage sludge as a dye (methylene blue) adsorbent and evaluated the toxicity of the resulting Co-HTC process water (PW) on *Daphnia magna*. Three hydrochars were produced by Co-HTC at 180, 215, and 250 °C and named H-180, H-215, and H-250, respectively. For methylene blue adsorption, H-180 and H-215 had a better performance than H-250. Both H-180 and H-215 presented a maximum adsorption capacity of approximately 70 $mg \cdot g^{-1}$, which was superior compared with the adsorption of methylene blue by other hydrochars in the literature. Moreover, the removal percentage obtained with H-180 remained satisfactory even after five cycles. Regarding the toxicological assays of the PWs, raising the Co-HTC temperature increased the variety of substances in the PW composition, resulting in higher toxicity to *D. magna*. The EC_{50} values of PW-180, PW-215, and PW-250 were 1.13%, 0.97%, and 0.51%, respectively. This highlights the importance of searching for the treatment and valorization of the PW. Instead of viewing this by-product as an effluent to be treated and disposed of, it is imperative to assess the potential of PWs for obtaining other higher added-value products.

Keywords: hydrothermal carbonization; adsorption; methylene blue; ecotoxicity; *Daphnia magna*; waste valorization; biochar

1. Introduction

The adverse effects of climate change have prompted discussions about the sustainability of current production systems. In this context, it is well-known that the transition from the current linear economic model to a circular one is necessary to achieve sustainable development. The circular economy also involves transforming waste into products, which can be understood as waste valorization approaches. Thus, research concerning waste valorization has been the subject of several studies about thermochemical processes, such as pyrolysis, gasification, and hydrothermal carbonization (HTC) [1–3].

Concerning HTC, its objective is to convert wastes into a solid, carbon-rich material named hydrochar. HTC has received great interest given the number of publications and established industries over the last decade [4]. In this process, waste is mixed with water at temperatures ranging from 180 to 350 °C under autogenous pressure conditions in sealed reactors [5]. The water between 100 and 374 °C becomes subcritical, acting not only as a solvent but also as a catalyst [6]. Advantageously, HTC uses water as the reaction medium, which is non-toxic, environmentally safe, and an economically cheap reagent [7]. Thus, HTC is convenient for wastes with high moisture content (e.g., sewage sludges), as it is carried out in an aqueous environment and does not require a pre-drying stage [8]. Furthermore, the temperatures used in HTC are considerably lower than those used in pyrolysis and gasification processes.

When more than one waste is employed in HTC, the process is named Co-HTC, which is an interesting approach for combining wet and dry wastes [9]. Thus, Co-HTC is advantageous in circumventing some drawbacks of single-feedstock HTC. It can avoid dependence on only one feedstock, provide lower demands for water when wet wastes are used, and mitigate the inconvenience of an undesired waste characteristic, such as a high inorganic content [3]. For example, by combining inorganic-rich (e.g., sewage sludge) and inorganic-poor (e.g., sawdust) wastes, it is possible to improve the hydrochar characteristics compared with a hydrochar obtained only from the waste with a high content of inorganics [10–12]. The application of hydrochar depends on its characteristics. Hydrochar has the potential to be used, for example, for agriculture and crop improvement, catalyst support, carbon sequestration, electrochemical devices, and pollutant adsorption [13]. Regarding pollutant adsorption, hydrochar has been proposed extensively to remove heavy metals, nutrients, emerging contaminants, and dyes from aquatic environments [3].

Dye removal is a pertinent issue from wastewater from the textile industries. Given the limited adsorption capacity of fabrics, 85% of the dye mixture is discharged from the dyeing process. It makes textile industries the primary source of dye effluents in the environment because of the high usage of dyes in the coloring process [14]. If not properly managed, this could have adverse environmental consequences due to the dye toxicity [15]. It is therefore essential to implement effective dye removal methods to prevent the contamination of water sources by dye effluents. Among the proposed solutions, adsorption has been suggested as the best alternative because of its advantages, including high efficiency and short reaction times required [14,16]. Thus, sustainable and cost-effective adsorbents, such as hydrochar, can be a viable alternative for removing dyes from wastewater.

In addition to hydrochar, HTC or Co-HTC produces a liquid fraction named process water (PW) that generally presents an acidic pH and high values of chemical oxygen demand (COD) [17]. According to the waste used in the HTC or Co-HTC, PW can present potentially toxic compounds, such as furfural, 5-hydroxymethylfurfural, phenol, cresol, catechol, resorcinol, and polycyclic aromatic hydrocarbons [18,19]. In the case of Co-HTC, PW can be more toxic due to the combination of different wastes. Despite that, studies evaluating the toxicity of the PW from Co-HTC processes are scarce in the literature [20,21]. It is important to understand the impact of HTC or Co-HTC PW in the environment to establish safe concentrations and design proper treatment and valorization processes.

Therefore, in light of the context mentioned above, the objectives of this study were (a) to evaluate the potential of hydrochar derived from the Co-HTC of sawdust and non-dewatered sewage sludge as a dye adsorbent and (b) to assess the ecotoxicity of the resulting PW.

2. Materials and Methods

2.1. Materials

The reagents employed in the experiments—$C_{16}H_{18}N_3ClS$ (methylene blue), sodium hydroxide (NaOH), hydrochloric acid (HCl), potassium dichromate ($K_2Cr_2O_7$), potassium hydroxide (KOH), and $C_4H_8O_2$ (ethyl acetate)—were of analytical grade (LABIMPORT, Piracicaba, Brazil).

Methylene blue was used as a model dye since it is a well-known dye employed by nearly all textile industries. This dye has found wide application in the dyeing of silk, wood, and cotton due to its absorbency and good fastness to materials. Methylene blue is a cationic and primary thiazine dye with a molecular weight of 319.85 $g \cdot mol^{-1}$, presenting a λ_{max} of 664 nm. It is highly water-soluble, forming a stable solution with water at room temperature [15,22–24].

2.2. Hydrochar Production

The hydrochars from Co-HTC of sawdust and non-dewatered sewage sludge were produced according to the best conditions found in a previous study [12]. Co-HTC used a sawdust-to-sludge mass ratio of 10 for 2 to 3 h at 180, 215, and 250 °C. After filtration and drying, three hydrochars—H-180, H-215, and H-250, respectively—were obtained. The PWs, named PW-180, PW-215, and PW-250, were properly stored for further utilization. Hydrochar yields and characteristics (e.g., volatile matter, ashes, and fixed carbon contents, O/C and H/C atomic ratios, zeta potential, BET surface area), as well as the pH, COD, and electrical conductivity of PW, were described in previous studies [12,25]. Additionally, the point of zero charge (pH_{PZC}) was estimated by measuring the zeta potential (Malvern Zeta Sizer—Nano-ZSP) (Malvern Panalytical Ltd., Malvern, UK) at pH 2, 4, 6, 8, 10, and 12 after mixing 20 mg of hydrochar with 20 mL of pH-adjusted deionized water at 20 °C for 24 h.

2.3. Adsorption of Methylene Blue

2.3.1. pH Effect

To evaluate the pH influence in the methylene blue adsorption by H-180, H-215, and H-250, dye solutions of 100 $mg \cdot L^{-1}$ had their pH adjusted with NaOH and HCl (0.1 $mol \cdot L^{-1}$) in the range from 2 to 10. The hydrochars and methylene blue solutions were mixed in Erlenmeyer flasks at a solid-to-liquid load of 3 $g \cdot L^{-1}$. The mixture was stirred (100 rpm) for 24 h at 30 °C before separating the hydrochars from the methylene blue solution by vacuum filtration. The experiments were carried out in triplicate for each pH in addition to blank tests (without hydrochar), and the mean, minimum, and maximum values were reported. The statistical analysis was conducted using the Origin® 2017 software.

2.3.2. Adsorption Isotherms

At the best pH determined in previous tests, isotherms were performed for H-180 and H-215. The hydrochar and methylene blue solutions (25, 50, 100, 200, 300 400, 600, and 800 $mg \cdot L^{-1}$) were mixed in Erlenmeyer flasks at a solid-to-liquid load of 3 $g \cdot L^{-1}$ The mixture was stirred (100 rpm) for 24 h at 30 °C before separating the hydrochars from the methylene blue solution by vacuum filtration. The experiments were carried out in triplicate for each isotherm point in addition to blank tests (without hydrochar), and the mean, minimum, and maximum values were reported. The suitability of the Langmuir (Equation (1)), Freundlich (Equation (2)), and Sips (Equation (3)) isotherm models for describing the data was evaluated [26]. The adjustment was performed in the software

Origin® using the function Nonlinear Curve Fit and assessed according to R^2 and mean relative deviation (MRD) [27,28]:

$$q_e = \frac{q_{max} \cdot K_L \cdot C_e}{1 + (K_L \cdot C_e)} \tag{1}$$

$$q_e = K_F \cdot C_e^{\frac{1}{n}} \tag{2}$$

$$q_e = \frac{q_{max} \cdot K_s \cdot C_e^{\,n}}{1 + (K_s \cdot C_e^{\,n})} \tag{3}$$

In Equations (1)–(3), q_e is the adsorption capacity ($mg \cdot g^{-1}$) at equilibrium, q_{max} is the maximum adsorption capacity ($mg \cdot g^{-1}$) at equilibrium, K_L is the Langmuir constant, C_e is the concentration at equilibrium ($mg \cdot L^{-1}$), K_F is the Freundlich constant, n is an empirical constant related to the heterogeneity of the adsorbent surface, and K_S is the Sips constant. In Origin® 2017 software, the 95% confidence intervals for the q_{max} values of the hydrochars were calculated using the two-tailed Student's t-test to assess the variability of the estimates and to check whether the intervals overlapped.

2.3.3. Adsorption Kinetics

Based on the previous tests, the adsorption kinetics were carried out for H-180. The hydrochar and methylene blue solutions (100 $mg \cdot L^{-1}$) were mixed in Erlenmeyer flasks at a solid-to-liquid load of 3 $g \cdot L^{-1}$. The mixture was stirred (100 rpm) at 30 °C, and sampled at 10, 15, 30, 45, 60, 90, 120, 180, 300, 480, 720, and 1440 min. After vacuum filtration, the concentration of the remaining methylene blue in the solution was determined. The experiments were carried out in triplicate for each point in addition to blank tests (without hydrochar), and the mean, minimum, and maximum values were reported. The suitability of the non-linear pseudo-first-order (PFO—Equation (4)), pseudo-second-order (PSO—Equation (5)), and Elovich (Equation (6)) models were tested [29,30]. The adjustment was performed in the software Origin® 2017 using the function Nonlinear Curve Fit and assessed according to R^2 and mean relative deviation (MRD) [27,28]:

$$q_t = q_e \cdot \left(1 - e^{(-K_1 \cdot t)}\right) \tag{4}$$

$$q_t = \frac{K_2 \cdot q_e^{\,2} \cdot t}{1 + (K_2 \cdot q_e \cdot t)} \tag{5}$$

$$q_t = \frac{1}{\beta} \cdot \ln(1 + \alpha \cdot \beta \cdot t) \tag{6}$$

In Equations (4)–(6), q_t and q_e are the adsorption capacity at time t and equilibrium, respectively ($mg \cdot g^{-1}$); t is the reaction time (min); K_1 (min^{-1}) and K_2 ($g \cdot mg^{-1} \cdot min^{-1}$) are the rate constants of PFO and PSO models, respectively; α is the initial adsorption rate ($mg \cdot g^{-1} \cdot min^{-1}$); and β is the desorption constant ($g \cdot mg^{-1}$).

2.3.4. Adsorption Thermodynamics

The adsorption thermodynamics was evaluated for H-180. The adsorption was carried out under different temperatures (30, 40, 50, and 60 °C). The hydrochar and methylene blue solutions (100 $mg \cdot L^{-1}$) were mixed in Erlenmeyer flasks at a solid-to-liquid load of 3 $g \cdot L^{-1}$ for 15 h. After vacuum filtration, the concentration of the remaining methylene blue in the solution was determined. The experiments were carried out in triplicate for each temperature, in addition to blank tests (without hydrochar). The changes in Gibbs free energy (ΔG, J·mol^{-1}), enthalpy (ΔH, J·mol^{-1}), and entropy (ΔS, J·mol^{-1}·K^{-1}) were obtained based on Equations (7)–(10) as described elsewhere [31]:

$$\Delta G = -RT \cdot \ln(K_c) \tag{7}$$

$$K_c = \frac{C_0 - C_e}{C_e} \tag{8}$$

$$\Delta G = \Delta H - T\Delta S \tag{9}$$

$$\ln(K_c) = \frac{-\Delta H}{R} \cdot \frac{1}{T} + \frac{\Delta S}{R} \tag{10}$$

2.3.5. Hydrochar Regeneration

The adsorbent regeneration was performed according to other studies to evaluate the hydrochar reusability [32–34]. A methylene blue solution of 100 mg·L^{-1} and pH 10 was mixed with hydrochar at a solid-to-liquid load of 3 g·L^{-1}. The mixture was stirred (100 rpm) for 15 h at 30 °C before separating the hydrochar from the methylene blue solution by vacuum filtration. The desorption of methylene blue from hydrochar was carried out with ethanol. The dye-contaminated hydrochar was mixed with ethanol, stirred (100 rpm) for 5 h at 25 °C, vacuum filtered, washed with water, dried at 80 °C for 24 h, and reused for another adsorption cycle. The experiment was performed in triplicate for each cycle, and the mean, minimum, and maximum values were reported.

2.3.6. Methylene Blue Quantification

The remaining methylene blue concentration in the solution was determined with a UV–Vis spectrophotometer (Bel Photonics—UV-M51) (BEL Engineering, Monza, Italy) at wavelength 664 nm based on a calibration curve (R^2 = 0.9960) previously prepared. The results were expressed as removal efficiency (%) and adsorption capacity ($mg_{dye} \cdot g_{hydrochar}{}^{-1}$) according to Equation (11) and Equation (12), respectively. C_i and C_f are the initial and final dye concentration (mg·L^{-1}), W is the hydrochar mass (g), and V is the volume of the dye solution (L):

$$RE = \frac{C_i - C_f}{C_i} \cdot 100 \tag{11}$$

$$q_e = \frac{\left(C_i - C_f\right) \cdot V}{W} \tag{12}$$

2.4. *Process Water (PW)*

2.4.1. Ecotoxicity of PW

Acute toxicity tests were carried out on the microcrustacean *Daphnia magna* in accordance with OECD guideline 202 (2004). The *D. magna* culture was maintained at 20 ± 2 °C with a 16:8 light cycle. The medium was refreshed every two days, and daphnids were fed with microalgae *Desmodesmus subspicatus*, three times weekly.

Based on preliminary tests, increased PW percentages of 1–1.4% for PW-180, 0.9–1.3% for PW-215, and 0.5–0.8% for PW-250 were tested, with no pH adjustment. A control with only dilution water was also included. Ten neonates (<24 h old), verified for sensitivity via $K_2Cr_2O_7$ exposure, were placed in 25 mL of each test dilution. After 48 h, immobility and concentrations that caused 50% effect (EC_{50}) values were assessed. The lower the EC_{50} value, the higher the inhibitory effect. Moreover, additional acute tests were conducted by exposing daphnids to the EC_{50} values of each PW and adjusting the pH of the exposure medium to 7 with NaOH, as recommended by OECD 202 (2004), to check the influence of pH on the toxicity of PWs.

Data analysis was conducted with GraphPad Prism® v6.0 and Statistica® v13.5.0.17, using the Kolmogorov–Smirnov and Bartlett's tests for normality and homogeneity. EC_{50} values with 95% confidence intervals were calculated from the logistic regression curves, and Dunnett's test ($p < 0.05$) was used to compare the treatment means to controls.

2.4.2. Characterization of PW

The analyses of COD and total nitrogen of PW-180, PW-215, and PW-250 followed the procedures 5220D and 4500N-C from the Standard Methods for the Examination of Water and Wastewater, respectively [35]. The effluents were also submitted to ionic liquid chromatography as well as gas chromatography coupled with mass spectrometry (GC-MS) analyses.

The ion chromatography analysis was performed to quantify the concentrations of acetate, chloride, nitrite, nitrate, sulphate, and phosphate. The analysis used an ion chromatography system (Thermo Scientific, Waltham, MA, USA—DIONEX ICS-5000) with a DIONEX IonPac AS19 capillary column (2 × 250 mm) (Sunnyvale, CA, USA). The eluent (EGC-KOH) was used in the following gradient mode: 10 mM KOH (0 to 10 min), 10 to 45 mM KOH (10 to 25 min), 45 mM KOH (25 to 30 min), and 10 mM KOH (30 to 35 min). The temperature was 30 °C, the flow rate was 0.250 $mL \cdot min^{-1}$, and the injection volume was 1.5 mL under 2000 psi. The detection was obtained using a conductivity detector with a suppressor module.

For the GC–MS analysis, the samples were acidified to pH 2 using HCl, and 5 mL of ethyl acetate was added. After vigorous agitation for 2 min, it was allowed to stand for 1 h. Then, 1 mL was collected from the organic phase and injected into the GC–MS equipment. GC–MS analysis used a 5975C GC MSD gas chromatograph mass spectrometry (Agilent Technologies, Santa Clara, CA, USA) equipped with an Agilent HP-5MS capillary column (30 m × 0.25 mm × 0.25 μm). Initially, the column temperature was held at 45 °C for 3 min, then programmed to 230 °C at a rate of 3 $°C \cdot min^{-1}$ with a hold time of 10 min, from 230 to 280 °C at a rate of 20 $°C \cdot min^{-1}$ and with a final hold time of 3 min. Helium was used as the carrier gas and the column head pressure was maintained at 83 kPa. The injector and detector temperatures were maintained at 250 and 300 °C, respectively, and the injection volume was 1 μL in splitless mode. The interface temperature was held at 250 °C. Mass spectra were scanned from 30 m/z to 500 m/z at a rate of 1.5 $scans \cdot s^{-1}$. The electron impact ionization energy was 70 eV [36].

3. Results and Discussion

3.1. Hydrochar Production

The yields and characteristics of H-180, H-215, and H-250 were reported in a previous study, which addressed hydrochar production and characterization [12]. Table 1 summarizes the H-180, H-215, and H-250 yield, volatile matter, ashes, fixed carbon, O/C and H/C atomic ratios, pH, zeta potential, and surface area.

Table 1. Yields and characteristics of hydrochar (H-180, H-215, and H-250) from Co-hydrothermal carbonization (Co-HTC) of sawdust and non-dewatered sewage sludge. Source: [12].

Hydrochar	Yield (%)	Volatile Matter (%) *	Ashes (%) *	Fixed Carbon (%) *	O/C	H/C	pH	BET Surface Area ($m^2 \cdot g^{-1}$)
H-180	87.3	78.5	10.4	11.1	0.5	1.3	5.4	14.8
H-215	72.7	74.4	8.0	17.6	0.4	1.2	5.2	9.6
H-250	64.1	59.3	18.4	22.3	0.2	0.9	5.8	22.7

* dry basis.

The yield of hydrochar was observed to decrease in proportion to the increase in temperature, while the fixed carbon content was enhanced. The O/C and H/C atomic ratios exhibited a reduction when Co-HTC was conducted at 250 °C, indicating an improved carbonization at elevated temperatures. This condition also resulted in a hydrochar with a superior surface area compared to those obtained at 180 and 215 °C. All hydrochars had a natural pH of around 5.5 [12]. The zeta potentials of H-180, H-215, and H-250 were −16, −17, and −21 mV, respectively [25].

The pH_{PZC} values of H-180, H-215, and H-250 were around 4. It means hydrochar is positively charged at a pH solution lower than 4 and negatively charged at a pH solution higher than 4. Therefore, at its natural pH, hydrochar presents a negatively charged surface, which could be useful for adsorbing cationic molecules, such as methylene blue.

3.2. Methylene Blue Adsorption

3.2.1. pH Effect

Figure 1 presents the removal efficiency of methylene blue with H-180, H-215, and H-250. The results from each pH were submitted to ANOVA and Tukey tests at a 95% confidence level to ascertain whether there were statistically significant differences in the removal efficiencies of H-180, H-215, and H-250. The superior performance of H-180 and H-215 compared with H-250 is clearly noticeable for all pHs. At pH 2, H-215 outperformed H-180 and H-250 ($p_{value} < 0.05$). Conversely, at pH 4, the removal efficiency of H-180 and H-250 were equal ($p_{value} > 0.05$), whereas H-250 presented the lowest removal (9.4%) of all assays. For pH values of 6, 8, and 10, all hydrochars promoted different removal efficiencies ($p_{value} < 0.05$), but H-250 remained the least effective adsorbent, irrespective of pH. The removal efficiency was found to exceed 60% at pH values of 6, 8, and 10 for both H-180 and H-215. However, H-250 exhibited a significantly lower adsorption capacity, with an efficiency of approximately 20% across all hydrochars at those pH values. Interestingly, at pH 10, the methylene blue removal was slightly higher with H-180 than with H-215 ($p_{value} < 0.05$). It is a different trend in comparison with lower pH values, in which H-215, except for pH 4, presented superior removal efficiencies than H-180 ($p_{value} < 0.05$).

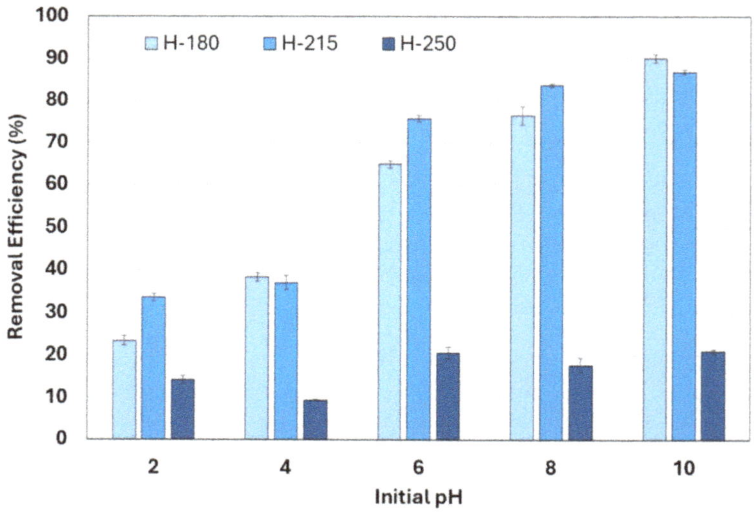

Figure 1. Influence of pH on methylene blue (100 mg·L^{-1}) adsorption (24 h at 30 °C and 100 rpm) by hydrochars (3 g·L^{-1}) from co-hydrothermal carbonization (Co-HTC) of sawdust and non-dewatered sewage sludge.

In aqueous solutions, methylene blue can exist as a cationic (MB^+) and as an undissociated molecule (MB^0). MB^0 predominates at pHs lower than 3.8, MB^0 (50%) and MB^+ (50%) coexisted equally at pH = 3.8, and MB^+ is the only species at pH > 6.0 [37]. Therefore, considering electrostatic interactions, one could explain the better adsorption performance of hydrochars at pH 6, 8, and 10 due to the hydrochars' pH_{PZC} and the pH-dependent methylene blue speciation. Even for H-250, it is possible to note a slight increase in the removal efficiency at pH 6, 8, and 10 compared with pH 2 and 4. Similarly, Khoshbouy and colleagues demonstrated that methylene blue removal was enhanced at basic pH due to the negative zeta potential of hydrochar, which was associated with elevated concentrations of OH^- ions and deprotonation of functional groups. This resulted in enhanced electrostatic attraction and hydrogen bond interactions between the hydrochar and dye [38].

The difference in removal efficiency of H-250 compared with H-180 and H-215 can be related to the surface functional groups of hydrochar. Hydrochars obtained from lignocellulosic materials can present oxygenated functional groups, such as phenolic, carboxylic, and lactonic, that can help with adsorption [39–41]. At low temperatures, Co-HTC can preserve more oxygenated functional groups in the hydrochar surface [12], which could explain why H-180 and H-250 outperformed H-250. Li and colleagues demonstrated that methylene blue adsorption was more effective for hydrochar derived from Co-HTC of bamboo and polyvinyl chloride than for hydrochar produced from a single feedstock. This finding was attributed to the presence of a greater number of hydroxyl and carboxyl groups on the surface of the Co-HTC-derived hydrochar, which improved the adsorption of methylene blue [11]. Therefore, given their better performance than H-250, only H-180 and H-215 were considered for the following tests, which were carried out at pH 10.

3.2.2. Adsorption Isotherms

Figure 2a,b shows the isotherm of methylene blue adsorption by H-180 and H-215, respectively. The q_{max} for both hydrochars was around 70 $mg \cdot g^{-1}$. Table 2 compares q_{max} for methylene blue of other hydrochars with that obtained in this study. The value obtained herein was similar to another hydrochar produced at 180 °C from citrus waste when the methylene blue adsorption was conducted at 4 °C, where q_{max} was 66 $mg \cdot g^{-1}$; however, when the adsorption temperature was increased to 20 and 36 °C, q_{max} reduced to 51 and 31 $mg \cdot g^{-1}$, respectively. In the same study, another hydrochar from winery waste, also obtained at 180 °C, presented q_{max} values between 29 and 37 $mg \cdot g^{-1}$ for adsorptions at 4, 20, and 36 °C, which are quite low compared with the results of this study. Additionally, it is worth noting that the BET surface area of citrus waste and winery waste hydrochars were 46 and 34 $m^2 \cdot g^{-1}$, respectively, while the surface area of H-180 was 14.8 $m^2 \cdot g^{-1}$ [42]. Even with a lower surface area, the hydrochar from Co-HTC of sawdust and sewage sludge at 180 °C produced herein had a higher adsorption capacity for methylene blue. Another study used only sewage sludge for hydrochar production at 190, 220, and 250 °C and showed q_{max} of 71, 54, and 38 $mg \cdot g^{-1}$, respectively, for methylene blue adsorption [43]. It is noteworthy that H-180 exhibits a comparable q_{max} value to the hydrochar derived from the 190 °C treatment in the aforementioned study. Additionally, the H-215 displays a higher q_{max} value than the hydrochar obtained from the 220 °C treatment in the aforementioned study. Therefore, it demonstrates that combining wastes, such as lignocellulosic materials and sewage sludge, in a Co-HTC process can improve the adsorption capacity of hydrochar.

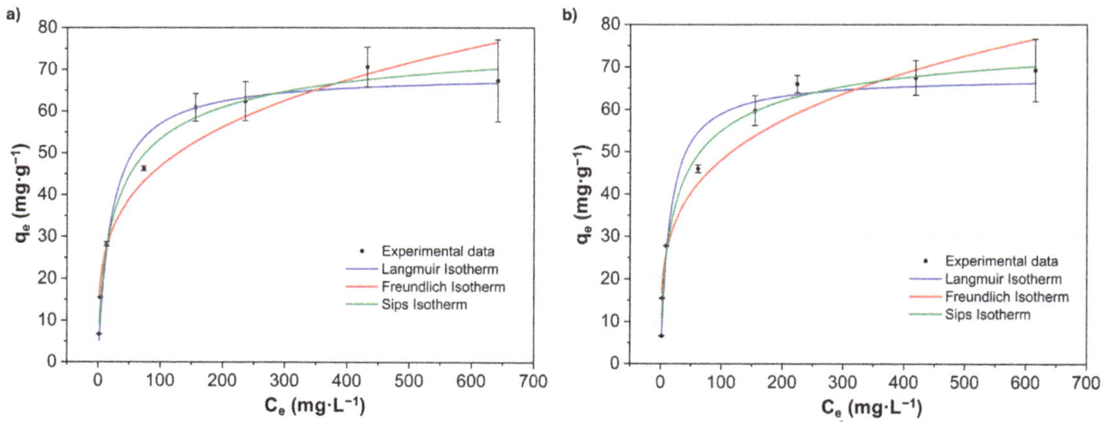

Figure 2. Isotherm models of methylene blue adsorption by H-180 (**a**) and H-215 (**b**) obtained at hydrochar load of 3 g·L^{-1} at pH 10, 30 °C, 100 rpm for 24 h.

Despite the q_{max} obtained for H-180 and H-215, one can mention hydrochar activation to increase the adsorption capacity, as performed in some works [30,44–46]. However, it is important to consider that an activation step would result in increasing the costs of adsorbent production, in addition to the generation of effluents when using wet routes. Furthermore, some studies that performed hydrochar activation demonstrated lower q_{max} values than those achieved for H-180 and H-215. For example, Lin and colleagues synthesized hydrochar from corn straw (HC), which was also N-doped with urea (UHC), melamine (MHC), and NH_4Cl (AHC) to increase methylene blue adsorption. Nevertheless, the q_{max} values of all hydrochars obtained were lower than those of H-180 and H-215 (HC: around 10 mg·g^{-1}; UHC: <20 mg·g^{-1}; MHC: <40 mg·g^{-1}; AHC: <60 mg·g^{-1}) [47]. In another study, an oxide–hydrochar composite produced from pomegranate peels was activated with $ZnCl_2$, but the q_{max} value of the activated material was lowered to 50 mg·g^{-1} [48].

Table 2. Maximum adsorption capacity (q_{max}) of different hydrochars for methylene blue.

Raw Material	Hydrochar Activation	q_{max} (mg·g^{-1})	Reference
Sawdust and sewage sludge	No	70	This study
Citrus waste	No	31–66	[42]
Winery waste	No	2–37	
Sewage sludge	No	38–71	[43]
Corn straw	Yes	20–60	[47]
Pomegranate peels	Yes	50	[48]

The q_{max} of a carbon-derived adsorbent for methylene blue can vary significantly depending on the feedstock and production conditions utilized. This phenomenon has been extensively discussed in several review articles. For instance, Kasuma et al. reported a range of q_{max} values from 1.33 to 300 mg·g^{-1}. Intriguingly, some activated carbons exhibited q_{max} values below 9 mg·g^{-1}, notably lower than the values observed for H-180 and H-215 [22]. In another study, q_{max} values of carbon-based adsorbents—activated carbon, graphite porous carbon, and carbon nanotubes—ranged from 0.003 to 2936 mg·g^{-1}, highlighting the variability in outcomes [49].

Figure 2a,b also presents the adjustment of the experimental data regarding H-180 and H-215, respectively, to Langmuir, Freundlich, and Sips isotherm models, for which the parameter values are reported in Table 3. The Langmuir isotherm model assumes monolayer adsorption, whereby the adsorption process occurs exclusively on a homogeneous adsorbent surface with identical adsorption sites, forming a single layer of adsorbate molecules. Once a site is occupied, no further adsorption can occur at that site. Furthermore, the model assumes that the adsorption–desorption process is reversible. On the other hand, the Freundlich isotherm model assumes multilayer adsorption, which occurs on a heterogeneous surface with adsorption sites of varying energy levels. The molecules can adsorb onto each other after the initial layer has formed. Regarding the Sips isotherm model, it is a hybrid model obtained from the combination of both the models aforementioned [50,51].

Table 3. Values of parameters related to Langmuir, Freundlich, and Sips isotherm models.

Isotherm Model	Parameter	H-180	H-215
Langmuir	q_{max} ($mg \cdot g^{-1}$)	69.01± 2.93 (56.40–81.61) *	67.78 ± 2.83 (55.60–79.96) *
	K_L	0.048 ± 0.012	0.068 ± 0.018
	R^2	0.9669	0.9636
	MRD (%)	11.56	9.89
Freundlich	K_F	13.92 ± 3.08	14.87 ± 3.09
	n	3.79 ± 0.57	3.92 ± 0.58
	R^2	0.9258	0.9289
	MRD (%)	25.54	27.68
Sips	q_{max} ($mg \cdot g^{-1}$)	80.35 ± 7.18 (49.45–111.25) *	79.96 ± 7.23 (48.85–111.07) *
	K_S	0.090 ± 0.020	0.112 ± 0.022
	n	0.672 ± 0.100	0.647 ± 0.099
	R^2	0.9852	0.9850
	MRD (%)	9.26	11.96

* 95% confidence interval.

Considering the R^2 and MRD values in Table 3, the Sips isotherm model is suggested for describing the methylene blue adsorption by both H-180 and H-215. This model indicated a q_{max} value of around 80 $mg \cdot g^{-1}$. Additionally, no significant difference was noticed for both hydrochars regarding their adsorption performance since the 95% confidence intervals for q_{max} values of H-180 and H-215 overlapped. Thus, only H-180 was used in the following tests considering energy-saving reasons. It is preferable to conduct Co-HTC at 180 °C rather than at 215 °C.

3.2.3. Adsorption Kinetics

Figure 3a presents the removal percentage of methylene blue with H-180 against the time. The adsorption increases rapidly in the first 3 h, reaching almost 60% of removal. From 13 h onwards, around 80% of methylene blue was removed and this removal percentage remained practically constant until 24 h. According to other studies that also evaluate methylene blue adsorption by hydrochar, a significant number of adsorption sites were occupied by MB molecules as the adsorption time increased, resulting in a gradual slowing down of the subsequent mass transfer and adsorption process until equilibrium was reached [46,47,52]. In general, it is possible to identify three stages: 10–60 min, 90–480 min, and 720–1440 min. Akbari and colleagues suggest that, in the first stage, the presence of numerous vacant active sites on the surface of the adsorbent facilitates the interaction between methylene blue molecules and active sites, thereby achieving a high adsorption rate. In the second stage, the rate of adsorption declines due to a reduction in the number of available active sites and the driving force, which is attributable to a decrease in the concentration of methylene blue. In the final stage, the reduction in the number of vacant active sites and dye concentration, coupled with the rise in electrostatic repulsion forces resulting from the adsorption of cationic molecules on the surface, might be the causes of the observed slow and steady adsorption rate, indicating that equilibrium was reached [53].

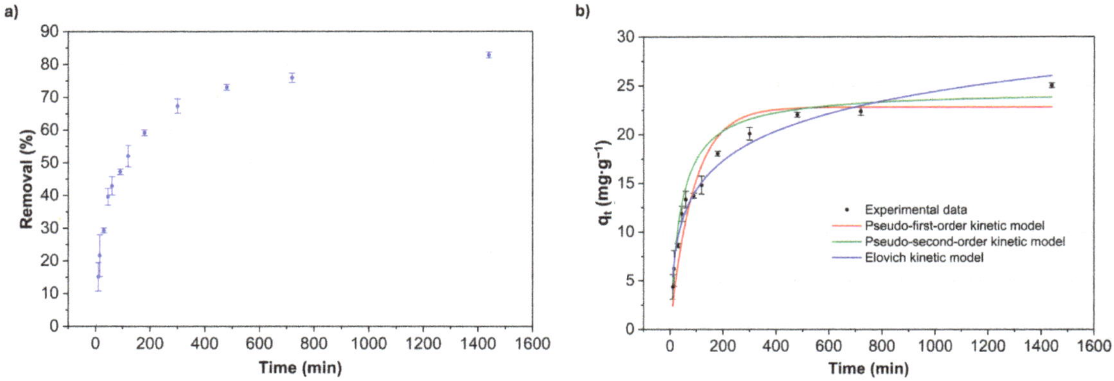

Figure 3. Removal percentage (**a**) and kinetic models (**b**) of methylene blue adsorption by hydrochar (H-180) obtained with an initial dye concentration of 100 mg·L^{-1} and a hydrochar load of 3 g·L^{-1} at pH 10, 30 °C, and 100 rpm.

In Figure 3b, kinetic models (PFO, PSO, and Elovich) were adjusted to the experimental data. As shown in Table 4, the Elovich kinetic model fits better than the others, considering the R^2 and MRD values. The suitability of the Elovich model compared with the PFO and PSO models was also reported in another study of methylene blue adsorption by hydrochar [30]. The Elovich equation is suitable for describing chemisorption in systems with heterogeneous adsorbing surfaces [30,54–56], suggesting therefore that this type of adsorption occurs between H-180 and methylene blue.

Table 4. Values of parameters related to pseudo-first-order (PFO), pseudo-second-order (PSO), and Elovich kinetic models.

Kinetic Model	Parameter	H-180
Pseudo-First-Order	K_1	0.011 ± 0.001
	q_e (mg·g^{-1})	22.85 ± 0.85
	R^2	0.9124
	MRD (%)	5.57
Pseudo-Second-Order	K_2	0.001 ± 0.0002
	q_e (mg·g^{-1})	24.54 ± 1.11
	R^2	0.8909
	MRD (%)	7.09
Elovich	β	0.225 ± 0.012
	α	1.08 ± 0.19
	R^2	0.9779
	MRD (%)	2.81

3.2.4. Adsorption Thermodynamics

Table 5 presents the thermodynamic parameters of the methylene blue adsorption by H-180, obtained by linearizing Equation (7) (Figure 4). The value of Gibbs free energy change (ΔG) tends to become more negative as the temperature increases from 30 to 60 °C. This suggests that increasing the temperature favors the spontaneity of the process. Also, the positive value of enthalpy change (ΔH) indicates that the process is endothermic. In this instance, the positive value of entropy change (ΔS) is advantageous for the spontaneity of adsorption, as it makes ΔG negative by compensating for ΔH, thereby facilitating the occurrence of adsorption. This corroborates with previous studies using other hydrochars for adsorbing methylene blue [30,42].

Table 5. Thermodynamic parameters of the methylene blue adsorption by hydrochar (H-180) from co-hydrothermal carbonization (Co-HTC) of sawdust and non-dewatered sewage sludge.

ΔG ($J \cdot mol^{-1}$)	ΔS ($J \cdot mol^{-1} \cdot K^{-1}$)	ΔH ($J \cdot mol^{-1}$)
−3845.12 (30 °C) −4128.52 (40 °C) −4411.92 (50 °C) −4695.32 (60 °C)	28.34	4746.13

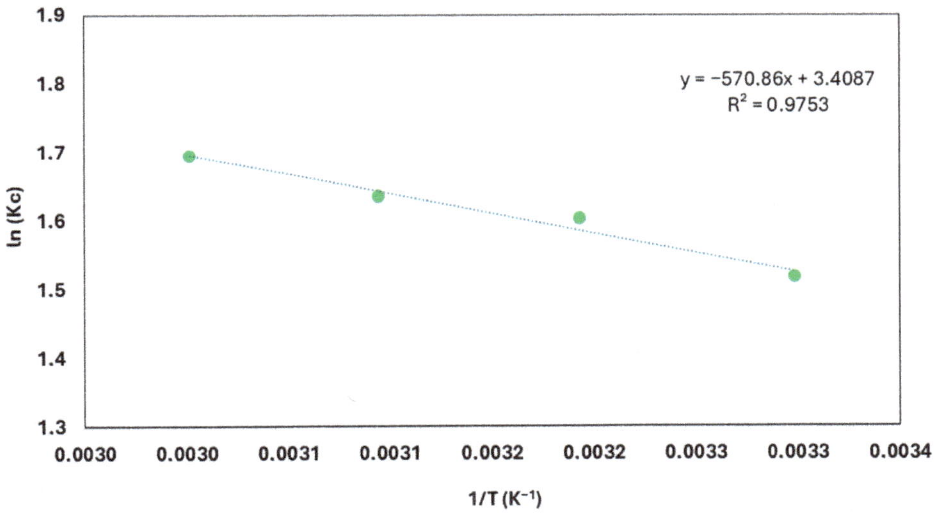

Figure 4. Linearization of Equation (7) to obtain the thermodynamic parameters of the methylene blue adsorption by hydrochar (H-180) from co-hydrothermal carbonization (Co-HTC) of sawdust and non-dewatered sewage sludge.

3.2.5. Hydrochar Regeneration

Five cycles of methylene blue adsorption were performed to evaluate the reusability of H-180. Figure 5 presents the removal percentage of methylene blue for each cycle. It is worth noting that after five cycles the removal percentage remained around 80%; it decreased from 92.5% (first cycle) to 79.2% (fifth cycle). This represents a reduction of only 14.4% in relation to the first cycle. For comparison purposes, Akbari and colleagues reported a removal percentage reduction from 85% (first cycle) to 77% (fourth cycle) in methylene blue adsorption by hydrochar from liquorice-root pulp. According to the authors, the changes to the adsorption sites on the adsorbent surface reduce the removal after each cycle. This occurs because some methylene blue molecules attach permanently to the surface, blocking the vacant pores and reducing dye adsorption efficiency [53].

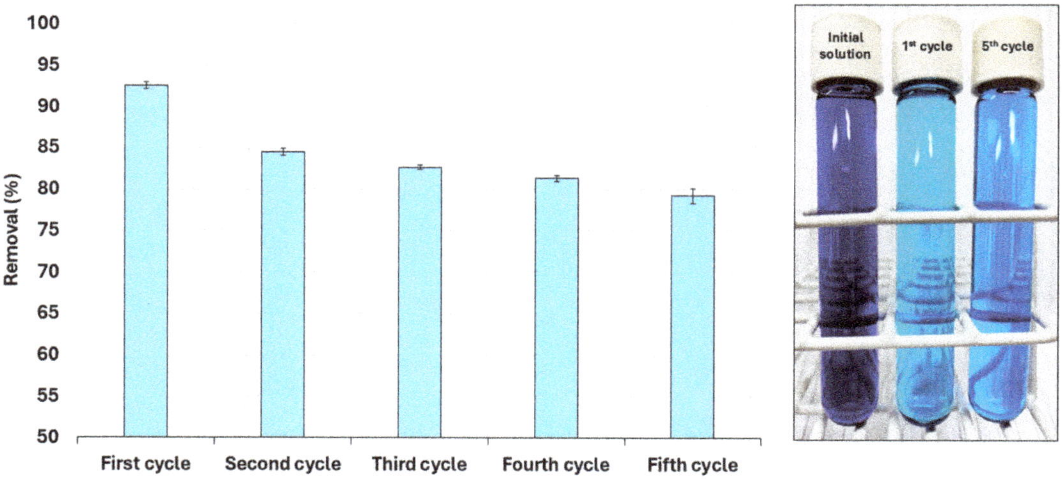

Figure 5. Removal percentage of methylene blue adsorption after five cycles by hydrochar (H-180) using an initial dye concentration of 100 mg·L^{-1} and a hydrochar load of 3 g·L^{-1} at pH 10, 30 °C, and 100 rpm for 16 h.

3.3. Process Water (PW)—Ecotoxicity and Characterization

Acute toxicity tests with PW-180, PW-215, and PW-250 were carried out on the microcrustacean *Daphnia magna* following OECD guideline 202 (2004) (Section 2.4.1). Exposure of *D. magna* to the PW-180, PW-215, and PW-250 showed high toxicity, with inhibitory effects observed after 48 h of exposure at the lowest percentage for all the PWs tested. The effect was enhanced by increasing the PW concentration in the medium (Figure 6).

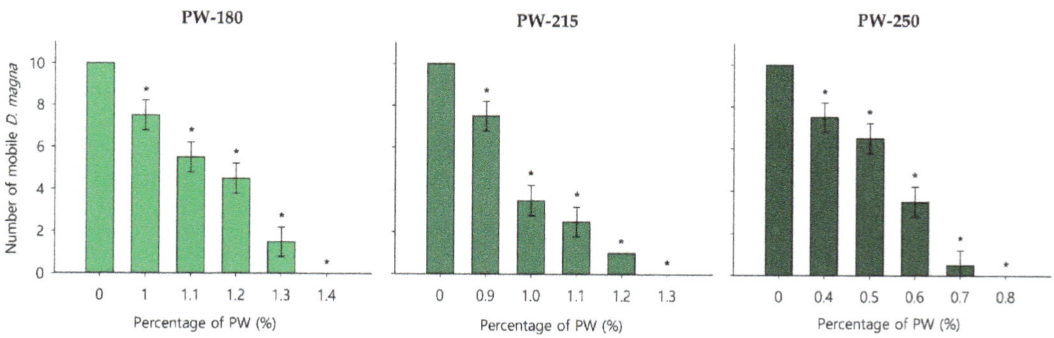

Figure 6. Average number of mobile *Daphnia magna* microcrustaceans (bars ± standard deviation, n = 6) after 48 h of exposure to process water (PW) from co-hydrothermal carbonization (Co-HTC) of sawdust and non-dewatered sewage sludge at 180, 215, and 250 °C. Asterisks (*) represent a significant increase in the number of immobile microcrustaceans compared with the respective control (Dunnett's test, $p \leq 0.05$).

PWs can be considered as complex mixtures of dissolved organic compounds and nutrients (Tables 5 and 6). The CODs of PW-180, PW-215, and PW-250 were 12.87 ± 0.55, 12.22 ± 0.35, and 17.33 ± 0.75 g·L^{-1} [12]. Regarding total nitrogen, the values for PW-180, PW-215, PW-250 were 1.02 ± 0.09, 0.92 ± 0.06, and 1.00 ± 0.11 g·L^{-1}, respectively. Both parameters align with other PWs reported in the literature, as presented in Table 6.

Table 6. Examples of chemical oxygen demand (COD) and total nitrogen values of process waters (PWs) from hydrothermal carbonization (HTC) of different biomass wastes.

Waste	Hydrothermal Carbonization Condition *	Chemical Oxygen Demand ($g \cdot L^{-1}$)	Total Nitrogen ($g \cdot L^{-1}$)	Reference
Agricultural residue digestate	T: 200; S/L: 1/5; t: 1	42.2	1.9	[57]
	T: 250; S/L: 1/5; t: 1	46.3	2.2	
Municipal solid waste digestate	T: 200; S/L: 1/5; t: 1	18.1	2.4	
	T: 250; S/L: 1/5; t: 1	16.4	1.7	
Sewage sludge digestate	T: 200; S/L: 1/5; t: 1	38.9	4.5	
	T: 250; S/L: 1/5; t: 1	43.6	4.7	
Oat husk	T: 219.2; S/L: 1/12.5; t: 0.5	13.2	1.8	[58]
Water hyacinth	T: 150; S/L: 1/10; t: 1	19.0	-	[59]
	T: 200; S/L: 1/10; t: 1	27.5	-	
	T: 250; S/L: 1/10; t: 1	31.4	-	
Grape Marc	T: 220; S/L: 1/10; t: 1	33.3		[60]
Grape Marc extracted		31.1	-	
Sewage sludge digestate	T: 160; S/L: 1/1; t: 0.5	12.6	-	[61]
	T: 220; S/L: 1/1; t: 0.5	12.9	-	
	T: 250; S/L: 1/1; t: 0.5	12.2	-	

* T: temperature (°C); S/L: solid-to-liquid ratio; t: time (h).

The presence and interaction of PW compounds in the exposure medium probably resulted in the toxic effects observed for *D. magna* [62,63]. Petrovič and colleagues showed the high toxicity of PW from Co-HTC of vegetable oil industry waste and sewage sludge to *D. magna* (EC_{50} = 0.5–2%) [64]. In another study, the PW from HTC of microalgae was toxic to *Allivibrio fischeri* (EC_{50} = 1.8%) [65]. For *Lemna minor*, the PW from sewage sludge was also harmful, causing complete decay at a 10% dilution [66]. In all the studies mentioned, the justification for the effects observed is consistent with our findings, which are also related to the complexity of PW composition (Tables 7 and 8).

Table 7. Content of acetate, chloride, nitrite, nitrate, sulphate, and phosphate in the process water (PW) from co-hydrothermal carbonization (Co-HTC) of sawdust and non-dewatered sewage sludge at three temperatures (180, 215, and 250 °C).

Ion	PW-180	PW-215	PW-250
Acetate ($mg \cdot L^{-1}$)	3333.67 ± 18.21	3600.67 ± 76.20	4584.00 ± 27.31
Chloride ($mg \cdot L^{-1}$)	118.94 ± 1.10	120.70 ± 1.44	124.70 ± 0.64
Nitrite ($mg \cdot L^{-1}$)	7.62 ± 0.00	1.99 ± 0.03	nd
Nitrate ($mg \cdot L^{-1}$)	13.80 ± 0.31	6.60 ± 0.14	nd
Sulfate ($mg \cdot L^{-1}$)	90.72 ± 0.39	74.87 ± 0.90	85.45 ± 14.32
Phosphate ($mg \cdot L^{-1}$)	441.93 ± 0.99	203.47 ± 0.53	194.32 ± 1.54

nd: not detected.

Table 8. Compounds identified by GC–MS in the process water (PW-180, PW-215, and PW-250) from co-hydrothermal carbonization (Co-HTC) of sawdust and non-dewatered sewage sludge.

Process Water	Retention Time (min.)	Peak Area (%)	Compound	CAS Number
PW-180	3.722	12.00	Isobutyl acetate	000110-19-0
	5.188	87.99	Furfural	000098-01-1
	6.120	2.32	Furfural	000098-01-1
	26.810	1.33	Cyclohexasiloxane, dodecamethyl-	000540-97-6
PW-215	3.657	4.15	Isobutyl acetate	000110-19-0
	5.135	5.60	Furfural	000098-01-1
	7.574	0.93	2-Cyclopenten-1-one, 2-methyl-	001120-73-6
	7.806	1.05	Ethanone, 1-(2-furanyl)-	001192-62-7
	10.049	2.33	2-Furancarboxaldehyde, 5-methyl-	000620-02-0
	12.661	7.74	1,2-Cyclopentanedione, 3-methyl-	000765-70-8
	15.510	14.07	Phenol, 2-methoxy-	000090-05-1
	17.071	3.05	2-Cyclopenten-1-one, 3-ethyl-2-hydroxy-	021835-01-8
	24.371	3.54	(Z)-4-Methyl-5-(2-oxopropylidene)-5H-furan-2-one	026474-45-3
	27.564	7.93	Phenol, 2,6-dimethoxy-	000091-10-1
	29.885	3.02	Vanillin	000121-33-5
	35.049	5.98	Homovanillyl alcohol	002380-78-1
PW-250	3.716	1.50	Isobutyl acetate	000110-19-0
	5.028	0.42	Pyrazine, methyl-	000109-08-0
	5.223	1.21	2-Cyclopenten-1-one	000930-30-3
	7.580	4.67	2-Cyclopenten-1-one, 2-methyl-	001120-73-6
	10.144	1.26	2-Cyclopenten-1-one, 3-methyl-	002758-18-1
	10.981	1.27	Phenol	000108-95-2
	11.467	1.99	2-Cyclopenten-1-one, 2,3-dimethyl-	001121-05-7
	12.773	6.22	1,2-Cyclopentanedione, 3-methyl-	000765-70-8
	13.248	1.39	2-Cyclopenten-1-one, 2,3-dimethyl-	001121-05-7
	15.551	21.46	Phenol, 2-methoxy-	000090-05-1
	20.240	0.42	Furan, 2,5-dihydro-2,5-dimethyl-	059242-27-2
	20.560	0.80	Creosol	000093-51-6
	21.724	5.04	Catechol	000120-80-9
	23.890	1.86	1,2-Benzenediol, 3-methoxy-	000934-00-9
	24.478	3.57	Phenol, 4-ethyl-2-methoxy-	002785-89-9
	26.804	1.31	Cyclohexasiloxane, dodecamethyl-	000540-97-6
	27.712	8.91	Phenol, 2,6-dimethoxy-	000091-10-1
	30.223	1.10	4-Hydroxy-2-methoxybenaldehyde	018278-34-7
	33.434	0.55	5-Hepten-3-yn-2-ol, 6-methyl-5-(1-methylethyl)-	063922-41-8
	33.933	0.51	Cycloheptasiloxane, tetradecamethyl-	000107-50-6
	35.102	3.86	Homovanillyl alcohol	002380-78-1
	39.589	1.12	Homovanillic acid	000306-08-1
	39.725	2.30	Homovanillic acid	000306-08-1

When comparing the toxicities of PWs from co-HTC at different temperatures, the EC_{50} values indicated the highest toxicity for PW-250 > PW-215 > PW-180, with values of 0.51% (0.48–0.54%), 0.97% (0.95–1.00%), and 1.13% (1.09–1.17%), respectively. These results suggest that the increase in Co-HTC temperature elevated the toxicity on *D. magna*. The observed increase in toxicity aligns with the rise in compound generation due to the temperature increase (Table 8). A possible explanation for this behavior could be the high level of acetate present in the PWs (Table 7). As lignocellulosic biomass, sawdust is composed mainly of cellulose, hemicellulose, and lignin [67]. During hydrothermal treatments, such as Co-HTC, the acetyl groups on the hemicellulose chains are cleaved and form acetic acid in the solution. Depending on the process conditions (e.g., temperature), cellulose, hemicellulose, and lignin can undergo partial or complete degradation. The more severe the hydrothermal treatment, the more degradation products are generated, as confirmed by PW compositions in Table 6. For instance, the degradation of cellulose and hemicellulose monomers forms hydroxymethylfurfural and furfural, while the lignin degradation produces phenolic compounds [68–70]. These molecules are intermediates in hydrochar formation [3]. Therefore, the increase in Co-HTC temperature favors the decomposition of organic biomass and catalyzes the formation of various other chemical substances that may be toxic [21,71]. Thus, PW became progressively more heterogeneous and toxic as the temperature rose.

Although no data were found in the literature regarding the toxicity of PW generated from Co-HTC of sawdust and sewage sludge, similar results were observed in a study exposing *D. magna* to dilutions of PW produced from Co-HTC of sewage sludge and whey at different temperatures. The study showed that the temperature increase affected the composition of PW; the toxicity of PW produced at 250 °C was approximately four times higher than that obtained at 200 °C for the same reaction time [21]. These findings reinforce that temperature significantly impacts both the composition and toxicity of PW.

When the daphnids were exposed to the previously obtained EC_{50} values with the pH of the exposure medium adjusted to neutral, there was no change in toxicity levels (Figure 7). This indicates that the toxicity effects observed for PWs at different Co-HTC temperatures were not influenced by the acidic pH of the exposure medium ($pH_{PW\text{-}180} = 5.8$; $pH_{PW\text{-}215} = 5.3$; $pH_{PW\text{-}250} = 5.1$), but probably by the specific composition of each PW and the interactions among their chemical constituents (Tables 7 and 8). However, while this conclusion appears evident, it should be interpreted with caution, as the toxicity of individual compounds and mixtures within each PW has not been tested.

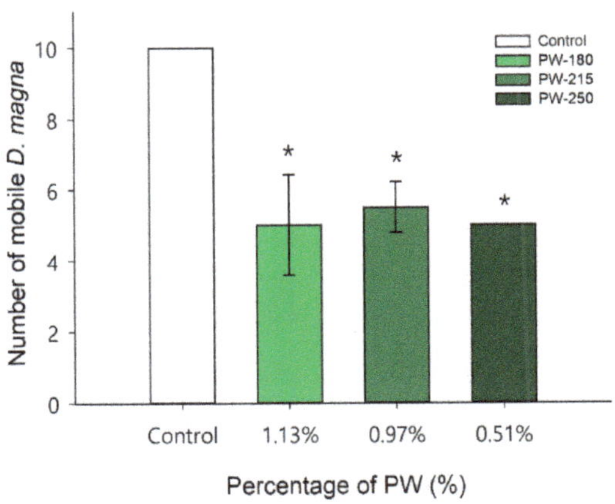

Figure 7. Average number of mobile *Daphnia magna* microcrustaceans (bars ± standard deviation, n = 6) after 48 h of exposure to effect concentration values for 50% of exposed organisms (EC_{50}) previously obtained for process water (PW) from the co-hydrothermal carbonization (Co-HTC) of sawdust and non-dewatered sewage sludge at 180, 215, and 250 °C with pH adjustment of the exposure medium with NaOH (pH = 7). The asterisks (*) represent a significant increase in the number of immobile microcrustaceans compared with the control test (Dunnett's test, $p \leq 0.05$).

These results underscore the necessity of identifying alternative methods for treating and, most crucially, valorization of PW to reduce its potential environmental impacts. Despite its toxicity, the PW from the Co-CHT of sawdust and sewage sludge contains molecules with commercial potential (e.g., furfural). Instead of viewing this by-product as an effluent to be treated and disposed of correctly in the environment, it is imperative to assess the potential of PW to obtain marketable products. For example, furfural, levulinic acid, vanillin, and catechol are considered platform chemicals, which are molecules that can serve as a substrate to produce a range of other higher value-added products [72–75].

4. Conclusions

This work assessed three hydrochars (H-180, H-215, and H-250) from Co-HTC of sawdust and non-dewatered sewage sludge as potential methylene blue adsorbents, and it also evaluated the ecotoxicological effects of the resulting PW. H-180 and H-215 had better adsorption performance than H-250. H-180 and H-215 presented a q_{max} value of around 70 mg·g^{-1}, which was superior compared with the adsorption of methylene blue by other hydrochars in the literature. The Sips isotherm model was better than the Langmuir and Freundlich models for describing the adsorption with H-180 and H-215. The Sips model indicated a q_{max} value of around 80 mg·g^{-1}. It is noteworthy that H-180 and H-215 were not subjected to modification by physical or chemical activation, which could enhance their adsorption capacity. However, an activation step would also raise the production cost of the adsorbent. Compared with the PFO and PSO models, the Elovich model presented a superior fit for describing the kinetics of methylene blue adsorption by H-180. Regarding the toxicological effects of the resulting PWs on *Daphnia magna*, the high toxicity, especially at higher Co-HTC temperatures, underscores the need for careful management and further study to assess their environmental impact and potential valorization approaches.

Author Contributions: Conceptualization, M.C.; methodology, M.C., T.B.H., N.L.J. and B.K.; formal analysis, M.C., T.B.H., N.L.J., B.K. and V.G.; investigation, M.C., T.B.H., N.L.J., B.K. and V.G.; resources, N.L.J., B.K., V.G., H.B., R.B., A.L.W., W.G.M. and A.B.d.C.J.; data curation, M.C., T.B.H., N.L.J. and B.K.; writing—original draft preparation, M.C. and T.B.H.; writing—review and editing, M.C., T.B.H., N.L.J., B.K., V.G., H.B., R.B., A.L.W., W.G.M. and A.B.d.C.J.; visualization, M.C. and T.B.H.; supervision, N.L.J., B.K., H.B., R.B., A.L.W., W.G.M. and A.B.d.C.J.; project administration, A.B.d.C.J.; funding acquisition, R.B., W.G.M., A.L.W. and A.B.d.C.J. All authors have read and agreed to the published version of the manuscript.

Funding: This research was funded by Brazilian Foundation Coordenação de Aperfeiçoamento de Pessoal de Nível Superior (CAPES)—Finance Code 001.

Institutional Review Board Statement: Not applicable.

Informed Consent Statement: Not applicable.

Data Availability Statement: The original contributions presented in this study are included in the article. Further inquiries can be directed to the corresponding author.

Acknowledgments: The authors thank the Brazilian foundation Coordenação de Aperfeiçoamento de Pessoal de Nível Superior (CAPES), the laboratory of the graduate program in Bioprocess Engineering and Biotechnology of the Federal University of Paraná, and the DEEP (Déchets Eaux Environnement Pollutions) Laboratory of the Lyon National Institute of Applied Sciences (INSA Lyon) for the supervision and the analyses carried out. This work was performed within the framework of the EUR H2O'Lyon (ANR-17-EURE-0018) of Université de Lyon (UdL), within the program "Investissements d'Avenir" operated by the French National Research Agency (ANR).

Conflicts of Interest: The authors declare no conflicts of interest.

References

1. Yu, C.; Chen, R.; Li, J.J.; Li, J.J.; Drahansky, M.; Paridah, M.; Moradbak, A.; Mohamed, A.Z.; Owolabi, F.H.; Abdulwahab, T.; et al. *Pyrolysis: A Sustainable Way to Generate Energy from Waste*; Intech: Vienna, Austria, 2012; p. 13.
2. You, S.; Sik, Y.; Chen, S.S.; Tsang, D.C.W.; Kwon, E.E.; Lee, J.; Wang, C. A Critical Review on Sustainable Biochar System through Gasification: Energy and Environmental Applications. *Bioresour. Technol.* **2017**, *246*, 242–253. [CrossRef] [PubMed]

3. Cavali, M.; Libardi Junior, N.; de Sena, J.D.; Woiciechowski, A.L.; Soccol, C.R.; Belli Filho, P.; Bayard, R.; Benbelkacem, H.; de Castilhos Junior, A.B. A Review on Hydrothermal Carbonization of Potential Biomass Wastes, Characterization and Environmental Applications of Hydrochar, and Biorefinery Perspectives of the Process. *Sci. Total Environ.* **2023**, *857*, 159627. [CrossRef] [PubMed]
4. Prochnow, F.D.; Cavali, M.; Dresch, A.P.; Belli, I.M.; Libardi, N.; de Castilhos, A.B. Biochar: From Laboratory to Industry Scale—An Overview of Scientific and Industrial Advances, Opportunities in the Brazilian Context, and Contributions to Sustainable Development. *Processes* **2024**, *12*, 1006. [CrossRef]
5. Parshetti, G.K.; Chowdhury, S.; Balasubramanian, R. Hydrothermal Conversion of Urban Food Waste to Chars for Removal of Textile Dyes from Contaminated Waters. *Bioresour. Technol.* **2014**, *161*, 310–319. [CrossRef]
6. Zhao, K.; Li, Y.; Zhou, Y.; Guo, W.; Jiang, H.; Xu, Q. Characterization of Hydrothermal Carbonization Products (Hydrochars and Spent Liquor) and Their Biomethane Production Performance. *Bioresour. Technol.* **2018**, *267*, 9–16. [CrossRef] [PubMed]
7. Langone, M.; Basso, D. Process Waters from Hydrothermal Carbonization of Sludge: Characteristics and Possible Valorization Pathways. *Int. J. Environ. Res. Public. Health* **2020**, *17*, 6618. [CrossRef] [PubMed]
8. Kambo, H.S.; Dutta, A. A Comparative Review of Biochar and Hydrochar in Terms of Production, Physico-Chemical Properties and Applications. *Renew. Sustain. Energy Rev.* **2015**, *45*, 359–378. [CrossRef]
9. Wang, Q.; Wu, S.; Cui, D.; Zhou, H.; Wu, D.; Pan, S.; Xu, F.; Wang, Z. Co-Hydrothermal Carbonization of Organic Solid Wastes to Hydrochar as Potential Fuel: A Review. *Sci. Total Environ.* **2022**, *850*, 158034. [CrossRef] [PubMed]
10. Shan, G.; Li, W.; Bao, S.; Li, Y.; Tan, W. Co-Hydrothermal Carbonization of Agricultural Waste and Sewage Sludge for Product Quality Improvement: Fuel Properties of Hydrochar and Fertilizer Quality of Aqueous Phase. *J. Environ. Manage* **2023**, *326*, 116781. [CrossRef] [PubMed]
11. Li, H.Z.; Zhang, Y.N.; Guo, J.Z.; Lv, J.Q.; Huan, W.W.; Li, B. Preparation of Hydrochar with High Adsorption Performance for Methylene Blue by Co-Hydrothermal Carbonization of Polyvinyl Chloride and Bamboo. *Bioresour. Technol.* **2021**, *337*, 125442. [CrossRef]
12. Cavali, M.; Benbelkacem, H.; Kim, B.; Bayard, R.; Libardi Junior, N.; Gonzaga Domingos, D.; Woiciechowski, A.L.; de Castilhos Junior, A.B. Co-Hydrothermal Carbonization of Pine Residual Sawdust and Non-Dewatered Sewage Sludge—Effect of Reaction Conditions on Hydrochar Characteristics. *J. Environ. Manage* **2023**, *340*, 117994. [CrossRef]
13. Masoumi, S.; Borugadda, V.B.; Nanda, S.; Dalai, A.K. Hydrochar: A Review on Its Production Technologies and Applications. *Catalysts* **2021**, *11*, 939. [CrossRef]
14. Katheresan, V.; Kansedo, J.; Lau, S.Y. Efficiency of Various Recent Wastewater Dye Removal Methods: A Review. *J. Environ. Chem. Eng.* **2018**, *6*, 4676–4697. [CrossRef]
15. Dutta, S.; Adhikary, S.; Bhattacharya, S.; Roy, D.; Chatterjee, S.; Chakraborty, A.; Banerjee, D.; Ganguly, A.; Nanda, S.; Rajak, P. Contamination of Textile Dyes in Aquatic Environment: Adverse Impacts on Aquatic Ecosystem and Human Health, and Its Management Using Bioremediation. *J. Environ. Manage* **2024**, *353*, 120103. [CrossRef]
16. Al-Tohamy, R.; Ali, S.S.; Li, F.; Okasha, K.M.; Mahmoud, Y.A.G.; Elsamahy, T.; Jiao, H.; Fu, Y.; Sun, J. A Critical Review on the Treatment of Dye-Containing Wastewater: Ecotoxicological and Health Concerns of Textile Dyes and Possible Remediation Approaches for Environmental Safety. *Ecotoxicol. Environ. Saf.* **2022**, *231*, 113160. [CrossRef] [PubMed]
17. Merzari, F.; Langone, M.; Andreottola, G.; Fiori, L. Methane Production from Process Water of Sewage Sludge Hydrothermal Carbonization. A Review. Valorising Sludge through Hydrothermal Carbonization. *Crit. Rev. Environ. Sci. Technol.* **2019**, *49*, 947–988. [CrossRef]
18. Wirth, B.; Reza, T.; Mumme, J. Influence of Digestion Temperature and Organic Loading Rate on the Continuous Anaerobic Treatment of Process Liquor from Hydrothermal Carbonization of Sewage Sludge. *Bioresour. Technol.* **2015**, *198*, 215–222. [CrossRef] [PubMed]
19. Liu, T.; Tian, L.; Liu, Z.; He, J.; Fu, H.; Huang, Q.; Xue, H.; Huang, Z. Distribution and Toxicity of Polycyclic Aromatic Hydrocarbons during CaO-Assisted Hydrothermal Carbonization of Sewage Sludge. *Waste Manag.* **2021**, *120*, 616–625. [CrossRef]
20. Fregolente, L.G.; Miguel, T.B.A.R.; de Castro Miguel, E.; de Almeida Melo, C.; Moreira, A.B.; Ferreira, O.P.; Bisinoti, M.C. Toxicity Evaluation of Process Water from Hydrothermal Carbonization of Sugarcane Industry By-Products. *Environ. Sci. Pollut. Res.* **2019**, *26*, 27579–27589. [CrossRef] [PubMed]
21. Petrovič, A.; Cenčič Predikaka, T.; Škodič, L.; Vohl, S.; Čuček, L. Hydrothermal Co-Carbonization of Sewage Sludge and Whey: Enhancement of Product Properties and Potential Application in Agriculture. *Fuel* **2023**, *350*, 128807. [CrossRef]
22. Kusuma, H.S.; Christa Jaya, D.E.; Illiyanasafa, N.; Ikawati, K.L.; Kurniasari, E.; Darmokoesoemo, H.; Amenaghawon, A.N. A Critical Review and Bibliometric Analysis of Methylene Blue Adsorption Using Leaves. *Chemosphere* **2024**, *356*, 141867. [CrossRef] [PubMed]
23. Khan, I.; Saeed, K.; Zekker, I.; Zhang, B.; Hendi, A.H.; Ahmad, A.; Ahmad, S.; Zada, N.; Ahmad, H.; Shah, L.A.; et al. Review on Methylene Blue: Its Properties, Uses, Toxicity and Photodegradation. *Water* **2022**, *14*, 242. [CrossRef]
24. Oladoye, P.O.; Ajiboye, T.O.; Omotola, E.O.; Oyewola, O.J. Methylene Blue Dye: Toxicity and Potential Elimination Technology from Wastewater. *Results Eng.* **2022**, *16*, 100678. [CrossRef]

25. Cavali, M.; Kim, B.; Tedoldi, D.; Benbelkacem, H.; Bayard, R.; Garnier, V.; Libardi, N.; Woiciechowski, A.L.; Borges de Castilhos, A. Hydrochar from Sawdust and Sewage Sludge—A Potential Media for Retaining Heavy Metals in Sustainable Drainage Systems (SuDS). *Environ. Technol.* **2024**. *online ahead of print*. [CrossRef] [PubMed]
26. Velinov, N.; Najdanović, S.; Vučić, R.; Mitrović, J.; Kostić, M.; Bojić, D.; Bojić, A. Biosorption of loperamide by lignocellulosic-Al_2O_3 hybrid: Optimization, kinetics, isothermal and thermodynamic studies. *Cellulose Chem. Technol.* **2019**, *53*, 175–189. [CrossRef]
27. Filipović, K.; Petrović, M.; Najdanović, S.; Velinov, N.; Hurt, A.; Bojić, A.; Kostić, M. Highly Efficient Nano Sorbent as a Superior Material for the Purification of Wastewater Contaminated with Anthraquinone Dye RB19. *J. Water Process Eng.* **2024**, *67*, 106118. [CrossRef]
28. Isquierdo, E.P.; Caldeira, D.S.A.; Siqueira, V.C.; Martins, E.A.S.; Quequeto, W.D. Fittings of Adsorption Isotherm Models and Thermodynamic Properties of Urunday Seeds. *Eng. Agric.* **2020**, *40*, 374–380. [CrossRef]
29. Velinov, N.; Radović Vučić, M.; Petrović, M.; Najdanović, S.; Kostić, M.; Mitrović, J.; Bojić, A. The Influence of Various Solvents Polarity in the Synthesis of Wood Biowaste Sorbent: Evaluation of Dye Sorption. *Biomass Convers. Biorefin* **2023**, *13*, 8139–8150. [CrossRef]
30. Chambers, C.; Saha, S.; Grimes, S.; Calhoun, J.; Reza, M.T. Physical and Morphological Alteration of Sargassum-Derived Ultra porous Superactivated Hydrochar with Remarkable Cationic Dye Adsorption. *Biomass Convers. Biorefin* **2023**, *14*, 29131–29144. [CrossRef]
31. Aktas, K.; Liu, H.; Eskicioglu, C. Treatment of Aqueous Phase from Hydrothermal Liquefaction of Municipal Sludge by Adsorption: Comparison of Biochar, Hydrochar, and Granular Activated Carbon. *J. Environ. Manage* **2024**, *356*, 120619. [CrossRef] [PubMed]
32. Alhawtali, S.; El-Harbawi, M.; Al-Awadi, A.S.; El Blidi, L.; Alrashed, M.M.; Yin, C.Y. Enhanced Adsorption of Methylene Blue Using Phosphoric Acid-Activated Hydrothermal Carbon Microspheres Synthesized from a Variety of Palm-Based Biowastes. *Coatings* **2023**, *13*, 1287. [CrossRef]
33. Genli, N.; Kutluay, S.; Baytar, O.; Şahin, Ö. Preparation and Characterization of Activated Carbon from Hydrochar by Hydrother mal Carbonization of Chickpea Stem: An Application in Methylene Blue Removal by RSM Optimization. *Int. J. Phytoremediat* **2022**, *24*, 88–100. [CrossRef]
34. Md. Munjur, H.; Hasan, M.N.; Awual, M.R.; Islam, M.M.; Shenashen, M.A.; Iqbal, J. Biodegradable Natural Carbohydrate Polymeric Sustainable Adsorbents for Efficient Toxic Dye Removal from Wastewater. *J. Mol. Liq.* **2020**, *319*, 114356. [CrossRef]
35. American Public Health Association. *Standard Methods for the Examination of Water and Wastewater*, 23rd ed.; Baird, R.B., Eaton, A.D., Rice, E.W., Eds.; American Public Health Association: Washington, DC, USA, 2017; ISBN 9780875532875.
36. Atallah, E.; Kwapinski, W.; Ahmad, M.N.; Leahy, J.J.; Al-Muhtaseb, A.H.; Zeaiter, J. Hydrothermal Carbonization of Olive Mill Wastewater: Liquid Phase Product Analysis. *J. Environ. Chem. Eng.* **2019**, *7*, 102833. [CrossRef]
37. Zhou, F.; Li, K.; Hang, F.; Zhang, Z.; Chen, P.; Wei, L.; Xie, C. Efficient Removal of Methylene Blue by Activated Hydrochar Prepared by Hydrothermal Carbonization and NaOH Activation of Sugarcane Bagasse and Phosphoric Acid. *RSC Adv.* **2022**, *12*, 1885–1896. [CrossRef] [PubMed]
38. Khoshbouy, R.; Takahashi, F.; Yoshikawa, K. Preparation of High Surface Area Sludge-Based Activated Hydrochar via Hydrother- mal Carbonization and Application in the Removal of Basic Dye. *Environ. Res.* **2019**, *175*, 457–467. [CrossRef] [PubMed]
39. Dhaouadi, F.; Sellaoui, L.; Hernández-Hernández, L.E.; Bonilla-Petriciolet, A.; Mendoza-Castillo, D.I.; Reynel-Ávila, H.E.; González-Ponce, H.A.; Taamalli, S.; Louis, F.; Lamine, A. Ben Preparation of an Avocado Seed Hydrochar and Its Application as Heavy Metal Adsorbent: Properties and Advanced Statistical Physics Modeling. *Chem. Eng. J.* **2021**, *419*, 129472. [CrossRef]
40. Zhou, N.; Chen, H.; Feng, Q.; Yao, D.; Chen, H.; Wang, H.; Zhou, Z.; Li, H.; Tian, Y.; Lu, X. Effect of Phosphoric Acid on the Surface Properties and Pb(II) Adsorption Mechanisms of Hydrochars Prepared from Fresh Banana Peels. *J. Clean. Prod.* **2017**, *165*, 221–230. [CrossRef]
41. Xue, Y.; Gao, B.; Yao, Y.; Inyang, M.; Zhang, M.; Zimmerman, A.R.; Ro, K.S. Hydrogen Peroxide Modification Enhances the Ability of Biochar (Hydrochar) Produced from Hydrothermal Carbonization of Peanut Hull to Remove Aqueous Heavy Metals: Batch and Column Tests. *Chem. Eng. J.* **2012**, *200–202*, 673–680. [CrossRef]
42. Saha, N.; Volpe, M.; Fiori, L.; Volpe, R.; Messineo, A.; Reza, M.T. Cationic Dye Adsorption on Hydrochars of Winery and Citrus Juice Industries Residues: Performance, Mechanism, and Thermodynamics. *Energies* **2020**, *13*, 4686. [CrossRef]
43. Ferrentino, R.; Ceccato, R.; Marchetti, V.; Andreottola, G.; Fiori, L. Sewage Sludge Hydrochar: An Option for Removal of Methylene Blue from Wastewater. *Appl. Sci.* **2020**, *10*, 3445. [CrossRef]
44. El Ouadrhiri, F.; Elyemni, M.; Lahkimi, A.; Lhassani, A.; Chaouch, M.; Taleb, M. Mesoporous Carbon from Optimized Date Stone Hydrochar by Catalytic Hydrothermal Carbonization Using Response Surface Methodology: Application to Dyes Adsorption. *Int. J. Chem. Eng.* **2021**, *2021*, 5555406. [CrossRef]
45. El Ouadrhiri, F.; Abdu Musad Saleh, E.; Husain, K.; Adachi, A.; Hmamou, A.; Hassan, I.; Mostafa Moharam, M.; Lahkimi, A. Acid Assisted-Hydrothermal Carbonization of Solid Waste from Essential Oils Industry: Optimization Using I-Optimal Experimental Design and Removal Dye Application. *Arab. J. Chem.* **2023**, *16*, 104872. [CrossRef]

46. Zhou, S.; Hu, A.; Jiang, J.; Tang, J.; Zhou, G.; Zhu, L.; Wang, S. Low-Temperature Synthesized Hierarchical Porous Carbon from Waste Hydrochar with Super Capacity for Dye Adsorption. *Biomass Bioenergy* **2023**, *177*, 106938. [CrossRef]
47. Lin, Z.; Wang, R.; Tan, S.; Zhang, K.; Yin, Q.; Zhao, Z.; Gao, P. Nitrogen-Doped Hydrochar Prepared by Biomass and Nitrogen-Containing Wastewater for Dye Adsorption: Effect of Nitrogen Source in Wastewater on the Adsorption Performance of Hydrochar. *J. Environ. Manage* **2023**, *334*, 117503. [CrossRef] [PubMed]
48. Hessien, M. Methylene Blue Dye Adsorption on Iron Oxide-Hydrochar Composite Synthesized via a Facile Microwave-Assisted Hydrothermal Carbonization of Pomegranate Peels' Waste. *Molecules* **2023**, *28*, 4526. [CrossRef] [PubMed]
49. Santoso, E.; Ediati, R.; Kusumawati, Y.; Bahruji, H.; Sulistiono, D.O.; Prasetyoko, D. Review on Recent Advances of Carbon Based Adsorbent for Methylene Blue Removal from Waste Water. *Mater. Today Chem.* **2020**, *16*, 100233. [CrossRef]
50. Mozaffari Majd, M.; Kordzadeh-Kermani, V.; Ghalandari, V.; Askari, A.; Sillanpää, M. Adsorption Isotherm Models: A Comprehensive and Systematic Review (2010−2020). *Sci. Total Environ.* **2022**, *812*, 151334. [CrossRef] [PubMed]
51. de Vargas Brião, G.; Hashim, M.A.; Chu, K.H. The Sips Isotherm Equation: Often Used and Sometimes Misused. *Sep. Sci. Technol.* **2023**, *58*, 884–892. [CrossRef]
52. Liu, J.L.; Qian, W.C.; Guo, J.Z.; Shen, Y.; Li, B. Selective Removal of Anionic and Cationic Dyes by Magnetic Fe_3O_4-Loaded Amine-Modified Hydrochar. *Bioresour. Technol.* **2021**, *320*, 124374. [CrossRef]
53. Akbari, A.; Peighambardoust, S.J.; Lotfi, M. Hydrochar Derived from Liquorice Root Pulp Utilizing Catalytic/Non-Catalytic Hydrothermal Carbonization: RSM Optimization and Cationic Dye Adsorption Assessment. *J. Water Process Eng.* **2023**, *55*, 104099. [CrossRef]
54. Kajjumba, G.W.; Emik, S.; Öngen, A.; Kurtulus Özcan, H.; Aydın, S. Modelling of Adsorption Kinetic Processes-Errors, Theory and Application. In *Advanced Sorption Process Applications*; Edebali, S., Ed.; IntechOpen: Rijeka, Croatia, 2019.
55. Wang, J.; Guo, X. Adsorption Kinetic Models: Physical Meanings, Applications, and Solving Methods. *J. Hazard. Mater.* **2020**, *390*, 122156. [CrossRef] [PubMed]
56. Wu, F.C.; Tseng, R.L.; Juang, R.S. Characteristics of Elovich Equation Used for the Analysis of Adsorption Kinetics in Dye-Chitosan Systems. *Chem. Eng. J.* **2009**, *150*, 366–373. [CrossRef]
57. Parmar, K.R.; Ross, A.B. Integration of Hydrothermal Carbonisation with Anaerobic Digestion—Opportunities for Valorisation of Digestate. *Energies* **2019**, *12*, 1586. [CrossRef]
58. Murillo, H.A.; Pagés-Díaz, J.; Díaz-Robles, L.A.; Vallejo, F.; Huiliñir, C. Valorization of Oat Husk by Hydrothermal Carbonization: Optimization of Process Parameters and Anaerobic Digestion of Spent Liquors. *Bioresour. Technol.* **2022**, *343*, 126112. [CrossRef] [PubMed]
59. Brown, A.E.; Adams, J.M.M.; Grasham, O.R.; Camargo-Valero, M.A.; Ross, A.B. An Assessment of Different Integration Strategies of Hydrothermal Carbonisation and Anaerobic Digestion of Water Hyacinth. *Energies* **2020**, *13*, 5983. [CrossRef]
60. Farru, G.; Cappai, G.; Carucci, A.; De Gioannis, G.; Asunis, F.; Milia, S.; Muntoni, A.; Perra, M.; Serpe, A. A Cascade Biorefinery for Grape Marc: Recovery of Materials and Energy through Thermochemical and Biochemical Processes. *Sci. Total Environ.* **2022**, *846*, 157464. [CrossRef] [PubMed]
61. Aragón-Briceño, C.; Ross, A.B.; Camargo-Valero, M.A. Evaluation and Comparison of Product Yields and Bio-Methane Potential in Sewage Digestate Following Hydrothermal Treatment. *Appl. Energy* **2017**, *208*, 1357–1369. [CrossRef]
62. Bamba, D.; Coulibaly, M.; Robert, D. Nitrogen-Containing Organic Compounds: Origins, Toxicity and Conditions of Their Photocatalytic Mineralization over TiO_2. *Sci. Total Environ.* **2017**, *580*, 1489–1504. [CrossRef] [PubMed]
63. Farru, G.; Dang, C.H.; Schultze, M.; Kern, J.; Cappai, G.; Libra, J.A. Benefits and Limitations of Using Hydrochars from Organic Residues as Replacement for Peat on Growing Media. *Horticulturae* **2022**, *8*, 325. [CrossRef]
64. Petrovič, A.; Cenčič Predikaka, T.; Parlov Vuković, J.; Jednačak, T.; Hribernik, S.; Vohl, S.; Urbancl, D.; Tišma, M.; Čuček, L. Sustainable Hydrothermal Co-Carbonization of Residues from the Vegetable Oil Industry and Sewage Sludge: Hydrochar Production and Liquid Fraction Valorisation. *Energy* **2024**, *307*, 132760. [CrossRef]
65. Mantovani, M.; Collina, E.; Marazzi, F.; Lasagni, M.; Mezzanotte, V. Microalgal Treatment of the Effluent from the Hydrothermal Carbonization of Microalgal Biomass. *J. Water Process Eng.* **2022**, *49*, 102976. [CrossRef]
66. Czerwińska, K.; Marszałek, A.; Kudlek, E.; Śliz, M.; Dudziak, M.; Wilk, M. The Treatment of Post-Processing Liquid from the Hydrothermal Carbonization of Sewage Sludge. *Sci. Total Environ.* **2023**, *885*, 163858. [CrossRef]
67. Cavali, M.; Ricardo Soccol, C.; Tavares, D.; Alberto Zevallos Torres, L.; Oliveira de Andrade Tanobe, V.; Zandoná Filho, A.; Lorenci Woiciechowski, A. Effect of Sequential Acid-Alkaline Treatment on Physical and Chemical Characteristics of Lignin and Cellulose from Pine (*Pinus* spp.) Residual Sawdust. *Bioresour. Technol.* **2020**, *316*, 123884. [CrossRef] [PubMed]
68. Cavali, M.; Soccol, C.R.; Tavares, D.; Zevallos Torres, L.A.; Oliveira de Andrade Tanobe, V.; Zandoná Filho, A.; Woiciechowski, A.L. Valorization of Lignin from Pine (*Pinus* spp.) Residual Sawdust: Antioxidant Activity and Application in the Green Synthesis of Silver Nanoparticles for Antibacterial Purpose. *Biomass Convers. Biorefin* **2021**, *13*, 10051–10063. [CrossRef]
69. Zhu, J.Y.; Pan, X. Efficient Sugar Production from Plant Biomass: Current Status, Challenges, and Future Directions. *Renew. Sustain. Energy Rev.* **2022**, *164*, 112583. [CrossRef]

70. Wang, Y.; Zhang, Y.; Cui, Q.; Feng, Y.; Xuan, J. Composition of Lignocellulose Hydrolysate in Different Biorefinery Strategies: Nutrients and Inhibitors. *Molecules* **2024**, *29*, 2275. [CrossRef] [PubMed]
71. Ischia, G.; Fiori, L. Hydrothermal Carbonization of Organic Waste and Biomass: A Review on Process, Reactor, and Plant Modeling. *Waste Biomass Valorization* **2021**, *12*, 2797–2824. [CrossRef]
72. Bozell, J.J.; Petersen, G.R. Technology Development for the Production of Biobased Products from Biorefinery Carbohydrates—The US Department of Energy's "Top 10" Revisited. *Green Chem.* **2010**, *12*, 539–555. [CrossRef]
73. Faure, E.; Falentin-Daudré, C.; Jérôme, C.; Lyskawa, J.; Fournier, D.; Woisel, P.; Detrembleur, C. Catechols as Versatile Platforms in Polymer Chemistry. *Prog. Polym. Sci.* **2013**, *38*, 236–270. [CrossRef]
74. Mariscal, R.; Maireles-Torres, P.; Ojeda, M.; Sádaba, I.; López Granados, M. Furfural: A Renewable and Versatile Platform Molecule for the Synthesis of Chemicals and Fuels. *Energy Environ. Sci.* **2016**, *9*, 1144–1189. [CrossRef]
75. Fache, M.; Darroman, E.; Besse, V.; Auvergne, R.; Caillol, S.; Boutevin, B. Vanillin, a Promising Biobased Building-Block for Monomer Synthesis. *Green Chem.* **2014**, *16*, 1987–1998. [CrossRef]

applied sciences

MDPI

Article

Exploring Circular Economy Strategies in Buildings: Evaluating Feasibility, Stakeholders Influence, and the Role of the Building Lifecycle in Effective Adoption

Abdulaziz AlJaber [1,2,*], Pedro Martinez-Vazquez [1] and Charalampos Baniotopoulos [1,*]

[1] Department of Civil Engineering, School of Engineering, University of Birmingham, Edgbaston, Birmingham B15 2TT, UK; p.vazquez@bham.ac.uk

[2] Department of Civil and Construction Engineering, College of Engineering, Imam Abdulrahman Bin Faisal University, Dammam 31451, Saudi Arabia

* Correspondence: asaljaber@iau.edu.sa (A.A.); c.baniotopoulos@bham.ac.uk (C.B.)

Abstract: The concept of circular economy (CE) has emerged as an effective strategy for addressing resource depletion, waste generation, and environmental challenges, offering a promising path towards a more sustainable future. In the building sector, adopting CE principles can significantly mitigate environmental impacts, minimize lifecycle costs, and promote sustainability throughout a building's lifecycle. Using a mixed-method approach via a pre-interview questionnaire and semi-structured interviews with 10 sustainability experts, this study analyses the significance of 15 CE strategies in building construction projects, assessing their importance and ranking their potential for adoption. Furthermore, this study evaluates the feasibility of applying CE principles to different building types, including storage, industrial, commercial, residential, business, and healthcare facilities. The role of lifecycle stages including initiation and planning, design, procurement, construction, operation and maintenance, and end of life is examined to identify phases with the highest potential for successfully embracing CE principles. The role of stakeholders in driving change is also analyzed. The outcomes of this study reveal that the most feasible strategies include the use of renewable energy, design for durability and longevity, prefabrication, and offsite construction. The study findings indicate that storage, industrial, and business (office) buildings are the most feasible for CE application, while the initiation and planning and design stages are identified as critical phases for embracing CE adoption. Owners and designers emerge as the stakeholders with the greatest influence on CE implementation. The results of this study provide a comprehensive overview of the feasibility of CE adoption in the building sector. These findings offer valuable insights that can inform the development of targeted strategies to support the effective adoption of CE principles.

Keywords: circular economy; building construction projects; building sector; circular building; construction industry; waste management

1. Introduction

The construction industry is a cornerstone of the global economy, accounting for approximately 13% of the world's gross domestic product (GDP) [1]. However, the sector also contributes significantly to environmental challenges, including natural resource depletion, high energy consumption, and waste generation [2,3]. The building sector accounts for 33% of global greenhouse gas emissions (GHGs), 40% of resource consumption, and 40% of waste production [4]. Furthermore, the traditional linear manufacturing processes that have dominated the construction industry for decades [5] only involve one direction of movement whereby raw materials transform into products that end up as waste after their service life. Globally, the construction industry involving building and infrastructure projects produces more than 10 billion tons of construction and demolition waste every year [6]. Around 10–15% of those building materials are wasted during construction [7], and many materials are discarded rather than reused or recycled at the end of their life. Only 20–30% of these materials are recycled or reused at the end-of-life phase [8].

A major obstacle to sustainable progress in the building sector is the substantial CO_2 emissions generated by the dependence on non-renewable energy sources during the planning, construction, and operation of buildings [9]. By 2050, construction activities are projected to account for nearly half of the total emissions associated with new buildings, a significant increase from the current contribution of 28% [10]. According to the Global Alliance for Buildings and Construction (GlobalABC) [11], the share of renewable energy in the building sector has only increased by 0.9% since 2015, reaching 5.9% in 2022, far below the target of 11%. If this trend persists, the renewable share will reach just 7% by 2030. To align with climate targets, the share of renewable energy needs to grow by 1.5 percentage points annually until 2030 [11]. The Ellen MacArthur Foundation [7] highlights that 20–40% of the energy used in existing buildings can be profitably conserved through enhanced efficiency measures and sustainable approaches.

The concept of Circular Economy (CE) has gained widespread attention as an innovative economic model aimed at decoupling economic growth from the reliance on finite resource consumption. It represents an alternative to the current linear economy of 'make-use-dispose'. According to Geissdoerfer et al. [12], CE can be defined as "*A regenerative system in which resource input and waste, emission, and energy leakage are minimised by slowing, closing, and narrowing material and energy loops. This can be achieved through long-lasting design, maintenance, repair, reuse, remanufacturing, refurbishing, and recycling*". The Ellen MacArthur Foundation highlighted that the adoption of a CE contributes to the preservation and restoration of natural resources, optimizes the utilization of renewable materials, and eliminates waste through proactive design [13]. The CE model represents a robust framework for fostering economic growth while advancing environmental sustainability [7,14]. As a result, CE is increasingly recognized as a pathway to addressing these persistent challenges in the building sector [15–17].

The application of a CE in buildings promotes building materials to undergo discretized stages of construction, use, deconstruction, and repurposed for reuse, recycling, and reintegration into further construction processes [18]. A circular building can be defined as "*A building that is designed, planned, built, operated, maintained, and deconstructed in a manner consistent with CE principles*". This means that every phase of its lifecycle is carefully considered to ensure materials and resources are continuously kept in a closed loop, supporting a regenerative system. Leising et al. [19] described the CE approach in buildings as a lifecycle framework that extends the building's useful life while incorporating end-of-life considerations during the design phase. They introduced the concept of buildings as material banks, where materials are temporarily stored for future recovery and reuse, promoting a shift from traditional models to a circular system that ensures resources remain accessible and are not wasted.

Materials in circular construction models are carefully selected for their durability, recyclability, reusability, and ability to support multiple services [20,21]. The implementation of CE strategies enhances material sustainability and significantly reduces the lifecycle costs (LCCs) of buildings by optimizing resource efficiency [22–24]. These strategies achieve cost savings and minimize environmental impacts by establishing systems that preserve resource value, ensure continuous circulation, and close the loops for materials, energy, and water [25]. Circular strategies in the built environment can reduce CO_2 emissions by 38% or 2 billion tons by 2050 due to the decreased demand for new materials like steel, aluminum, cement, and plastic [26].

Past studies have proposed various strategies for incorporating CE principles into buildings. For example, Askar et al. [27] explored how design for adaptability (DfA) serves as a critical strategy to extend building lifespans and enable material recovery within CE practice. Akanbi et al. [28] developed a disassembly and deconstruction analytics system using Building Information Modeling (BIM) to evaluate and enhance the end-of-life performance of building designs. Bertino et al. [29] proposed a systematic approach to building deconstruction as a sustainable alternative to demolition focusing on material reuse align with CE goals. Munaro et al. [30] explored how eco-design methodologies can be applied in the construction sector to facilitate building deconstruction and promote material recovery. Other studies focus on the use of modular, prefabricated, and standardized building materials and components to as circular strategies to promote efficiency and sustainability. Wuni and Shen [31] identified 23 critical success factors for integrating circular modular construction projects. Neaves [32] highlighted the role of standardization in the adoption of CE in the construction sector. Li et al. [33] developed a framework to account for the lifecycle carbon footprint of prefabricated buildings within the CE principles, whereas Machado and Morioka [34] conducted a systematic literature review to identify how modularity can contribute to the CE.

However, circular buildings remain rare despite the acknowledged environmental and economic benefits of the CE approach [16,35,36]. Achieving circularity in the building sector demands a holistic approach that considers the building's full lifecycle and emphasizes collaboration and engagement among stakeholders [37,38]. Akhimien et al. [39] emphasized that the full potential of a CE in the construction of buildings cannot be achieved without integrating its principles throughout all stages of a building's lifecycle. Furthermore, according to Giorgi et al. [40], most studies focus on a single stakeholder perspective without examining the diverse viewpoints across the building value chain.

This study builds on these insights to evaluate the significance and feasibility of CE strategies within the building sector. This includes identifying key CE strategies, analyzing their applicability across various building types and lifecycle stages, and assessing the roles of stakeholders in influencing CE adoption. By addressing these points, this study seeks to provide actionable insights and targeted recommendations to support the effective integration of CE principles in building construction projects. Policymakers can use these findings to develop targeted incentive schemes to encourage industry stakeholders to align their practices with CE principles. Moreover, the findings provide key stakeholders, including construction managers, project planners, designers, developers, and investors, with actionable insights to prioritize the most impactful strategies for CE adoption.

Addressing the stated objectives, the following research questions were established:

1. Which CE strategies are most feasible and impactful for the wider adoption of circular practices in the building sector?
2. How does the type of building influence the feasibility of adopting CE strategies?
3. What roles do different stages of a building's project lifecycle play in the effective implementation of CE strategies?
4. How do various stakeholders influence the adoption of CE strategies in the building sector?

2. Materials and Methods

2.1. Research Design

This study investigates the potential of applying CE principles within the building sector by analyzing their implementation across various building types, assessing the influence of lifecycle stages, and evaluating the role of stakeholders in the process. A systematic literature review was conducted to identify key CE strategies within the building sector. These identified strategies were then incorporated into the pre-interview questionnaire (PIQ) and further explored through semi-structured interviews as part of a mixed-method approach.

Mixed research methods allow researchers to use a variety of techniques to solve research questions, without being limited to a single method [41]. Denzin [42] argued that employing mixed methods can significantly enhance the validity and reliability of research findings. Consequently, adopting a mixed-methods approach is considered the most suitable strategy to achieve the objectives of this study. Figure 1 illustrates the research methodology for this study.

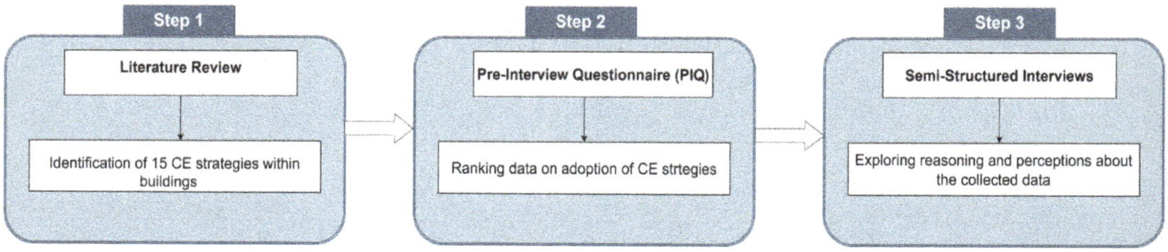

Figure 1. The research framework for the study.

2.2. Data Collection

2.2.1. Systematic Literature Review

The literature review enabled the identification of CE strategies and exploration of their relevance to each stage of the building lifecycle. The authors adopted Scopus as the primary search engine, selected for its extensive coverage of peer-reviewed journals and conference proceedings relevant to the fields of engineering, environmental science, and energy. To ensure a comprehensive review, the search focused on journal articles and conference papers published between 2012 and 2024. Additionally, Google Scholar was utilized to access key industry reports, such as those from the Ellen MacArthur Foundation, to integrate practical insights and frameworks into the review.

The search query used in Scopus was as follows: TITLE ("circular economy" OR "circular buildings" OR "circular construction" OR "circular economy strategies" OR "circularity in buildings") AND (ABS ("built environment" OR "construction sector" OR "construction industry" OR "building sector" OR "buildings" OR "circular strategies") OR KEY ("built environment" OR "construction sector" OR "construction industry" OR "building sector" OR "buildings" OR "circular strategies")) AND PUBYEAR > 2011 AND PUBYEAR < 2025 AND (LIMIT-TO (SUBJAREA, "ENGI") OR LIMIT-TO (SUBJAREA, "ENVI") OR LIMIT-TO (SUBJAREA, "ENER")) AND (LIMIT-TO (LANGUAGE, "English")) AND (LIMIT-TO (SRCTYPE, "j") OR LIMIT-TO (SRCTYPE, "p")).

This search query enabled a systematic exploration of a wide range of studies addressing CE strategies, their implementation, and potential applications within the building sector. Through this initial search, a total of 1056 articles were identified and screened. During the first screening phase, based on titles, articles that did not align with the research scope were excluded. This process left 238 articles for abstract review. After evaluating the abstracts, many were deemed irrelevant to the research questions, further narrowing the selection. Finally, the full-text review of the remaining articles resulted in the inclusion of 47 studies that provided valuable insights into CE strategies in the construction industry.

The process of identification, screening, eligibility, and inclusion is illustrated in the PRISMA flowchart as shown in Figure 2. This provides an overview of the systematic review process and the rationale for article inclusion and exclusion. Furthermore, Table 1 presents the list of identified CE strategies along with their corresponding references, providing a comprehensive overview of the key strategies derived from the literature review.

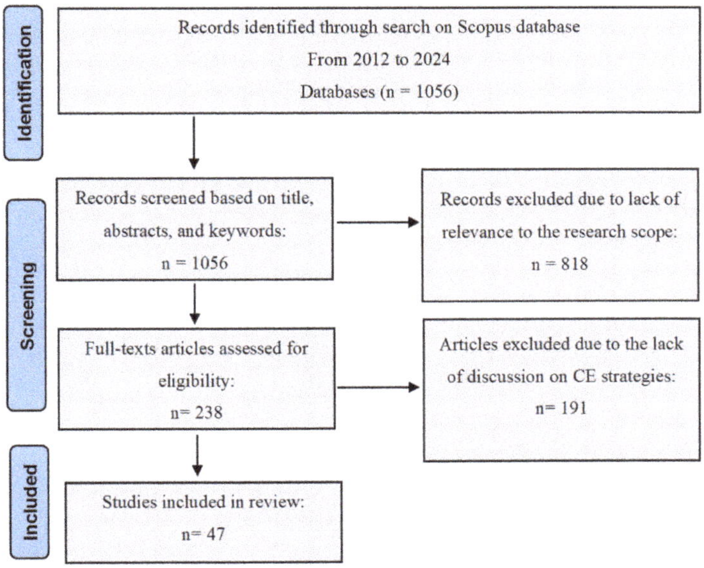

Figure 2. Flow diagram of literature selection for review.

Table 1. CE Strategies related to the building sector found in the literature.

Lifecycle Stage	CE Strategy	References
Design	Design for Deconstruction/Disassembly	[18,29,43–53]
	Design for Adaptability and Flexibility.	[18,21,27,46–48,50–52,54–56]
	Design for Standardization.	[32,46,51,55–57]
	Design for Durability and Longevity.	[18,20,25,46,58,59]
	Design for Modularity.	[31,34,46,57,60–62]
Procurement and Construction	Prefabrication and Offsite Construction.	[31,33,46,48,51–53,63]
	Procuring Reused and Recycled Materials.	[46,51–53,64,65]
	Procuring Bio-based Materials.	[8,18,59]
Operation and Maintenance	Renewable Energy Integration.	[18,21,52,66–70]
	Adaptive Reuse.	[29,49,52,71–73]
End of Life	Reuse of Materials and Products.	[45,46,48,49,51,53,61,64,74]
	Closed-Loop Recycling (up-cycling).	[29,46,48,49,51,53,58,65,74,75]
	Open-Loop Recycling (down-cycling).	[29,46,53,58,75]
	Selective Demolition.	[46,51,53]
	Deconstruction.	[18,29,43–53]

2.2.2. Pre-Interview Questionnaire (PIQ)

The pre-interview questionnaire (PIQ) was distributed to participants before the interviews to collect background information and preliminary insights on key aspects of CE principles and their adoption in the sector. The PIQ specifically aimed to evaluate the importance and feasibility of CE strategies, assess the feasibility of applying CE principles to various building types, and examine the role of lifecycle stages in defining sub-phases with the highest potential for successfully embracing CE principles.

Furthermore, PIQ analyzed the influence of stakeholders, highlighting their roles in driving CE. Participants were asked to rank these points listed in Table 2, providing structured data that informed the development of targeted interview questions. These questions enabled a more in-depth exploration of participants' reasoning and perceptions.

Table 2. Pre-interview questionnaire questions.

Section	Question	Format
Interviewee Background	Name	Open-ended (Free Text)
	Organization	Open-ended (Free Text)
	Years of Experience	Open-ended (Free Text)
	Level of Education	Open-ended (Free Text)
Circular Economy Strategies	Importance of CE strategies for fostering circular economy across the building sector (Design, Procurement, Operation, and End-of-Life Phases)	Likert Scale (1–5)
	Feasibility of adopting CE strategies in the Saudi Arabian building sector (Design, Procurement, Operation, and End-of-Life Phases)	Likert Scale (1–5)
Building Types	Feasibility of applying CE principles to building types (Residential, Commercial, Business, Educational, Healthcare, Industrial, and Storage)	Ranking (1 = Highest; 7 = Lowest)
Building Life Cycle	Identification of critical lifecycle phases for embracing CE principles (Initiation, Procurement, Design, Construction, Operation, and End-of-Life)	Ranking (1 = Highest; 6 = Lowest)
Stakeholders	Prioritization of stakeholder impact on CE adoption (Owner, Manufacturer, Contractor, Consultant, Sub-contractor, Designer, and Demolition Contractor)	Ranking (1 = Highest; 7 = Lowest

2.2.3. Semi-Structured Interview

Semi-structured interviews are a valuable qualitative research method for collecting rich data and conducting in-depth explorations of topics without being constrained by a fixed number of questions or answers [76]. These interviews can take various forms, ranging from structured formats resembling questionnaires with opportunities for follow-up probing to broader discussions focused on key topic areas, capturing respondents' detailed perspectives [77].

Following the completion of the PIQ, face-to-face semi-structured interviews were conducted with 10 sustainability experts representing different areas within the building sector. A purposive sampling approach was employed to select participants, ensuring that individuals with the most relevant expertise were included. The selection criteria required participants to have at least eight years of experience in roles related to sustainability practices in the building sector. Stakeholder diversity was ensured by including representatives such as owners, contractors, consultants, designers, and academics to offer research-based perspectives and theoretical insights. Participants were chosen from organizations with significant construction projects and academic institutions engaged in sustainability and CE-related research.

The semi-structured interview allowed participants to expand on their PIQ responses, providing deeper insights into their reasoning and perspectives. The questions in the interview focused on gaining a deeper understanding of participants' reasoning behind their rankings of key aspects of CE adoption in building construction. For example, participants were asked, "Could you please explain the reasoning behind your ranking of importance and feasibility of implementing various CE strategies in the building sector?" and "Could you please explain the reasoning behind your ranking of the top three building types for the application of CE principles?" Other questions included, "Can you please provide an explanation for your ranking of the three most critical building project lifecycle phases for the successful adoption of CE principles?" To explore stakeholder influence, participants were asked, "Could you please explain the reasoning behind your ranking of the top three stakeholder influences in implementing CE principles?" These open-ended questions encouraged participants to elaborate on their insights, providing richer and more detailed responses that aligned with the study's objectives and to enhance both the validity and richness of the data collected.

The interviews were conducted individually, each lasting approximately 55 min. The flexibility of the semi-structured format enabled focused discussions while allowing participants to address additional topics as needed. To ensure the most relevant data were collected, a purposive sampling approach was used, targeting individuals best positioned to address the research questions. Table 3 shows the details of the ten participants involved in the semi-structured interviews. The participant names were coded to protect their identity but agreed to reveal the type of stakeholder they are, positions within their organizations, and years of experience in the building sector.

Table 3. Participants' demographic details.

No.	ID	Type	Position	Years of Experience	Level of Education
1	A1	Academia	Lecturer/Circular B Member	10	PhD
2	A2	Academia	Research Associate	16	PhD
3	C1	Client	Project Manager	19	PhD
4	C2	Client	Project Manager	11	MSc
5	C3	Client	Portfolio Manager	14	PhD
6	CN1	Contractor	Environmental Engineer	13	BSc
7	CN2	Contractor	Procurement Manager	25	MSc
8	CT1	Consultant	Sustainability Manager	21	MSc
9	CT2	Consultant	Sustainability Senior Director	20	PhD
10	D1	Designer	Senior Development Architect	12	MSc

2.3. Data Analysis

2.3.1. Relative Importance Index (RII)

The RII was employed in this study as an effective tool for prioritizing and ranking CE strategies based on their perceived importance and feasibility. The RII is a common method used extensively within construction industry research to measure attitude, level of importance, and level of practice [78]. The feedback from respondents was analyzed using Microsoft Excel, and the RII was then applied using the following formula:

$$\frac{\sum w}{A \times N} = \frac{1n_1 + 2n_2 + 3n_3 + 4n_4 + 5n_5}{A \times N}; (0 \leq \text{RII} \leq 1) \quad (1)$$

w = The weighting given to each strategy by the respondent, ranging from 1 to 5 where n_5 is "Very important" and n_1 is "Not important";

A = Highest weight (in this study: 5);

N = Overall number of respondents (in this study: 10).

2.3.2. Thematic Analysis

Thematic analysis is a method that focuses on identifying, examining, and interpreting patterns (themes) within qualitative data [79]. The method is widely recognized for its flexibility, making it suitable for analyzing diverse forms of qualitative data across various research contexts such as interview transcripts and focus group discussions [80]. Thematic analysis enables researchers to explore and understand participants' perspectives and experiences in-depth, offering valuable and comprehensive insights [79–82].

The data collected from the interviews were analyzed using thematic analysis. The six phases of thematic analysis outlined by Kiger and Varpio [81] were systematically followed in this study: (1) familiarization with data; (2) generating initial codes; (3) searching for themes; (4) reviewing themes; (5) defining and naming themes; and (6) producing the report. Audio recordings were carefully transcribed into Word documents, enabling a detailed review of the participants' responses. Repeated listening the recordings ensured transcription accuracy and helped uncover additional themes that might have been initially overlooked.

3. Results and Discussion

3.1. Assessment of Circular Economy Strategies

This section presents an integrated analysis of participants' perceptions regarding the importance and feasibility of implementing various CE strategies in the building sector. These strategies are evaluated across four distinct construction project lifecycle phases: design, procurement and construction, operation and maintenance, and end of life. The respondents rated the strategies for importance on a five-point Likert scale ranging from "(1) not important" to "(5) very important", with rankings determined based on the relative importance index (RII) scores calculated for this.

Similarly, the feasibility of implementing these strategies in the Saudi Arabian building sector was assessed using a five-point Likert scale ranging from "(1) not feasible" to "(5) very feasible", reflecting participants' views on the practicality of adopting these strategies within the sector. This method allowed to prioritize the options, with higher RII scores reflecting higher importance and feasibility for implementation. The RII values are divided into five levels: High (H) ($0.8 \leq \text{RII} \leq 1$); High–Medium (H–M) ($0.6 \leq \text{RII} < 0.8$); Medium (M) ($0.4 \leq \text{RII} < 0.6$); Medium–Low (M–L) ($0.2 \leq \text{RII} < 0.4$); and Low (L) ($0 \leq \text{RII} < 0.2$) [83].

The analysis of the 15 CE strategies, as shown in Figure 3 and based on the mean scores of a five-point Likert scale, reveals significant insights into their perceived importance and feasibility for implementation across various project lifecycle phases. In addition, Tables 4 and 5 present the detailed results concerning the respondents' views on the importance and feasibility of the 15 CE strategies included in the study.

Table 4. Ranking of the importance of CE strategies.

Project Phase	CE Strategy	Participant Responses										Mean	RII	Rank	Importance Level
		A1	A2	C1	C2	C3	CN1	CN2	CT1	CT2	D1				
Design	Design for Deconstruction/Disassembly	5	5	5	5	5	5	4	5	4	5	4.80	0.96	2	High
	Design for Adaptability and Flexibility.	5	5	5	4	5	5	3	5	3	5	4.50	0.90	7	High
	Design for Standardization.	5	4	4	5	4	4	2	4	2	4	3.80	0.76	11	High–Medium
	Design for Durability and Longevity.	5	2	5	5	5	4	5	5	5	5	4.60	0.92	6	High
	Design for Modularity.	4	3	4	4	5	3	3	4	3	4	3.70	0.74	13	High–Medium
Procurement and Construction	Prefabrication and Offsite Construction.	4	2	4	4	4	4	4	5	3	4	3.80	0.76	11	High–Medium
	Procuring Reused and Recycled Materials.	5	5	4	4	5	5	5	5	5	5	4.80	0.96	2	High
	Procuring Bio–based Materials.	3	4	4	4	5	3	4	3	4	3	3.70	0.74	13	High–Medium
Operation and Maintenance	Renewable Energy Integration.	5	4	5	5	5	5	4	5	5	5	4.80	0.96	2	High
	Adaptive Reuse.	4	5	5	4	5	4	5	5	5	5	4.70	0.94	5	High
End of Life	Reuse of Materials and Products.	5	5	5	4	5	5	5	5	5	5	4.90	0.98	1	High
	Closed–Loop Recycling (up–cycling).	5	5	4	4	5	5	4	5	4	4	4.50	0.90	7	High
	Open–Loop Recycling (down–cycling).	3	3	3	4	5	4	4	3	4	3	3.60	0.72	15	High–Medium
	Selective Demolition.	5	3	4	3	5	4	4	4	3	4	3.90	0.78	10	High–Medium
	Deconstruction.	5	4	4	4	5	4	4	5	3	5	4.30	0.86	9	High

Table 5. Ranking of the feasibility of implementing CE strategies in the Saudi Arabian building sector.

Project Phase	CE Strategy	Participant Responses										Mean	RII	Rank	Feasibility Level
		A1	A2	C1	C2	C3	CN1	CN2	CT1	CT2	D1				
Design	Design for Deconstruction/Disassembly	2	3	3	2	3	3	3	3	3	2	2.70	0.54	10	Medium
	Design for Adaptability and Flexibility.	3	3	3	2	3	3	3	3	3	3	2.90	0.58	6	Medium
	Design for Standardization.	3	3	2	3	3	3	2	2	3	3	2.70	0.54	10	Medium
	Design for Durability and Longevity.	3	3	5	4	4	4	4	5	5	5	4.20	0.84	2	High
	Design for Modularity.	2	3	3	2	4	3	3	2	3	3	2.80	0.56	8	Medium
Procurement and Construction	Prefabrication and Offsite Construction.	3	5	4	4	2	3	3	3	4	4	3.50	0.70	3	High–Medium
	Procuring Reused and Recycled Materials.	2	2	2	1	1	2	2	2	3	2	1.90	0.38	13	Medium–Low
	Procuring Bio–based Materials.	1	2	3	1	1	1	2	2	3	2	1.80	0.36	15	Medium–Low
Operation and Maintenance	Renewable Energy Integration.	5	5	4	4	4	5	4	5	5	5	4.60	0.92	1	High
	Adaptive Reuse.	3	3	3	2	2	3	3	3	4	2	2.80	0.56	8	Medium
End of Life	Reuse of Materials and Products.	2	3	3	2	3	2	3	2	4	3	2.70	0.54	10	Medium
	Closed-Loop Recycling (up-cycling).	2	3	3	2	3	3	3	3	4	3	2.90	0.58	6	Medium
	Open-Loop Recycling (down-cycling).	2	4	3	2	3	4	3	3	4	4	3.20	0.64	4	High–Medium
	Selective Demolition.	2	4	3	3	3	3	2	3	4	3	3.00	0.60	5	High–Medium
	Deconstruction.	1	2	2	2	2	1	2	2	3	2	1.90	0.38	13	Medium–Low

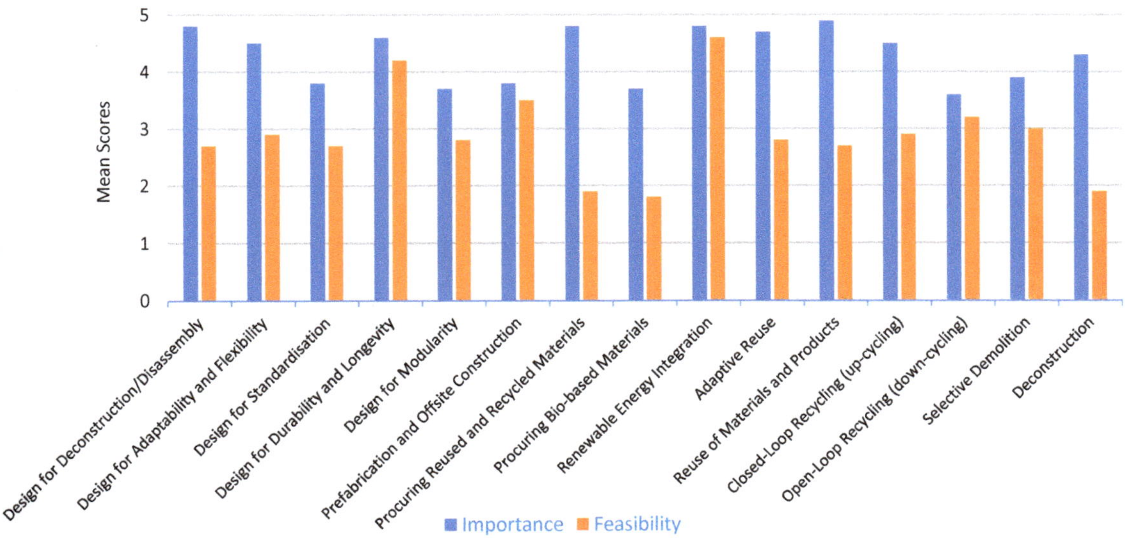

Figure 3. Analysis of circular economy strategies rankings.

The top five strategies in terms of importance were as follows: "Reuse of Materials and Products" addressing the end-of-life phase (RII = 0.98); followed by three strategies with the same ranking (RII = 0.96): "Design for Disassembly", "Procuring Reused and Recycled Materials", and "Renewable Energy Integration" applicable at the operation and maintenance phase, and, and lastly "Adaptive Reuse" at the operation and maintenance phase (RII = 0.94).

When considering the feasibility of these strategies, as shown in Table 5, there were noticeable differences in implementation potential. For instance, "Renewable Energy Integration", and " Design for Durability and Longevity" were not only considered highly important but also stood out as the only two strategies rated as very feasible to implement with RIIs of 0.92 and 0.84, respectively, indicating strong alignment between perceived importance and practical application. Conversely, although "Procuring Reused and Recycled Materials" and "Deconstruction" were ranked high in importance, they faced significant challenges in feasibility, receiving "Medium-Low" feasibility scores (RII = 0.38), which highlights barriers to their adoption. In addition, "Procuring Bio-based Materials" reveals a complex perception among participants regarding its importance and feasibility within the CE framework. It received an RII of 0.74, categorizing it as "High-Medium" in importance, indicating recognition of its value in promoting circularity in buildings.

However, when it comes to feasibility, this strategy faced challenges, as indicated by its lower scores in the feasibility ranking (RII = 0.36), which suggested it is not as easily implementable as other strategies. Overall, nine out of the fifteen strategies were rated as highly important, while the feasibility scores showed a more varied distribution, reflecting the complexities and challenges of implementing these strategies in the Saudi Arabian building sector. It is important to note that feasibility may vary across different countries due to variations in regulatory frameworks, market conditions, and resource availability.

3.1.1. Design for Deconstruction/Disassembly

Design for deconstruction/disassembly was rated highly important, with RII of 0.96, which places it at the top of strategic action to enforce. This high score also reflects the growing recognition to disassembly for improving the circularity of buildings. Yet, its feasibility score was notably lower, with an RII of 0.54, ranking it much lower compared to its importance. In this regard, participant A1 explained why the construction industry is not well-prepared to embrace this design. They commented "*We lack skilled labours, established concepts, market conditions, and appropriate regulations or policies. There are also no incentives for construction sector stakeholders to adopt this methodology. Even if we design buildings for disassembly, the most important consideration is that there are no established methods for using and recycling the resulting materials after disassembly. This raises the question of why we should invest in this design when we lack ways to recover, reuse, and recycle the materials*". Similarly, D1 commented "*In my view, the feasibility of this design is limited due to the continued dominance of traditional construction methods. Additionally, there may be a lack of specialized expertise and experience among designers and contractors necessary for the successful implementation of such designs.*"

3.1.2. Design for Adaptability and Flexibility

Design for Adaptability and Flexibility ranked seventh in importance with a high RII of 0.90, reflecting its prominence. However, its feasibility was rated at a medium level, with an RII of 0.58, positioning it sixth in feasibility. Respondents regard adaptability and flexibility as critical for extending building lifespan and offering solutions that can accommodate evolving needs. Participant C3 emphasized its importance and feasibility, stating "*In my view, the long-term benefits really make it an interesting design option. With flexibility in design societal needs can be met easily as there is no need to make significant changes in original design*". On the other hand, some respondents expressed concerns about the practical challenges of implementing this strategy. D1 who rated feasibility as low noted that "*Designing adaptable spaces often requires higher upfront costs for materials and structural elements that allow for future modifications, which can strain project budgets, especially if clients are focused on immediate costs rather than long-term savings. Moreover, adaptable designs add a level of complexity to the planning and construction process. Creating flexible layouts that can support multiple future uses often requires additional advanced planning, which can make the design process more challenging and time-consuming*". Similarly, C2 added "*Clients might consider that the benefits of adaptability are too far in the future to justify the current investment, particularly if short-term budgets and timelines are prioritized*".

3.1.3. Design for Standardization

Design for Standardization received a high–medium importance score (RII = 0.76) and ranked 10th in feasibility with a score of RII = 0.54 which places it in the medium level of feasibility. Respondents recognize that standardization plays a critical role in simplifying construction processes, enabling easier recycling, and facilitating the reuse of building materials. A1 elaborated on the significance of standardization, emphasizing its ability to streamline design processes, enhance recycling efficiency, and drive economic profitability. The participant commented "*Standardizing materials provides conditions for more efficient recycling, particularly regarding narrowing the loop. By eliminating toxic materials and selecting mono-materials or materials with clear compositions, we create appropriate conditions for recycling. At the component level, standardization simplifies the design process, while standardizing connections makes overall design management more efficient without requiring standardization of individual elements. I think standardization would also be moderately feasible because it can create financial profit. It can be viewed this way: it is profitable for some companies, not necessarily with the intention of helping the environment, but with the intention of being economically profitable. Therefore, it can be followed more easily than design for disassembly*". Additionally, C1 acknowledged the advantages of standardization, highlighting how compatibility simplifies the construction process and facilitates easier maintenance and upgrades throughout the building's lifecycle. They also pointed out that "*While the current building code does not mandate standardisation, it promotes an environment that encourages the adoption of standardised components and construction methods*". CN2 believed that resistance from building materials manufacturers hinder the adoption of standardized systems. C1 added that "*They're often hesitant to change their organisational structures or business routines because it requires significant investment and can disrupt their established processes. This hesitation slows the progress toward broader adoption of standardisation*".

3.1.4. Design for Durability and Longevity

Design for Durability and Longevity received a high importance score (RII = 0.92) and ranked second in feasibility (RII = 0.84), reflecting its recognized value and strong potential for implementation. This strong consensus reflects the recognition that durability is a key factor in creating sustainable buildings. By ensuring that materials and structures last longer, the need for repairs and replacements is minimized, which directly supports CE principles. While most of the respondents believe that durable designs can be practically implemented, one respondent likely sees barriers to feasibility. A2 who rated the feasibility of this design as slightly feasible expressed concerns regarding the upfront costs of using more durable materials and advanced construction techniques. They commented "*My rating is based on a cost prospective. In my opinion, designing for durability and longevity often requires using higher-quality or high-strength materials, which increase the initial investment in the project*". On the other hand, C1, who rated the feasibility of this design as highly feasible, commented that "*I believe designing for durability and longevity is a highly feasible approach, especially when considering the long-term benefits through life cycle cost (LCC). If you factor in long-term savings, such as reduced maintenance and an extended lifespan for the building, it starts to make financial sense. In the long run, this approach helps keep the project more cost-efficient and sustainable*". CT1 added, "*Design for durability is a great strategy as it reduces waste along the concrete life cycle*".

3.1.5. Design for Modularity

Design for Modularity received a high–medium importance score (RII = 0.74) but ranked eighth in feasibility with a medium score (RII = 0.56), reflecting its recognized value for flexible design but highlighting the practical challenges of implementation. C2, who rated the feasibility of design for modularity slightly feasible, pointed out that "*The adoption of modular construction is still in its early stages and would require industry adaptation. To enhance the feasibility of design for modularity, increased investment in logistics and supply chains is essential, particularly for handling and transporting modular units efficiently. Specialized equipment and skilled labours are required to load, secure, and unload modules properly*". Similarly, CN1 added "*The market currently prefers the traditional designs, and the awareness or demand for modular buildings is still in its early stages. Shifting the mindset of stakeholders towards modularity will require time, education, and investment*". Furthermore, A1 shared some thoughts on why design for modularity might be a low-importance strategy. A1 commented "*I chose modularity to be a bit less important because this is not applicable to all types of buildings. In some cases, we cannot apply modular solutions to certain types of buildings, and other strategies like standardization can likely compensate for this aspect*".

3.1.6. Prefabrication and Offsite Construction

Prefabrication and offsite construction received a high–medium importance score (RII = 0.76), indicating that respondents value its potential to reduce waste and increase efficiency in construction projects. In terms of feasibility, this strategy ranked third with a high–medium score (RII = 0.7), showing that it is considered one of the more feasible strategies for implementation. The controlled manufacturing environment in prefabrication enables better quality control and more efficient use of materials which align with CE goals. CT1 who ranked this strategy as very important but moderately feasible commented that "*Prefabrication and off-site construction allow for stricter quality control measures compared to on-site construction which ensure the project meets the highest quality standards. I rated it as moderately feasible due to its limitation in design flexibility.* In contrast, A2 and D1 expressed optimism about the growing adoption of this strategy. They pointed out that public projects, such as schools and social housing, are increasingly utilizing prefabrication and off-site construction methods, indicating a shift towards broader implementation. Furthermore, C1 highlighted the potential of off-site construction to encourage project developers to adopt this approach, emphasizing its significant advantages. According to C1, workflow in off-site construction can reduce construction schedules by 30% to 45% compared to traditional methods. For investors and developers, this accelerated timeline translates into earlier occupancy and the potential for faster returns on investment, making it an attractive option for modern construction projects.

3.1.7. Procuring Reused and Recycled Materials

The procurement of reused and recycled materials in construction is seen by industry professionals as highly important (RII = 0.96), with a strong consensus that this strategy plays a crucial role in promoting CE principles. Most participants gave it the highest possible importance score, highlighting their recognition of its value in reducing waste and conserving resources. Despite the high importance rating, the feasibility of procuring reused and recycled materials construction projects was rated much lower, with a feasibility score of RII = 0.38, placing it in the medium–low range. This indicates significant doubts about the potential of procuring reused and recycled materials in building construction projects. The participants shared similar concerns regarding the use of recycled and reused materials, particularly around their quality, supply chain limitations, and social acceptance. Others expressed doubts about whether these materials can consistently meet the necessary standards for building code. CN2, who is a procurement manager, highlighted that "*Recycled and reused materials often raise questions about their quality and adequacy for construction. While it's not impossible to use them, it is slightly feasible due to the mentality of the construction sector. There is social hesitation to adopt these kinds of materials, making them less feasible*". C1, who rated the strategy as moderately feasible, commented that "*I believe there is less demand for such materials as a result there are logistical difficulties associated with sourcing them. In addition, stakeholders are often concerned that reused materials won't be as durable or reliable as new ones*".

3.1.8. Procuring Bio-Based Materials

The strategy received a high–medium importance score (RII = 0.74) but ranked the lowest in feasibility with a medium–low feasibility score (RII = 0.36), suggesting significant challenges in its practical implementation. These challenges include the limited availability of bio-based materials, quality conditions, and supply chain constraints, which make it less feasible for widespread adoption in the near future. In this regard, C2 commented "*I believe there is a limited supply chain for procuring bio-based materials not like the conventional materials such as concrete and steel. In addition, bio-based materials may struggle to perform effectively in regions with extreme hot climate which make their feasibility very low due to quality performance*". CT1 discussed how adopting bio-based materials would require changes to building regulations and certifications, which can take time and add complexity to the approval process. They added that "*Biodegradable materials break down over time while building codes and regulations require that all materials used in construction meet strict safety, durability, and performance standards*". Furthermore, A1 emphasized that bio-based materials are not always readily available and might not be suitable for all construction applications, also pointing out that "*It's not always feasible to find bio-based materials, and they are not adequate for all types of uses. They might only be suitable for certain elements. Of course, bio-based materials are important. However, if we ensure that we are closing the loop by procuring reused and recycled materials, bio-based materials become a secondary priority. They cannot fully substitute the need for technical cycle materials.*"

3.1.9. Renewable Energy Integration

Renewable energy integration was rated as a highly important strategy by the respondents, coming in as the second-most important strategy (RII = 0.96) and the highest in feasibility (RII = 0.92). There was a strong agreement among participants about both the importance and feasibility of renewable energy integration. All the participants agreed that incorporating renewable energy sources like solar panels is essential for making buildings more sustainable. They see it as a way to reduce carbon emissions and improve energy efficiency, which aligns perfectly with CE goals. In this regard, A1 commented that "*It's one of the main principles of the CE to rely on renewable resources, as reported by the Ellen MacArthur Foundation. During the use phase of a building, we may not have as many strategies compared to the design and end-of-life phases, but we cannot ignore its significance. The use phase is the longest in the building's life cycle. Therefore, using renewable energies can save a significant amount of carbon emissions and energy, referencing the overall life cycle analysis*". CN1 highlighted that "*There is a high potential of solar energy in the region. It is very feasible to harvest solar energy, and its feasibility makes it a viable option. In addition, solar panels and energy storage systems are much more affordable making renewable energy not just an environmentally responsible choice, but a financially smart one as well*". A2, CN2, and CT1 further emphasized that the solar energy market in the region is experiencing substantial growth, supported by government initiatives and significant investments in renewable energy.

3.1.10. Adaptive Reuse

Adaptive reuse received a relatively high score (RII = 0.94), ranking it among the top five strategies in this study. Most participants rated adaptive reuse highly, with most giving it a score of 5 and a few rating it a 4, indicating strong agreement on its importance. This reflects the respondents' recognition of the potential value in repurposing existing buildings rather than demolishing them, thus conserving resources. However, when considering feasibility, the strategy was rated slightly lower, with a feasibility score of RII = 0.56, placing it in the medium range. Participant A1, who rated the feasibility as moderately feasible, commented, "*I believe adaptive reuse is important, but probably not the most important strategy. If we design buildings for disassembly, we end up with two options: adapt the use and extend the building's life cycle or disassemble the building and reuse its components elsewhere. Adaptive reuse is a strategy for slowing the loop, but there is an alternative. While it is important, the existence of this alternative makes adaptive reuse slightly less critical in my view*". Additionally, participant D1, who also rated the strategy as slightly feasible, highlighted some challenges to its wider adoption. D1 stated that "*Older buildings may have deteriorated, requiring extensive renovations to meet current standards. Integrating modern systems into old structures can be technically challenging. Moreover, a significant challenge in adaptive reuse projects lies in the discrepancies between the original as-built drawings and the existing building layout due to poor quality and incomplete technical information*". C3 added that there are legal challenges involved in meeting the necessary requirements and securing municipal approval for the proposed new use. C2 discussed the financial challenges stating that adaptive reuse could be costly when extensive reforms are needed and predicting the economic feasibility of the new use is also among the financial challenges of adaptive reuse projects.

3.1.11. Reuse of Materials and Products

The reuse of materials and products at the end-of-life phase was rated as the most important strategy by respondents, with a mean score of 4.9 and an RII of 0.98. The reasoning behind this high rating was consistent across participants, as they all agreed that reuse aligns with the core principles of a CE, where maximizing the lifespan of materials is key to reducing resource extraction and waste. A1 commented, "*Reuse of materials is a preferred strategy over recycling, whether it be upcycling or downcycling, because it's at the top of the waste hierarchy. It extends the life of products and materials without the need for additional energy or resources that recycling requires. However, sometimes it is not possible to reuse certain materials, so we must opt for recycling*". Despite its importance, the feasibility of reusing materials and products received a medium feasibility score (RII = 0.54). This indicates that, while respondents recognize the importance of material and product reuse in sustainable building practices, they see challenges in implementing it effectively. D1 shared the perspective that to achieve better material recovery at the end of life, buildings should be designed with material recovery in mind from the design phase. CN1 argued that "*The construction sector often favors rapid demolition techniques that prioritize speed over material recovery.*" Moreover, C2 highlighted the need for more infrastructure for storing and distributing reused materials, as this is necessary to enhance material recovery. Similarly, A2 argued that "*It is recommended to address the complexity of a building's end-of-life phase by enhancing the efficiency and functionality of local collection schemes to improve material recovery*".

3.1.12. Closed-Loop Recycling (Up-Cycling)

Closed-loop recycling was rated high in importance, with an RII of 0.89, placing it eighth in importance among the CE strategies. Most participants rated this strategy as very important, with 9 out of 11 giving it a top importance score. Despite this high level of perceived importance, feasibility scored lower, with an RII of 0.58, ranking closed-loop recycling 6th in feasibility. One significant barrier cited by respondents is the difficulty of applying closed-loop recycling to certain materials, particularly concrete. Participants, including A1, C2, CN2, and D1, expressed the view that, while upcycling is generally preferable to downcycling, not all materials lend themselves to this approach. Concrete, for example, presents unique challenges. A1 explained "*Upcycling is more preferred than downcycling, and this applies to concrete as well. However, concrete is a very complex material and is irreversible. Once you make concrete, you cannot revert it to its original state. Due to its irreversible nature, it's challenging to find established methods to reuse concrete as it is*". C2 also pointed out the regional shortage of recycling facilities, stating, "*There are currently only around 8–10 authorised metal recycling facilities, and these are primarily located in major cities*". Conversely, CT2, who rated the strategy as feasible, believed that the growing global emphasis on resource conservation and the Saudi Vision 2030 agenda could support the development of a reliable system for reclaiming used materials from demolition and construction sites. This perspective highlights a sense of optimism that, with sufficient focus and infrastructure investment, closed-loop recycling could become a more viable strategy in the region, especially as sustainability becomes an increasing priority.

3.1.13. Open-Loop Recycling (Down-Cycling)

Closed-loop recycling was rated moderately important strategy with an RII of 0.65, placing it in the high–medium importance category region but with the lowest rank of Importance. The strategy was considered relatively feasible and ranked fourth in feasibility (RII 0.64). The lower quality requirements of down-cycled materials and their suitability for non-structural applications places it higher in feasibility. Both contractors (CN1 and CN2) shared a similar view about concrete being so dominant in the region's construction practices; therefore, open-loop recycling is a highly feasible option. They pointed out that concrete can generally only be down-cycled, making this approach practical for managing construction waste in the region. However, another important consideration was highlighted by A1, C1, and C2 is the lack of recycling facilities. For instance, C1 claimed that *"There is a need for more materials storage and recycling facilities to have more advanced recycling practices. Therefore, Investment in recycling facilities support the development of a circular economy, aligning with Vision 2030's sustainability goals"*.

3.1.14. Selective Demolition

Selective demolition allows for careful removal of materials in good condition, which can then be reused or recycled, reducing landfill waste. The strategy received a high-medium level of importance and feasibility, ranking tenth in importance with an RII of 0.78 and fifth in feasibility with an RII of 0.60 among the 15 CE strategies evaluated. A2, who rated the strategy as moderately important, saw selective demolition as reclaiming only specific materials, with the rest likely ending up in landfills, which led him to rate its importance slightly lower. In terms of feasibility, he believed that the strategy is feasible as it is relatively straightforward process, requiring fewer specialized skills and less time than deconstruction. In contrast, A1, C1, and CN2 shared the view that a lack of awareness about the sustainable benefits of material recovery contributes to conventional demolition methods. These methods are less time-consuming, less labor-intensive, and more familiar to local demolition contractors. In addition, CN1 highlighted that *"I believe there is a need for more specialized demolition professionals that trained to carefully recover valuable materials of the building. This shortage of skilled labours might slow the progress toward Vision 2030 sustainability goals"*. CN2, who rated the strategy as feasible, believes that the growing emphasis on sustainability promoted by Vision 2030 creates potential for a shift toward selective demolition among contractors.

3.1.15. Deconstruction

Deconstruction at a building's end-of-life phase allows for the recovery of nearly all materials for reuse, making it a highly valuable strategy. This approach ranked ninth in importance, with a RII of 0.86. However, its feasibility ranked second-lowest, with a medium–low score of 0.38, as most participants rated its feasibility between 1 and 2, with only one rating it as 3. This consensus highlights the significant barriers to implementing deconstruction effectively. D1, who rated the strategy with low feasibility, noted that deconstruction's success is limited if it has not been considered from the design phase. He pointed out that current building designs in the region are not optimized for deconstruction, making it challenging to reclaim materials efficiently and fully realize its benefits. The value of materials at the end of life was discussed by A2, CN1, CN2, and CT1. They believe that most materials may only be suitable for down-cycling rather than reuse due to quality limitations. A1 claimed that deconstruction requires specialized skills and tools, and significant time to dismantle structures carefully, which currently makes it less practical compared to simpler demolition methods. They added that additional time required to carefully remove, sort, and store materials can lead to higher budget cost. Finally, C1 shared the perspective that limited recycling infrastructure further complicates the feasibility of deconstruction as without sufficient recycling facilities, even the materials that are recovered may not be effectively processed or reused.

3.2. *Feasibility of Applying Circular Economy Strategies to Different Building Types*

This section assesses the feasibility of applying CE strategies to different building types, as presented in Table 6. The participants were asked to evaluate the feasibility of implementing CE principles across various building types using a scoring system. The scoring method classifies feasibility into four levels: high feasibility (scores of 1 and 2), medium feasibility (scores of 3 and 4), low feasibility (scores of 5 and 6), and very low feasibility (score of 7). In the post-processing of this exercise, for each response highly feasible building are awarded 3 points, moderately feasible receive 2 points, less feasible earn 1 point, and least feasible receive 0 points. The total points for each building type are calculated based on these weighted responses, providing a clear measure of the overall feasibility. See second last column on Table 6. The feasibility levels are then ranked according to the total score, offering valuable insights into which building types are most suitable for the application of CE principles. This method highlights the building types where CE practices can be most effectively implemented, enabling focused efforts to drive circularity in the building sector.

Table 6. Ranking of the feasibility of applying CE strategies to different building types.

Building Type	Participant Responses										Total	Rank
	A1	A2	C1	C2	C3	CN1	CN2	CT1	CT2	D1		
Storage	1	1	6	1	1	1	1	1	1	1	28	1
Industrial	6	2	7	2	3	2	2	2	2	2	24	2
Business (offices)	2	5	3	6	2	3	3	3	4	3	20	3
Educational	4	3	4	3	4	5	4	4	5	4	18	4
Commercial	3	6	1	5	5	4	5	5	3	5	15	5
Healthcare	7	4	5	4	6	6	6	6	7	6	10	6
Residential	5	7	2	7	7	7	7	7	6	7	5	7

3.2.1. Storage Buildings

Storage buildings were ranked first in the feasibility of applying CE principles, with a total score of 28 indicating a strong consensus on their suitability for incorporating CE strategies. The primary rationale behind this high-ranking relates to the simplicity and flexibility of storage facility designs. Participant A1 noted that "*Storage buildings are highly feasible for circular strategies because their layouts are very simple. They are easier to convert or adapt to other uses due to their open-plan design. Even reused materials can be utilized within them. I believe storage buildings are the simplest type of buildings, and their simplicity makes them more feasible for adopting circular strategies. That's why I consider storage buildings to be the most feasible option*". This was supported by C3, who mentioned that "*In my view, less complicated designs are usually attached to storage buildings, which means that standard materials can be used, as well as reused/recyclable ones. Simple designs can easily facilitate the adoption of CE principles, such as reused materials and modular construction. Adaptability can be achieved easily, as the structure can be used for different purposes over time, which meets the requirements of the CE agenda*". Additionally, D1 commented that "*The straightforward design of storage facilities reduces complexities when disassembling and repurposing materials. Also, the lack of complex internal finishes in storage buildings simplifies the incorporation of recyclable and reusable materials*".

3.2.2. Industrial Buildings

Industrial buildings were ranked second in the feasibility of implementing CE principles, with a total score of 24. Out of the ten participants, seven rated industrial buildings as having high feasibility. The majority who rated industrial buildings highly emphasized the simplicity and flexibility in their design. Industrial buildings are often constructed with standardized components, making them ideal for disassembly and reuse. C3 highlighted their structural advantages by stating that "*Industrial buildings are usually constructed with portal frames and large open spans, which makes them easy to adapt for different uses and deconstructed easily"*. Moreover, D1 highlighted that industrial buildings usually have simpler design requirements focused on functionality rather than appearance. This makes it easier to integrate reused/recycled materials, which might not satisfy the esthetic standards required in commercial or residential buildings. The potential for integrating sustainable energy solutions was another advantage mentioned by CN1. They commented *"Industrial buildings often have large, flat roofs that are perfect for installing solar panels. This makes it easier to integrate on-site renewable energy solutions which enhance the generation of clean energy"*. In contrast, C1, who rated the potential as low, believed that in some industrial sectors, there may be a perception that CE principles are not essential for operations, especially if the focus is on immediate profitability and efficiency.

3.2.3. Business Buildings (Offices)

Office buildings were ranked third in the feasibility of implementing CE principles, with a total score of 20. Only one respondent rated business buildings with a low potential. The respondents identified several factors that enhance the potential of business buildings for CE implementation. For instance, A1 pointed out that *"Office buildings are feasible for circular strategies because they typically feature open spaces or light separations. Instead of concrete walls, offices often use glass partitions or movable wall separations, which are easier to manage, adapt, or disassemble. This is why I chose office buildings as the second most feasible option"*. Similarly, C3 commented that *"Business office can be renovated on a regular basis. Therefore, it would be easy to implement CE principals through refurbishment programmes and reuse of various materials"*. C1, CT1, and CT2 shared a similar perspective, stating that there is strong potential for incorporating CE in business buildings due to the compelling business case. This view aligns with the priorities of stakeholders who focus on financial returns. On the other hand, A2, who rated the feasibility of implementing CE principles as low, believed that the upfront costs could pose a significant barrier. A2 suggested that some owners in business may prioritize short-term gains and favor lower initial expenses, potentially overlooking the long-term benefits that CE practices could provide.

3.2.4. Educational Buildings

Educational buildings were ranked fourth in feasibility for applying CE principles, with a score of 18, as shown in Table 6. Many respondents rated the feasibility as medium, with only one participant rating it as having low potential. Respondent C2 believed that the adoption of CE principles in educational buildings could be higher due to public sector ownership of many such buildings. C2 noted that this ownership is a potential driver, as public sector clients often have environmental commitments and a greater responsibility to embed CE principles. Participants A2 and D1 highlighted that prefabrication and offsite construction practices are commonly used in educational buildings, aligning well with CE principles. CN1 claimed that educational institutions are increasingly obligated to enhance their role as leaders in environmental responsibility, as this contributes positively to their public image. By adopting CE practices, educational institutions demonstrate a commitment to sustainability, positioning themselves as role models for the community.

3.2.5. Commercial Buildings

Commercial buildings were ranked fifth in feasibility for applying CE principles, with a total score of 15. Respondents generally rated the feasibility as low to medium, indicating a moderate potential for CE practices. Respondents A1 and C1 discussed how the high rate of fit-outs and refurbishments creates opportunities for CE principles. They believe that, due to the regular tenant turnover in retail spaces, stores could be designed with deconstruction in mind. This approach would allow modular interior elements, like partitions, to be easily dismantled, reused, or refurbished whenever there is a change in tenant. Conversely, CN2, who rated the feasibility of CE principles in commercial buildings as low, argued that commercial spaces especially retail often have short lease terms, which makes it challenging to establish a long-term CE strategy. According to CN2, owners and tenants may be reluctant to invest in materials that extend beyond the current lease. Similarly, D1 noted that the rapid changes in retail spaces often lead to frequent, cost-driven fit-outs that prioritize fast, low-cost solutions over sustainable, durable ones. CN1, who rated commercial buildings with medium feasibility, emphasized that the long-term savings in lifecycle costs, particularly through reduced maintenance and operational expenses, make a strong case for adopting CE principles among shopping mall owners.

3.2.6. Healthcare Buildings

Healthcare buildings were ranked sixth in the feasibility of CE principles, with a total score of 10, reflecting a consensus among participants on the challenges associated with incorporating CE strategies. Several participants who scored healthcare buildings lower highlighted their highly specialized nature, where strict standards are required which limit the circularity principles. On the other hand, C2 believed that long-term ownership incentivizes healthcare clients to think about building whole life cycle and adopt strategies that provide financial and operational efficiency such as energy consumption, which directly contribute to lowering long-term costs. In addition, A2 who rated it with medium feasibility pointed out that many healthcare facilities are publicly funded and are expected to lead by example in sustainability.

3.2.7. Residential Buildings

Residential buildings ranked seventh in feasibility for applying CE principles, with a total score of 5. This rank places residential buildings at the lowest level for CE feasibility among the various building types examined in this study. Most interviewees who rated residential buildings low provided a similar rationale for their scores. The reason was residential properties are often privately owned by individual homeowners who lack awareness of CE concepts and tend to prioritize esthetics over sustainability. Additionally, many homeowners favor cost-effective solutions with immediate benefits, rather than investing in long-term sustainable practices. Demand for reused or recycled materials is also limited due to concerns about perceived quality and esthetic preferences. Conversely, one interviewee (C1), who ranked residential buildings highly, highlighted the potential for modular and offsite construction in social housing.

3.3. The Impact of Building's Project Lifecycle on the Adoption of CE Principles in Construction Projects

This section evaluates the perceived impact of different building's project lifecycle on the adoption of CE principles, as shown in Table 7. Participants were asked to assess the significance of each lifecycle stage on the application of CE principles, using a ranking system. The lifecycle stages were categorized into three levels: high (scores of 1 and 2), medium (scores of 3 and 4), and low (scores of 5 and 6). High-impact stages were then awarded 3 points per response, medium-impact stages receive 2 points, and low-impact stages earn 1 point. The total points for each lifecycle stage are calculated from these weighted responses, providing a comprehensive measure of its overall influence. The stages are then ranked by total score, identifying the critical phases where CE principles can be most effectively adopted.

Table 7. Ranking of the impact of building's project lifecycle on the adoption of CE.

Lifecycle Stage	Participant Responses										Total	Rank
	A1	A2	C1	C2	C3	CN1	CN2	CT1	CT2	D1		
Initiation and Planning	2	2	2	2	1	2	1	2	1	2	30	1
Design	1	1	1	1	2	1	2	1	2	1	30	1
Procurement	5	4	3	4	4	3	3	3	3	3	19	3
End of life	3	3	6	5	6	4	4	4	5	4	16	4
Construction	6	6	4	3	3	6	5	5	4	5	14	5
Operation and Maintenance	4	5	5	6	5	5	6	6	6	6	11	6

3.3.1. Initiation and Planning Stage

As shown in Table 7, the initiation and planning stage were ranked as the highest lifecycle stage with a total score of 30. All the participants ranked the stage as highly impactful for the successful adoption of CE principles. Several participants shared a similar rationale, noting that this stage establishes the framework for integrating CE principles across all subsequent phases. For instance, C3 argued that *"All decision-making processes are made during this phase so if CE principal were highlighted earlier so it would be carried out in the following process. Moreover, within this phase all goals are defined and aligned with organization or any strategies goals. So, it's critical to link CE principals with goals to ensure its implementation throughout the project lifecycle. In general, business case, and project management benefit plan are conducted in this phase"*. Similarly, CN1 and CT2 argued that the initiation and planning phase is the most influential stage for CE adoption. They emphasized that this phase is critical because it is where the project's CE principles are defined and prioritized. Decisions made during this stage set the foundation for the entire project, establishing its scope and objectives while influencing the direction of CE integration.

3.3.2. Design Stage

The design phase was ranked first, equal to the initiation and planning phase, with a score of 30. All participants identified this phase as highly impactful for the adoption of CE principles in construction projects. According to A1, A2, C2 CN1, and D1 this stage has the most significant influence on the entire lifecycle of the building, as key decisions are made regarding material selection, building layout, and construction methods, which directly affect the adoption of CE strategies. D1 commented that *"Early design decisions influence the entire building lifecycle. Circular strategies are identified during this phase, which will shape the whole lifecycle of the building while ensuring cost-effective solutions"* CT2 added that *"The design phase is where theoretical commitments to CE are translated into specific, actionable project specifications"*. CT1 argued that *"Without planning during the design stage, achieving a sustainable and efficient end-of-life outcome would be significantly more challenging"*. CN1 emphasized that contractors' detailed design specifications directly influence their ability to execute CE strategies effectively on-site.

3.3.3. Procurement Stage

Procurement had the third greatest impact based on the interview responses with a score of 19 with no high ratings but a majority of medium scores. Participants A1, A2, C2, C3, and D1 emphasized that the procurement phase heavily relies on market availability and pre-established design specifications. If CE principles are not prioritized early, procurement has a limited ability to drive circularity. CN2, who is a procurement manager, stated that *"The availability of circular materials often comes down to whether local suppliers are around. Without a local supply chain, it can be really challenging to source materials that meet CE standards. On top of that, cost pressures at this stage can lead to tough choices, where cheaper, less circular options get picked over more than CE-aligned materials"*. D1 highlighted that a well-developed circular supply chain is essential for the availability of materials and products, which is critical for delivering CE projects. They commented that, *"Without a well-developed circular supply chain, designers are limited in their ability to create sustainable solutions, as the availability of CE-compliant materials and products, directly shapes the feasibility of their designs"*. CT2 highlighted that the procurement phase is crucial for engaging suppliers who understand and support CE principles and further emphasized that decisions about suppliers and availability of circular material specifications during this stage can have a significant impact on the building's overall sustainability.

3.3.4. End-of-Life Stage

The end-of-life stage was ranked fourth with a total score of 16. Participants acknowledged this stage as an important opportunity to recover value from materials and components through strategies like recycling, reuse, and repurposing. However, responses varied, with higher ratings contingent on how effectively CE principles were implemented in earlier stages. For example, A2 highlighted that *"The end of life phase is where the building's circularity potential is truly tested. If the structure was not designed for disassembly or reuse, recovering materials becomes much more challenging and expensive."* CN1 added that successful CE implementation at this stage depends significantly on decisions made during the design and planning phases. They stated, *"Buildings that were not designed with circularity in mind often result in more waste and fewer opportunities for material recovery"*. C1, C3, and CT2 expressed concerns about the challenges of implementing CE principles at the end-of-life stage, with three believing that most of the product value is already lost by this phase. They acknowledged that recovering significant value can be particularly difficult for certain building materials and components. C1 argued that current practices, such as downcycling, often dominate at this stage and can be hard to move away from due to technical and economic constraints, and finally commented, *"This may not align with higher-value circular strategies, it remains a practical solution in many cases"*.

3.3.5. Construction Stage

The construction stage, ranked fifth with a total score of 14, reflects a moderate level of impact on the adoption of CE principles. Participants expressed mixed views, with four participants rating it as medium and six rating it as low. These results indicate that, while construction plays a role in CE implementation, its influence is limited compared to the earlier stages in the chain. The rationale provided by respondents who rated the construction stage as low was consistent. They believe that the construction phase primarily focuses on executing the plans and specifications developed during the design phase. The stage depends heavily on how well CE principles were embedded in the design. In contrast, C3 emphasized that this stage plays a critical role in the application of CE principles by actively reducing waste through enhanced resource management and efficient use of materials. C3 also noted that minimizing on-site errors decreases material wastage and optimizes project efficiency which is aligned with CE goals.

3.3.6. Operation and Maintenance Stage

The operation and maintenance stage received the lowest score (11), ranking it as the phase with the least influence on adopting CE principles. Only one participant rated the stage with a medium impact on the adoption while the rest rated it as low. While interviewees acknowledged that this stage provides opportunities to implement CE principles such as energy management and water efficiency, they emphasized that its effectiveness heavily relies on decisions made during the earlier phases. Several interviewees shared the perspective that the regular maintenance of mechanical and electrical systems plays a crucial role in supporting CE principles in this stage. They emphasized that consistent upkeep ensures these components remain in good condition throughout their operational life and retains its value at the end of life. In this regard, A1 pointed out that "*Having a proper maintenance programme in place is so important at this stage. For example, well-maintained HVAC systems can often upgrade or reuse them rather than having to replace them entirely. That way, they can be refurbished, resold, or repurposed, which fits perfectly with CE goals*". C3 and CN1 shared a similar perspective, emphasizing building management teams and facility operators play essential roles in promoting operational practices with CE goals among users, such as recycling programs and energy efficiency initiatives.

3.4. Stakeholder's Influence in Implementing CE Principles

This section assesses the influence of different stakeholders in the implementation of CE principles in the building sector, as presented in Table 8. The participants were asked to rate the level of influence of stakeholders in implementing CE principles by providing a scoring system. The scoring method categorizes stakeholder influence into four levels: high (scores of 1 and 2), medium (scores of 3 and 4), low (scores of 5 and 6), and very low (score of 7). Stakeholders with high influence are awarded 3 points for each response, medium influence receives 2 points, low influence earns 1 point, and very low receives 0 points. The total points for each stakeholder are calculated based on these weighted responses, providing a clear measure of the overall influence. The level of influence is then ranked according to the total score, offering insights into who plays a critical role in applying CE principles. This approach highlights key players in the adoption of CE principles, enabling better targeted efforts to drive circular practices within the building sector.

Table 8. Ranking of the stakeholder's influence in implementing CE principles.

Stakeholder	Participant Responses										Total	Rank
	A1	A2	C1	C2	C3	CN1	CN2	CT1	CT2	D1		
Owner	3	1	1	1	1	1	1	1	1	1	29	1
Designer	1	3	4	2	2	2	3	2	2	2	27	2
Consultant	2	2	3	3	3	3	4	3	3	3	22	3
Manufacturer	4	5	5	4	4	4	2	4	5	4	18	4
Contractor	6	4	2	5	5	5	5	5	4	5	14	5
Sub-contractor	7	6	6	7	6	7	7	7	6	6	5	6
Demolition contractor	5	7	7	6	7	6	6	6	7	7	5	6

3.4.1. Owner

The respondents ranked the owner as the most influential stakeholder in implementing CE principles, with the highest total points as shown in Table 8. The reasoning behind owner having the greatest influence was consistent across all interviewees. These respondents see the owner as the primary decision-maker who has the authority to set project goals, allocate financial resources, and drive the adoption of CE strategies. In this regard, CT2 commented that "*The project owners have the ultimate control over a project, including setting the budget, scope, and priorities. Their commitment to CE principles is crucial because it directly influences the overall direction of the project. If the owner prioritizes sustainability and circular practices, these principles are likely to be embedded from the very start*". Similarly, C1 heightened that "*The owner is the one who will pay and invest in the project. If the owner is not convinced that the short-term or long-term benefits justify the investment, they will be hesitant to invest in circular solutions. Circular solutions can be costly from an immediate perspective, as their benefits are realized over the long term. Therefore, the owner needs to be convinced of the value. To achieve this, stakeholders need to first persuade the owner. Once the owner is convinced of the benefits, they are more likely to invest in circular solutions."* C3 also discussed the significant role of the client by stating *"As the client has the full authority over the project so, they should have more power to influence the adoption of CE principals by documenting all requirements in the project charter, business case and mapping all requirements in requirements documentations. If the client supports the adoption of CE at all levels of an organization, so sustainable practices can be achieved, as it will be a habit of all employees, which would foster an organization's commitment to CE principles"*.

3.4.2. Designer

The designer was ranked as the second most influential stakeholder in implementing CE principles, with a total of 27 points. This high ranking highlights the critical role designers play in embedding CE strategies into construction projects. Respondents view designers as key influencers because they are responsible for making decisions early in the project lifecycle, decisions that have a lasting impact on the building's functional and environmental performance throughout its lifecycle. A1, who ranked the designer as the most influential stakeholder, commented that *"Designers can have a full life cycle perspective, unlike owners who may not have the experience to see the complete picture. If the designer can effectively communicate this perspective to the owner, it can significantly influence the owner's decisions regarding investment and the long-term benefits of design choices. Design choices have implications for all subsequent lifecycle stages. If the designer makes the right decisions from the beginning, it will positively impact the entire lifecycle"*. Additionally, C3 commented that *"It could be noticed and understood that designer plays a vital role in defining and documenting how products should be created, used and finally disposed. Implementation of CE principles influenced directly by designer decisions as it could be influencing selected materials, adoption of recyclable materials which can improve the Tribal Bottom Line achieving social needs, environmentally friendly materials and reducing cost through the use of modular design which supports reuse concept as well as extend lifecycle"*. D1 also emphasized that designers who are well-knowledge in BIM can play a pivotal role in convincing owners of the benefits of LCC and LCA. By leveraging BIM, designers can effectively demonstrate the long-term financial and environmental advantages of adopting CE strategies, making it easier for owners to understand and commit to CE practices.

3.4.3. Consultant

The interviewees identified the consultant as the third most influential stakeholder in implementing CE principles with a total of 22 points. There was a consensus of the consultant's significant role in providing technical expertise and guidance throughout the project lifecycle. Consultants can also support designers by highlighting the environmental and economic benefits of design choices. CT1 pointed out that *"Consultants are generally trusted by owners because they do not have a direct economic interest in misleading the client. When a client hires a consultant, the consultant is paid regardless of what they recommend, so they are more likely to provide honest advice. This trust in consultants is why they play a valuable role in influencing design decisions."* C3 discussed the role of consultant by highlighting that *"A client sometimes does not have required skills to determine their needs so consultants or project management team can use their expertise to introduce, implement all best practices related to CE. Consultants or PMT could easily identify opportunities and threat which can help to move to sustainable practices. As Knowledge transfer is a unique process to ensure effective knowledge sharing about CE strategies. Therefore, consultants play a vital role in this regard"*.

3.4.4. Manufacturer

The product manufacturer was regarded as the fourth most influential stakeholder in implementing CE principles, with a total of 18 points. This ranking highlights the moderate influence manufacturers hold in promoting CE, as they are responsible for producing and supplying the materials and components used in construction. Respondents believe that manufacturers play a crucial role in shaping the material supply chain, and their ability to produce products that align with CE principles is essential for wider adoption. CN1 commented, *"If manufacturers do not invest in creating circular products, it becomes challenging for the rest of the supply chain to implement CE practices"*. Furthermore, C2 emphasized the important role of product manufacturers, stating, *"If manufacturers can supply circular products at affordable prices, it would remove a key obstacle for project owners. It becomes much easier for owners and other stakeholders to justify the investment in circular solutions"*. Although manufacturers are seen as significant contributors to the implementation of CE principles, A2 believes that their influence is somewhat limited. Their ability to drive the adoption of CE practices depends on market demand for circular products, which is largely set by higher-ranking stakeholders, such as owners and designers.

3.4.5. Contractor

Contractors were ranked as the fifth most influential stakeholder in implementing CE principles, with a total of 14 points. Respondents who rated the contractor's influence as low shared a common rationale: they believe that contractors primarily follow the specifications and guidelines established by the owner and design team. In this view, contractors are seen as executors of the project rather than key decision-makers, especially when it comes to setting or driving CE objectives. However, one respondent (C1), who ranked contractors as having high influence, offered a different perspective. C1 argued that contractors can have a significant impact on the implementation of CE principles through their choices regarding construction methods, material procurement, and on-site practices, such as waste management. According to C1, contractors play a pivotal role in shaping the practical aspects of how circular economy strategies are realized during the construction phase.

3.4.6. Sub-Contractor and Demolition Contractor

The sub-contractor and demolition contractor were both seen as having limited influence in implementing CE principles, tying for the lowest ranking among all stakeholders. Respondents view these roles as more focused on carrying out specific tasks, rather than being involved in the bigger strategic decisions related to CE. For sub-contractors, their job is typically to handle parts of the project as directed by the main contractor. Because they are focused on specialized tasks, their ability to influence broader CE initiatives is seen as minimal. They are carrying out the work rather than shaping the sustainability vision. Similarly, the demolition contractor' role comes at the end of the building's life, mainly tasked with deconstructing or demolishing structures. Even though this phase can be important for recycling and reusing materials which are key to CE principles, respondents feel that the demolition contractor's influence is limited. As by the time they step in, most of the critical CE decisions have already been made during the design phase. Their role is more about executing the end-of-life stage, rather than influencing the project's CE approach from the start. However, A1 believed that demolition contractors can influence in deciding which materials can be reused or recycled.

4. Recommendations

To foster the wider adoption of CE principles within the building sector, a multi-faceted approach is essential. Policymakers should establish a robust regulatory framework that incentivizes CE practices, such as tax reductions for using recycled materials, grants for renewable energy integration, and penalties for excessive construction waste. Additionally, policies mandating LCA and pre-demolition audits can help maximize material recovery and reuse. Setting clear national standards for CE practices, including certifications and benchmarks, can ensure consistency and alignment across projects while encouraging compliance. Governments can drive the adoption of circular construction by funding and supporting pilot projects that demonstrate how CE principles can be successfully applied in real-world scenarios. These projects can serve as practical examples for the industry, showing the benefits of CE strategies and inspiring wider adoption.

Owners and developers should be encouraged to integrate CE principles at the earliest stages of a project, as decisions made during initiation and planning significantly influence the project's overall sustainability. This can be supported by developing clear CE guidelines and standards to ensure alignment across projects. Strengthening the focus on CE principles during the initiation and planning stages is vital for embedding CE throughout the building lifecycle. This phase sets the foundation for all subsequent decisions, making it the ideal stage to prioritize CE objectives. Project briefs should explicitly include CE strategies such as material reuse, renewable energy integration, and adaptability. Additionally, incorporating lifecycle assessments (LCAs) and lifecycle costing (LCC) during planning can ensure a holistic view of environmental and economic impacts, enabling better decision-making. Early stakeholder engagement is also critical, as it fosters collaboration, aligns priorities, and ensures that all parties understand the benefits of CE principles.

Designers play a critical role and should be equipped with tools and training to prioritize CE strategies, such as design for disassembly, modularity, and durability. Collaborative platforms that bring together designers, consultants, and contractors can help in integrating CE objectives across the project lifecycle. Design guidelines and mandates should explicitly incorporate CE strategies such as design for disassembly, adaptability, and durability to ensure that buildings are designed with their entire life cycle in mind. Collaboration between designers and consultants is critical to integrating lifecycle thinking into material selection, building layouts, and construction methods. This collaborative approach helps identify opportunities for resource efficiency and material reuse early in the process and evaluate design options using circular metrics. Additionally, investing in advanced design tools and technologies, such as BIM, can support the simulation and optimization of circular outcomes, enabling teams to predict the environmental impact and economic benefits of their design choices.

Supply chain manufacturers must be encouraged to adopt CE principles by developing circular products and components, supported by incentives for adopting circular practices. This can be further enhanced by improving local supply chain infrastructure to ensure the availability of CE-compliant materials, reducing logistical and cost barriers. Establishing and supporting local supply chains for circular materials, such as reused, recycled, and bio-based, is fundamental to ensuring their availability and accessibility. Procurement policies should be updated to prioritize suppliers and materials aligned with CE principles, offering incentives to contractors who opt for sustainable choices. Strong partnerships with suppliers can enhance the reliability and consistency of CE-compliant materials, creating a more resilient supply chain.

Contractors and sub-contractors should adopt advanced construction practices that minimize waste, such as prefabrication and on-site material recovery processes, supported by training and capacity-building programs. It is essential to provide targeted training for contractors and workers, equipping them with the knowledge and skills needed to implement circular construction methods effectively. Demolition contractors should be equipped with specialized training and tools to recover valuable materials efficiently, minimizing waste sent to landfills. Additionally, establishing partnerships between demolition contractors and material recovery facilities can ensure that salvaged materials are reintegrated into the construction supply chain. Governments and industry stakeholders can further support this transition by introducing regulations and incentives that prioritize material recovery during demolition projects, thereby enhancing the role of demolition contractors as key players in the circular construction ecosystem.

Developing key performance indicators (KPIs) and metrics to measure the effectiveness of CE strategies is essential for tracking progress and ensuring measurable project outcomes. KPIs, such as the percentage of reused/recycled materials, waste and CO_2 emissions reduction rates, and LCC savings, can provide measurable benchmarks for assessing the impact of CE practices. These metrics should be applied across all project phases, enabling stakeholders to evaluate success and identify opportunities for improvement. Aligning these KPIs with recognized credible tool for the assessment of product circularity such as Material Circularity Indicators (MCIs), ensures consistency and comparability, driving wider adoption of CE strategies.

Increasing investment in research and development for innovative materials and technologies, such as advanced recycling methods, will enhance the feasibility of implementing CE strategies. Leveraging digital tools like BIM can improve material tracking, lifecycle data analysis, and efficient resource recovery, supporting data-driven decision-making throughout a building's lifecycle. In addition, public awareness campaigns highlighting the economic and environmental benefits of CE adoption are crucial to shifting stakeholder mindsets. Incorporating stakeholder education into professional certification programs can address the lack of awareness of CE benefits.

5. Conclusions

This study provides a comprehensive examination of the factors influencing the adoption of CE principles within the building sector, drawing on both quantitative and qualitative insights from 10 sustainability experts. By assessing the importance and feasibility of 15 CE strategies, evaluating their applicability to diverse building types, exploring role of multiple project lifecycle stages on effective implementation, and identifying the stakeholders most capable of driving CE adoption. This is performed by addressing four research questions: (1) Which CE strategies are most feasible and impactful for the wider adoption of circular practices in the building sector? (2) How does the type of building influence the feasibility of adopting CE strategies? (3) What roles do different stages of a building's project lifecycle play in the effective implementation of CE strategies? (4) How do various stakeholders influence the adoption of CE strategies in the building sector?

In this investigation, we identified and evaluated 15 CE strategic actions throughout the entire building lifecycle, encompassing design, procurement and construction, operation and maintenance, and ultimately the end-of-life phase. The results reveal that, among these, renewable energy integration, design for durability and longevity, and prefabrication and offsite construction stand out as the most feasible options. Their strong alignment between perceived importance and practical applicability underscores their potential to be readily implemented within the building sector. Conversely, procuring reused/recycled and bio-based materials and facilitating building deconstruction at the end of life, were acknowledged as critical for advancing circularity, however exposed to more significant feasibility hurdles.

Furthermore, this study indicates that building type and project phase critically influence CE feasibility. Storage and industrial buildings emerged as especially conducive environments for adopting CE principles, likely due to their simpler designs and flexible layouts. They can serve as pilot projects for CE adoption, given their high feasibility scores. Lessons learned from these projects can inform broader implementation across other building types. In contrast, residential and healthcare facilities, with their stringent requirements and stakeholder preferences, present more formidable hurdles.

Through examination of the building lifecycle stages, we noted the varying levels of potential for integrating CE principles across different phases. Key findings reveal that the briefing and design stages hold the greatest potential for incorporating CE practices. These early stages are pivotal for defining the project's vision, strategy, and sustainability goals, allowing stakeholders to embed circularity principles effectively. Decisions made during these stages, such as material selection and construction methods, have long-lasting implications for the entire building lifecycle.

The results obtained underscore the critical role of stakeholder collaboration throughout the lifecycle. Owners and designers are identified as key drivers during the initial stages, shaping the project's CE framework. Owners, as the primary decision-makers that set the project's direction and influencing the project stakeholders to align with CE goals. Designers influence the integration of CE principles through material selection, architectural choices, and shape the owner's brief into actionable design solutions. Moreover, designers and consultants have a crucial role in ensuring that materials and components are selected for their recyclability, reusability, and longevity, contributing to a circular framework that enhances resource efficiency and reduces waste.

While the findings are broadly applicable, the regional focus limits their direct generalizability to contexts with different economic conditions, cultural attitudes, and environmental priorities. Future studies should aim to include a more diverse range of participants, such as subcontractors, material suppliers, and policymakers, to provide a more holistic understanding of the barriers and enablers of CE adoption across the value chain. Expanding the geographical scope of research and incorporating comparative analyses across regions will further enrich the global discourse on circularity in the building sector.

Author Contributions: Conceptualization, A.A., C.B. and P.M.-V.; Data curation, A.A.; Formal analysis, A.A.; Investigation, AA., C.B. and P.M.-V.; Methodology, A.A.; Project administration, A.A., C.B. and P.M.-V.; Supervision, A.A., C.B. and P.M.-V.; Validation, A.A.; Visualization, A.A., C.B. and P.M.-V.; Writing—original draft, A.A.; Writing—review and editing, A.A., C.B. and P.M.-V. All authors have read and agreed to the published version of the manuscript.

Funding: This research received no external funding.

Institutional Review Board Statement: Not applicable.

Informed Consent Statement: Informed consent was obtained from all subjects involved in the study.

Data Availability Statement: This study was conducted using data sets and analyses that are reasonably accessible to the corresponding author.

Acknowledgments: The first author is grateful for the support of his research project at the University of Birmingham by the Imam Abdulrahman Bin Faisal University. All authors consented to these Acknowledgments.

Conflicts of Interest: The authors declare no conflicts of interest.

References

1. World Economic Forum Engineering and Construction. Available online: https://www.weforum.org/communities/engineering-and-construction (accessed on 8 May 2024).
2. Global Alliance for Buildings and Construction (GlobalABC). *2018 Global Status Report: Towards a Zero-Emission, Efficient and Resilient Buildings and Construction Sector*; International Energy Agency and United Nations Environment Programme: Nairobi, Kenya, 2018.
3. Zuo, J.; Zhao, Z.-Y. Green Building Research–Current Status and Future Agenda: A Review. *Renew. Sustain. Energy Rev.* **2014**, *30*, 271–281. [CrossRef]
4. Ness, D.A.; Xing, K. Toward a Resource-Efficient Built Environment: A Literature Review and Conceptual Model. *J. Ind. Ecol.* **2017**, *21*, 572–592. [CrossRef]
5. Neves, S.A.; Marques, A.C. Drivers and Barriers in the Transition from a Linear Economy to a Circular Economy. *J. Clean. Prod.* **2022**, *341*, 130865. [CrossRef]
6. Wu, H.; Zuo, J.; Zillante, G.; Wang, J.; Yuan, H. Status Quo and Future Directions of Construction and Demolition Waste Research: A Critical Review. *J. Clean. Prod.* **2019**, *240*, 118163. [CrossRef]
7. Ellen MacArthur Foundation Growth Within: A Circular Economy Vision for a Competitive Europe. Available online: https://ellenmacarthurfoundation.org/growth-within-a-circular-economy-vision-for-a-competitive-europe (accessed on 5 May 2023).
8. Ellen MacArthur Foundation. Towards the Circular Economy: Accelerating the Scale-Up Across Global Supply Chains. Available online: https://ellenmacarthurfoundation.org/towards-the-circular-economy-vol-3-accelerating-the-scale-up-across-global (accessed on 4 January 2023).
9. Huang, L.; Krigsvoll, G.; Johansen, F.; Liu, Y.; Zhang, X. Carbon Emission of Global Construction Sector. *Renew. Sustain. Energy Rev.* **2018**, *81*, 1906–1916. [CrossRef]
10. Architecture 2030. The 2030 Challenge for Embodied Carbon. Available online: https://www.architecture2030.org/2030_challenges/embodied/ (accessed on 17 September 2024).
11. Global Alliance for Buildings and Construction (GlobalABC). *Global Status Report for Buildings and Construction: Beyond Foundations: Mainstreaming Sustainable Solutions to Cut Emissions from the Buildings Sector*; UN Environment Programme (UNEP) and the Global Alliance for Buildings and Construction (GlobalABC): Nairobi, Kenya, 2024.
12. Geissdoerfer, M.; Savaget, P.; Bocken, N.M.P.; Hultink, E.J. The Circular Economy—A New Sustainability Paradigm? *J. Clean. Prod.* **2017**, *143*, 757–768. [CrossRef]
13. Ellen MacArthur Foundation. Towards the Circular Economy: Economic and Business Rationale for an Accelerated Transition. Available online: https://ellenmacarthurfoundation.org/towards-the-circular-economy-vol-1-an-economic-and-business-rationale-for-an (accessed on 1 March 2023).
14. Stahel, W.R. The Circular Economy. *Nature* **2016**, *531*, 435–438. [CrossRef] [PubMed]
15. Sariatli, F. Linear Economy Versus Circular Economy: A Comparative and Analyzer Study for Optimization of Economy for Sustainability. *Visegr. J. Bioecon. Sustain. Dev.* **2017**, *6*, 31–34. [CrossRef]
16. Charef, R.; Lu, W. Factor Dynamics to Facilitate Circular Economy Adoption in Construction. *J. Clean. Prod.* **2021**, *319*, 128639. [CrossRef]
17. AlJaber, A.; Martinez-Vazquez, P.; Baniotopoulos, C. Circular Economy in the Building Sector: Investigating Awareness, Attitudes, Barriers, and Enablers through a Case Study in Saudi Arabia. *Sustainability* **2024**, *16*, 1296. [CrossRef]
18. Rahla, K.M.; Mateus, R.; Bragança, L. Implementing Circular Economy Strategies in Buildings—From Theory to Practice. *Appl. Syst. Innov.* **2021**, *4*, 26. [CrossRef]
19. Leising, E.; Quist, J.; Bocken, N. Circular Economy in the Building Sector: Three Cases and a Collaboration Tool. *J. Clean. Prod.* **2018**, *176*, 976–989. [CrossRef]
20. Rahla, K.M.; Mateus, R.; Bragança, L. Selection Criteria for Building Materials and Components in Line with the Circular Economy Principles in the Built Environment—A Review of Current Trends. *Infrastructures* **2021**, *6*, 49. [CrossRef]
21. Rahla, K.M.; Mateus, R.; Bragança, L. Practical Strategies for Embodying the Circular Economy Concept in the Construction Sector. In *Sustainable Urban Development: Topics, Trends and Solutions*; Bragança, L., de Alvarez, C.E., Cabeza, L.F., Eds.; IOP Publishing Ltd.: Bristol, UK, 2021; p. 15-1.
22. AlJaber, A.; Alasmari, E.; Martinez-Vazquez, P.; Baniotopoulos, C. Life Cycle Cost in Circular Economy of Buildings by Applying Building Information Modeling (BIM): A State of the Art. *Buildings* **2023**, *13*, 1858. [CrossRef]
23. Jansen, B.W.; van Stijn, A.; Gruis, V.; van Bortel, G. A Circular Economy Life Cycle Costing Model (CE-LCC) for Building Components. *Resour. Conserv. Recycl.* **2020**, *161*, 104857. [CrossRef]

24. Alasmari, E.; Martinez-Vazquez, P.; Baniotopoulos, C. Utilising BIM on LCC to Enhance the Sustainability of Saudi Residential Projects Through Simulation. A Case Study at the Kingdom of Saudi Arabia. In *Proceedings of the 4th International Conference "Coordinating Engineering for Sustainability and Resilience" & Midterm Conference of CircularB "Implementation of Circular Economy in the Built Environment", Timisoara, Romania, 29–31 May 2024*; Ungureanu, V., Bragança, L., Baniotopoulos, C., Abdalla, K.M., Eds. Springer: Cham, Switzerland, 2024; pp. 659–668. ISBN 9783031577994.
25. Mercader-Moyano, P.; Esquivias, P.M. Decarbonization and Circular Economy in the Sustainable Development and Renovation of Buildings and Neighbourhoods. *Sustainability* **2020**, *12*, 7914. [CrossRef]
26. Ellen MacArthur Foundation. Completing the Picture: How the Circular Economy Tackles Climate Change. Available online: https://www.ellenmacarthurfoundation.org/completing-the-picture (accessed on 18 November 2024).
27. Askar, R.; Bragança, L.; Gervásio, H. Design for Adaptability (DfA)—Frameworks and Assessment Models for Enhanced Circularity in Buildings. *Appl. Syst. Innov.* **2022**, *5*, 24. [CrossRef]
28. Akanbi, L.A.; Oyedele, L.O.; Omoteso, K.; Bilal, M.; Akinade, O.O.; Ajayi, A.O.; Delgado, J.M.D.; Owolabi, H.A. Disassembly and Deconstruction Analytics System (D-DAS) for Construction in a Circular Economy. *J. Clean. Prod.* **2019**, *223*, 386–396. [CrossRef]
29. Bertino, G.; Kisser, J.; Zeilinger, J.; Langergraber, G.; Fischer, T.; Österreicher, D. Fundamentals of Building Deconstruction as a Circular Economy Strategy for the Reuse of Construction Materials. *Appl. Sci.* **2021**, *11*, 939. [CrossRef]
30. Munaro, M.R.; Tavares, S.F.; Bragança, L. The Ecodesign Methodologies to Achieve Buildings' Deconstruction: A Review and Framework. *Sustain. Prod. Consum.* **2022**, *30*, 566–583. [CrossRef]
31. Wuni, I.Y.; Shen, G.Q. Developing Critical Success Factors for Integrating Circular Economy into Modular Construction Projects in Hong Kong. *Sustain. Prod. Consum.* **2022**, *29*, 574–587. [CrossRef]
32. Neaves, M. *Circular Economy for the Built Environment, Research and Practice*; Taylor & Francis: Abingdon, UK, 2024; pp. 88–112. [CrossRef]
33. Li, C.Z.; Tam, V.W.Y.; Lai, X.; Zhou, Y.; Guo, S. Carbon Footprint Accounting of Prefabricated Buildings: A Circular Economy Perspective. *Build. Environ.* **2024**, *258*, 111602. [CrossRef]
34. Machado, N.; Morioka, S.N. Contributions of Modularity to the Circular Economy: A Systematic Review of Literature. *J. Build. Eng.* **2021**, *44*, 103322. [CrossRef]
35. Hossain, M.U.; Ng, S.T.; Antwi-Afari, P.; Amor, B. Circular Economy and the Construction Industry: Existing Trends, Challenges and Prospective Framework for Sustainable Construction. *Renew. Sustain. Energy Rev.* **2020**, *130*, 109948. [CrossRef]
36. AlJaber, A.; Martinez-Vazquez, P.; Baniotopoulos, C. Barriers and Enablers to the Adoption of Circular Economy Concept in the Building Sector: A Systematic Literature Review. *Buildings* **2023**, *13*, 2778. [CrossRef]
37. Zimmann, R.; O'Brien, H.; Hargrave, J.; Morrell, M. *The Circular Economy in the Built Environment*; Arup: London, UK, 2016.
38. AlJaber, A.; Martinez-Vazquez, P.; Baniotopoulos, C. Developing Critical Success Factors for Implementing Circular Economy in Building Construction Projects. *Buildings* **2024**, *14*, 2319. [CrossRef]
39. Akhimien, N.G.; Latif, E.; Hou, S.S. Application of Circular Economy Principles in Buildings: A Systematic Review. *J. Build. Eng.* **2021**, *38*, 102041. [CrossRef]
40. Giorgi, S.; Lavagna, M.; Wang, K.; Osmani, M.; Liu, G.; Campioli, A. Drivers and Barriers towards Circular Economy in the Building Sector: Stakeholder Interviews and Analysis of Five European Countries Policies and Practices. *J. Clean. Prod.* **2022**, *336*, 130395. [CrossRef]
41. Creswell; Clark, V.P. *Designing and Conducting Mixed Methods Research*; Sage: Thousand Oaks, CA, USA, 2007.
42. Denzin, N.K. *The Research Act: A Theoretical Introduction to Sociological Methods*; Transaction Publishers: New York, NY, USA, 1978.
43. Akinade, O.; Oyedele, L.; Oyedele, A.; Delgado, J.M.D.; Bilal, M.; Akanbi, L.; Ajayi, A.; Owolabi, H. Design for Deconstruction Using a Circular Economy Approach: Barriers and Strategies for Improvement. *Prod. Plan. Control* **2020**, *31*, 829–840. [CrossRef]
44. O'Grady, T.; Minunno, R.; Chong, H.-Y.; Morrison, G.M. Design for Disassembly, Deconstruction and Resilience: A Circular Economy Index for the Built Environment. *Resour. Conserv. Recycl.* **2021**, *175*, 105847. [CrossRef]
45. Eberhardt, L.C.M.; Birgisdottir, H.; Birkved, M. Potential of Circular Economy in Sustainable Buildings. *IOP Conf. Ser. Mater. Sci. Eng.* **2019**, *471*, 092051. [CrossRef]
46. Adams, K.T.; Osmani, M.; Thorpe, T.; Thornback, J. Circular Economy in Construction: Current Awareness, Challenges and Enablers. *Proc. Inst. Civ. Eng.-Waste Resour. Manag.* **2017**, *170*, 1–11. [CrossRef]
47. Cheshire, D. *Building Revolutions, Applying the Circular Economy to the Built Environment*; RIBA publishing: London, UK, 2019. [CrossRef]
48. Minunno, R.; O'Grady, T.; Morrison, G.M.; Gruner, R.L.; Colling, M. Strategies for Applying the Circular Economy to Prefabricated Buildings. *Build* **2018**, *8*, 125. [CrossRef]
49. Joensuu, T.; Edelman, H.; Saari, A. Circular Economy Practices in the Built Environment. *J. Clean. Prod.* **2020**, *276*, 124215. [CrossRef]
50. Tsoka, S.; Tsikaloudaki, K. *Circular Economy Design and Management in the Built Environment, A Critical Review of the State of the Art*; Springer: Cham, Switzerland, 2024; pp. 257–272. [CrossRef]

51. Charef, R.; Lu, W.; Hall, D. The Transition to the Circular Economy of the Construction Industry: Insights into Sustainable Approaches to Improve the Understanding. *J. Clean. Prod.* **2022**, *364*, 132421. [CrossRef]
52. Mangialardo, A.; Micelli, E. Smart and Sustainable Planning for Cities and Regions, Results of SSPCR 2017. In *Green Energy and Technology*; Springer: Cham, Switzerland, 2018; pp. 333–344. [CrossRef]
53. Guerra, B.C.; Leite, F. Circular Economy in the Construction Industry: An Overview of United States Stakeholders' Awareness, Major Challenges, and Enablers. *Resour. Conserv. Recycl.* **2021**, *170*, 105617. [CrossRef]
54. Askar, R.; Bragança, L.; Gervásio, H. Adaptability of Buildings: A Critical Review on the Concept Evolution. *Appl. Sci.* **2021**, *11*, 4483. [CrossRef]
55. Eberhardt, L.C.M.; Birkved, M.; Birgisdottir, H. Building Design and Construction Strategies for a Circular Economy. *Archit. Eng. Des Manag.* **2022**, *18*, 93–113. [CrossRef]
56. Geldermans, R.J. Design for Change and Circularity—Accommodating Circular Material & Product Flows in Construction. *Energy Procedia* **2016**, *96*, 301–311. [CrossRef]
57. Dewagoda, K.G.; Ng, S.T.; Kumaraswamy, M.M.; Chen, J. Design for Circular Manufacturing and Assembly (DfCMA): Synergising Circularity and Modularity in the Building Construction Industry. *Sustainability* **2024**, *16*, 9192. [CrossRef]
58. Marsh, A.T.M.; Velenturf, A.P.M.; Bernal, S.A. Circular Economy Strategies for Concrete: Implementation and Integration. *J. Clean. Prod.* **2022**, *362*, 132486. [CrossRef]
59. Eberhardt, L.C.M.; van Stijn, A.; Stranddorf, L.K.; Birkved, M.; Birgisdottir, H. Environmental Design Guidelines for Circular Building Components: The Case of the Circular Building Structure. *Sustainability* **2021**, *13*, 5621. [CrossRef]
60. Mignacca, B.; Locatelli, G.; Velenturf, A. Modularisation as Enabler of Circular Economy in Energy Infrastructure. *Energy Polic.* **2020**, *139*, 111371. [CrossRef]
61. Minunno, R.; O'Grady, T.; Morrison, G.M.; Gruner, R.L. Exploring Environmental Benefits of Reuse and Recycle Practices: A Circular Economy Case Study of a Modular Building. *Resour. Conserv. Recycl.* **2020**, *160*, 104855. [CrossRef]
62. Kyrö, R.; Jylhä, T.; Peltokorpi, A. Embodying Circularity through Usable Relocatable Modular Buildings. *Facilities* **2019**, *37*, 75–90. [CrossRef]
63. Zairul, M. The Recent Trends on Prefabricated Buildings with Circular Economy (CE) Approach. *Clean. Eng. Technol.* **2021**, *4*, 100239. [CrossRef]
64. Arora, M.; Raspall, F.; Cheah, L.; Silva, A. Buildings and the Circular Economy: Estimating Urban Mining, Recovery and Reuse Potential of Building Components. *Resour. Conserv. Recycl.* **2020**, *154*, 104581. [CrossRef]
65. Nuñez-Cacho, P.; Górecki, J.; Molina-Moreno, V.; Corpas-Iglesias, F.A. What Gets Measured, Gets Done: Development of a Circular Economy Measurement Scale for Building Industry. *Sustainability* **2018**, *10*, 2340. [CrossRef]
66. Patil, R.A.; Shabunko, V.; Ramakrishna, S. An Introduction to Circular Economy. In *An Introduction to Circular Economy*; Springer: Cham, Switzerland, 2020; pp. 201–218. [CrossRef]
67. Klemeš, J.J.; Foley, A.; You, F.; Aviso, K.; Su, R.; Bokhari, A. Sustainable Energy Integration within the Circular Economy. *Renew. Sustain. Energy Rev.* **2023**, *177*, 113143. [CrossRef]
68. Olabi, A.G. Circular Economy and Renewable Energy. *Energy* **2019**, *181*, 450–454. [CrossRef]
69. Goh, K.C.; Kurniawan, T.A.; Goh, H.H.; Zhang, D.; Jiang, M.; Dai, W.; Khan, M.I.; Othman, M.H.D.; Aziz, F.; Anouzla, A.; et al. Harvesting Valuable Elements from Solar Panels as Alternative Construction Materials: A New Approach of Waste Valorization and Recycling in Circular Economy for Building Climate Resilience. *Sustain. Mater. Technol.* **2024**, *41*, e01030. [CrossRef]
70. Ghufran, M.; Khan, K.I.A.; Ullah, F.; Nasir, A.R.; Alahmadi, A.A.A.; Alzaed, A.N.; Alwetaishi, M. Circular Economy in the Construction Industry: A Step towards Sustainable Development. *Buildings* **2022**, *12*, 1004. [CrossRef]
71. Marika, G.; Beatrice, M.; Francesca, A. Adaptive Reuse and Sustainability Protocols in Italy: Relationship with Circular Economy. *Sustainability* **2021**, *13*, 8077. [CrossRef]
72. Foster, G. Circular Economy Strategies for Adaptive Reuse of Cultural Heritage Buildings to Reduce Environmental Impacts. *Resour. Conserv. Recycl.* **2020**, *152*, 104507. [CrossRef]
73. Foster, G.; Kreinin, H.; Stagl, S. The Future of Circular Environmental Impact Indicators for Cultural Heritage Buildings in Europe. *Environ. Sci. Eur.* **2020**, *32*, 141. [CrossRef]
74. Sanchez, B.; Haas, C. Capital Project Planning for a Circular Economy. *Constr. Manag. Econ.* **2018**, *36*, 303–312. [CrossRef]
75. Huysveld, S.; Hubo, S.; Ragaert, K.; Dewulf, J. Advancing Circular Economy Benefit Indicators and Application on Open-Loop Recycling of Mixed and Contaminated Plastic Waste Fractions. *J. Clean. Prod.* **2019**, *211*, 1–13. [CrossRef]
76. Petrescu, S.H.; Lazar, A.; Cioban, C.; Doroftei, I. Semi-Structured Interview. In *Qualitative Research in Regional Geography: A Methodological Approach*; Ilovan, O.R., Doroftei, I., Eds.; Presa Universitară Clujeană: Cluj-Napoca, Romania, 2017; ISBN 9786063703140.
77. Fellows, R.; Liu, A. Data Collection. In *Research Methods for Construction*; Wiley Blackwell: Oxford, UK, 2015.
78. Chang, R.-D.; Zuo, J.; Zhao, Z.-Y.; Soebarto, V.; Lu, Y.; Zillante, G.; Gan, X.-L. Sustainability Attitude and Performance of Construction Enterprises: A China Study. *J. Clean. Prod.* **2018**, *172*, 1440–1451. [CrossRef]

79. Clarke, V.; Braun, V. Thematic Analysis. *J. Posit. Psychol.* **2017**, *12*, 297–298. [CrossRef]
80. Terry, G.; Hayfield, N.; Clarke, V.; Braun, V. Thematic Analysis. In *The SAGE Handbook of Qualitative Research in Psychology*; Willig C., Roger, W.S., Eds.; Sage: London, UK, 2017.
81. Kiger, M.E.; Varpio, L. Thematic Analysis of Qualitative Data: AMEE Guide No. 131. *Méd. Teach.* **2020**, *42*, 846–854. [CrossRef]
82. Nowell, L.S.; Norris, J.M.; White, D.E.; Moules, N.J. Thematic Analysis. *Int. J. Qual. Methods* **2017**, *16*, 1609406917733847 [CrossRef]
83. Akadiri, P.O.; Olomolaiye, P.O.; Chinyio, E.A. Multi-Criteria Evaluation Model for the Selection of Sustainable Materials for Building Projects. *Autom. Constr.* **2013**, *30*, 113–125. [CrossRef]

applied sciences

MDPI

Article

Furniture Wood Waste Management Towards the Circular Economy

Giusilene Costa de Souza Pinho [1,*], João Luiz Calmon [1], Diego Lima Medeiros [2] and Clarisse Pinho [3]

1 Department of Environmental Engineering, Federal University of Espírito Santo (UFES), Vitória Campus, Fernando Ferrari Highway, 514, Vitória 29075-910, ES, Brazil; calmonbarcelona@gmail.com
2 Clean Technologies Network (TECLIM), Federal University of Maranhão (UFMA), Balsas Campus, MA-140 Highway, km 4, Balsas 65800-000, MA, Brazil; diego.lm@ufma.br
3 Department of Civil Engineering, Federal University of Espírito Santo (UFES), Vitória Campus, Fernando Ferrari Highway, 514, Vitória 29075-910, ES, Brazil; clarisse.pinho@edu.ufes.br
* Correspondence: giusilene.pinho@ifes.edu.br

Abstract: Wood waste (WW) from the furniture industry is a material that deserves special attention, given that it is still relatively unexplored in manufacturing industries. Therefore, this study aims to elaborate guidelines for decision-making in the furniture industry's wood waste, focusing on the north–central region of the state of Espírito Santo, Brazil, as a spatial outline of the research. The SWOT matrix is the main methodology for this stage of analysis, based on the previous study of Life Cycle Assessments (LCAs), the calculation of the Material Circularity Index (MCI), and the field research conducted in the industries. The obtained results show promising recycling scenarios over energy recovery, and the studied guidelines indicate this path. The three basic guidelines include discouraging the disposal of wood waste potentially usable as raw material into landfills; encouraging the use of wood waste based on the concept of industrial symbiosis; and promoting the creation of a market for the use of WW as a raw material in new products. Finally, with the proposed guidelines, it will be possible to contribute to the decision-making process on the management of wood waste from the furniture industry, reduce the amount of wood waste discarded in landfills, and increase the recycling rate. All of this being covered by a market created and maintained for the adequate management of wood waste. Furthermore, despite focusing on a Brazilian region for analysis, this paper intends to contribute to future research in the field, to be conducted in other locations.

Keywords: LCA; industrial symbiosis; SWOT matrix; circularity index; sustainable management

1. Introduction

The history of humanity has undergone major changes with the phenomenon of economic growth, which began in the middle of the 18th century [1]. Thus, the day-to-day life of people was modified. For a long time, the linear model of production, in which the depletion of natural resources and the accumulation of waste are typically ignored, was determinant in the global economy [2,3].

The waste management systems adopted by the majority of communities result from a vision that perceives natural resources as endless. However, the era of this production model is coming to an end, necessitating a shift towards a circular model that incorporates a comprehensive understanding of all stages of the product life cycle [4]. Therefore, for an adequate transition towards the Circular Economy (CE), public and private managers need to make assertive decisions about waste management.

https://doi.org/10.3390/app15031360

Within this context, the wood waste from the furniture industry is a material that deserves special attention, considering that it is still relatively unexplored in manufacturing industries, despite the existence of many proven studies guaranteeing a wide portfolio of products for incorporating said material as a raw material [5]. This unexpressive use occurs mainly due to the lack of efficiency in classifying these materials [6,7]. In Brazil, the vast majority of furniture industries utilize this wood waste either by burning it in ovens for heat and energy production or by sending it to landfills. Therefore, to better understand the dynamic of this sector, this research focused on the north–central region of the state of Espírito Santo, where timber residue is also often used for energy production [8].

Regarding the direct destination for burning, the cascade use of wood, which is an extremely widespread concept in the European Union, encourages the utilization of wood waste as a raw material at all possible stages, to be later used for energy recovery through burning. Thonemann and Schumann [9] highlighted that, in a multi-stage cascade, wood is processed into a product and this product is used at least once more in material form, before disposal or recovery for energy purposes. Adding to this, the cascade use and multiple uses of materials and resources, before their conversion into energy, has been identified as a strategy to improve resource efficiency and contribute to the Circular Economy, extending the lifespan of the wood resource [10].

In the study conducted by Pinho et al. [8], the recycling scenarios, mainly for the manufacturing of panels, were quite promising regarding the results of the comparative Life Cycle Assessment (LCA), conducted in the five proposed scenarios in the north–central regions of the state of Espírito Santo, Brazil. This also corroborates the concepts of the cascading use of wood. For a more effective contribution within this context, stakeholders must be guided on the best ways to dispose of the wood waste from the furniture industry. This can be accomplished through guidelines, strategies, and goals to achieve an adequate management towards the Circular Economy.

As far as it is known, to date, no other work has addressed the creation of guidelines for decision-making on the management of wood waste in the furniture industry, having as its main basis an LCA study of the projected scenarios. Furthermore, this research steps forward by studying the circularity index of the proposed products.

In this sense, there exists a scientific gap of studies that associate the LCA results with the concepts of Circular Economy, transforming them into guidelines to assist decision-making on the management of furniture waste.

In this way, acknowledging the importance of applying LCA combined with the concepts of CE for adequate waste management, this study aimed to develop guidelines for decision-making on the management of wood waste from the furniture industry in the north–central region of the state of Espírito Santo. Furthermore, despite adopting a region for analysis, this paper intends to contribute to future research in the field, to be developed in other locations.

2. Materials and Methods

Figure 1 presents the methodologies applied to achieve the proposed objective in this research. A previous study, which included a bibliographic review on Life Cycle Assessment applied to wood waste management systems, was the starting point in the first stage [5].

Subsequently, field research was conducted using a field journal, which served as a fundamental tool for recording relevant observations [11]. The field journal can be utilized by researchers to document activities, events, and conversations, capturing ideas, strategies, and thoughts [12]. Marconi and Lakatos [13] highlighted that field research aims to "obtain information and/or knowledge about a problem, for which an answer is sought, or a hypostasis that one wants to prove, or even discover new phenomena or relations between them." The technical visits took place in Colatina, Linhares, Pinheiros, São Roque do Canaã, and in Grande Vitória, Espírito Santo, Brazil, in the years 2019, 2020, 2022, and 2023.

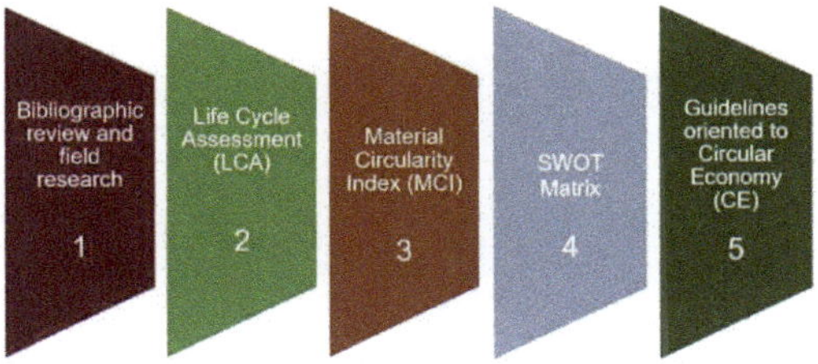

Figure 1. Methodological stages in the creation of guidelines for managing wood waste from the furniture industry.

In the second stage, the environmental performance of the proposed scenarios was evaluated based on the Life Cycle Assessment study carried out by Pinho et al. [8]. The attributional LCA conducted in the previous study was based on ISO 14040 [14] and ISO 14044 [15]. The use and final destination phases are not included within the scope of this study due to difficulties in tracking the use and final destination of the product. The adopted system boundary was from cradle-to-gate.

Additionally, the circularity (MCI) and linearity (LFI) indicators calculated for the proposed products in the third stage. These results were used to construct comparative tables, followed by the application of the SWOT matrix technique. The fifth stage was the development of guidelines for decision-making on the disposal of wood waste from the furniture industry.

2.1. Proposed Scenarios

A comparative LCA was conducted in Pinho et al. [8], in which was projected 5 scenarios for the disposal of wood waste from the furniture industry in the Linhares and Colatina furniture hubs, in the state of Espírito Santo, Brazil. The studied scenarios include the use of uncoated boards of Medium-Density Fiberboard (MDF), produced with 20% *w*/*w* of wood waste; uncoated boards of Medium-Density Particleboard (MDP), with 20% *w*/*w* of wood waste incorporation; production of solid ceramic brick with 11% *w*/*w* wood of waste incorporation; and heat production in ceramic industries (current practice) and disposal in landfills. Furthermore, the products and the energy generation were compared with the predominant scenarios in the market that utilizes virgin sources.

Regarding the percentages of recycled materials, some studies have adopted up to 100% recycled material for the manufacture of MDF [16] and MDP [17,18]. However, other authors have stated that for MDF, a replacement of 33% of fresh fibers by recycled fibers resulted in a decrease in the physical and mechanical properties of the boards [19,20]. The Brazilian study by Teixeira [21] highlights that with a 20% proportion of recycled material, new MDP boards can be produced without changing technology or compromising their quality.

Pinho et al. [8] found that scenarios that included wood waste as a raw material for manufacturing products appeared promising. Therefore, this study focused on 3 projected scenarios for the production of MDF, MDP, and ceramic brick. Subsequently, the calculation of the circularity indicator for the products was performed for the Scenarios 1, 2, and 3, which were analyzed using the SWOT matrix. Figure 2 illustrates the products considered for the insertion of wood waste.

Figure 2. Scenarios 1–3 were studied for the disposal of wood waste from the furniture industry in the state of Espírito Santo.

2.2. Circularity Indicator

The Material Circularity Indicator (MCI) was developed to quantify the circularity of products individually. The MCI expresses the extent to which the linear flow is reduced and the restorative flow is maximized for the materials comprising it. It also considers the duration and intensity of its use compared to similar products manufactured on an industrial scale [2].

The MCI was utilized to quantify the circularity of the proposed products in Scenarios 1, 2, and 3. The MCI takes into account the amount of virgin raw material (V) used in manufacturing, the amount of irrecoverable waste (W) associated with the product and the utility factor (X) which considers the duration (lifespan) and intensity of use (functional units) [2]. Table 1 presents the parameters for calculating the MCI, and Table 2 shows the input data for computing the circularity and linearity indices of the products from Scenarios 1–3.

Table 1. Parameters for calculating the MCI.

Symbol	Definition
M	Mass of a product
FR	Fraction of mass of a product's feedstock from recycled sources
FU	Fraction of mass of a product's feedstock from reused sources
FS	Fraction of a product's biological feedstock from Sustained Production. Biological material that is recycled or reused is captured as recycled or reused material, not biological feedstock.
V	Material that is not from reuse, recycling or, for the purpose of this methodology, biological materials from Sustained Production.
CC	Fractionof mass of a product being collected to go into a composting process
Cg	Fraction of mass of a product being collected for energy recovery where the materials satisfies the requeriments of inclusion.
CR	Fraction of mass of a product being collected to go into a recycling process
CU	Fraction of mass of a product going into component reuse
EC	Efficiency of the recycling process used for the portion of a product collected for recycling
EE	Efficiency of the energy recovery process for biological materials satisfying the requirements for inclusion
EF	Efficiency of the recycling process used to produce recycled feedstock for a product
BC	The carbon content of a biological material, by defaut a value of 45% is used unless supported evidence to the contrary
W	Mass of unrecoverable waste associated with a product

Table 1. *Cont.*

Symbol	Definition
WO	Mass of unrecoverable waste through a product material going into landfill, waste to energy and any other type of process where the materials are no longer recoverable
WC	Mass of unrecoverable waste generated in the process of recycling parts of a product
WF	Mass of unrecoverable waste generated when producing recycled feedstock for a product
LFI	Linear Flow Index
F(X)	Utility factor built as a function of the utility X of a product
X	Utility of a product
L	Actual average lifetime of a product
Lav	Average lifetime of an industry-average product of the same type
ER	Recovery energy
HHV	Higher heating value
MB	Mass of biological material
U	Actual average number of functional units achieved during the use phase of a product
Uav	Average number of functional units achieved during the use phase of an industry-averege product of the same type
MCIp	Material Circularity Indicator of a product

Table 2. Input data for computing the circularity indicator from Scenarios 1–3.

Input Data for MCI				
Factor	**Unity**	**MDF**	**MDP**	**Brick**
M	kg	690	630	1
FR	%	0.172	0.170	0.1045
FU	%	0.00	0.00	0.00
Fs	%	0.688	0.680	0.00
Cc	%	0.00	0.00	0.00
CE	%	0.05	0.05	0.12
CR	%	0.95	0.95	1.00
CU	%	0.00	0.00	0.00
EC	Factor	0.95	0.95	0.95
EF	Factor	0.95	0.95	0.95
Bc	%	0.45	0.45	0.00
MB	kg	593.4	535.5	0.110
F(x)	Factor	0.9	0.9	0.9
L	years	20	20	40
LAV	years	20	20	40
U	qtt	0.80	0.80	1.00
UAV	qtt	0.80	0.80	1.00

The calculation of the Linear Flow Index (LFI) and MCI was conducted using the software Microsoft Excel 2016, following the script presented in *"Circularity Indicators Methodology"*, published by the Ellen MacArthur Foundation [2].

Equation (1) of the LFI quantifies how much material comes from virgin raw material and ends as irrecoverable waste. LFI presents a value between 0 and 1, in which 0 is a circular flow and 1 is totally linear. The same is calculated using the following equation:

$$LFI = \frac{V + W}{2M + \frac{Wf - Wc}{2}} \quad (1)$$

Equation (2) of the MCI considers the LFI and an F(X) factor, which is constructed as an F function with X utility. It is calculated using the following equation:

$$\boldsymbol{MCI = 1 - LFI \cdot F(X)} \quad (2)$$

The data collected for the LCA studies conducted in Scenarios 1 to 3 in Pinho et al. [8] served as the basis for composing the input parameters for calculating the MCI, which are presented in Table 2. The mass (M) represents the weight of each finished product, while FR represents the percentage of wood waste incorporated into MDF, MDP, and ceramic brick. FS denotes the fraction of biological raw material originating from sustainable sources. In the panels, FR and the FS were calculated considering only the final wood incorporated. It was possible to exclude the value of other raw materials contained in the product; thus, V = 148.35 kg for MDF and V = 113.40 kg for MDP. The values of 108.33 kg (20% w/w) and 433.32 kg (80% w/w) are associated with MDF, respectively, for FR and FS. Regarding MDP, 113.4 kg (20% w/w) is related to FR and 413.28 kg (80% w/w) to FS. The FR of the ceramic brick corresponds to 0.111 kg (11% w/w). For the ceramic brick, FS = 0, while clay accounts for 0.890 kg (89% w/w) of the composition. None of the products studied used raw materials from reused sources; hence, FU = 0.

At the end of its lifespan, no waste is collected for composting, so CC = 0 in the three scenarios analyzed. Given that the adopted methodology focuses on the materials present in the final product, the biological materials constituting the panels and ceramic bricks do not satisfy all the conditions required for energy recovery, and thus CE = 0 [2]. At the end of their lifespan, the products are collected for recycling. CR = 0.95 was adopted for panels and CR = 0.80 for bricks [22]. Since there is no reuse of components in any of the products, CU = 0. The recycling process was considered 95% efficient, resulting in EC = 0.95. Similarly, EF = 0.95 [23].

The utility factor X was considered to be 1, adopting the recommended average utility of the industries [2]. A duration of 20 years was assumed for the panel's lifespan and 40 years for bricks [24,25]. For the LAV referring to a similar product life, the same lifetime was considered. An average value of 0.80 for functional unit "U" was adopted for panels and similar products, accounting for losses in the furniture industries, while the value of 1.0 was used for ceramic bricks [22].

2.3. SWOT Matrix

The SWOT matrix (strengths, weaknesses, opportunities, and threats) is a strategic planning tool that aims to diagnose the strengths, weaknesses, opportunities, and threats of the object under analysis [26]. Initially formulated for the field of administration, this matrix has been applied in various fields, including research in the field of solid waste management [27,28].

The tool is typically employed to conduct scenario analysis, aiming to compile all information into a matrix and thus facilitate the visualization of the characteristics involved in the study [29]. In this research, Scenarios 1, 2, and 3 were studied, which were intended for the incorporation of WW as a raw material in the manufacturing processes of MDF, MDP, and ceramic brick. The strengths and weaknesses in the internal environment of the panel factory and the ceramic industry, which will utilize WW as a raw material, are analyzed. The strengths are variables that provide favorable conditions for integrating the product into the company's portfolio. The opportunities are listed, considering external factors that can enhance the industrialization of the proposed product. External conditions, understood as difficulties, are considered threats to the manufacturing of the studied products. Figure 3 depicts the basic design of the SWOT matrix.

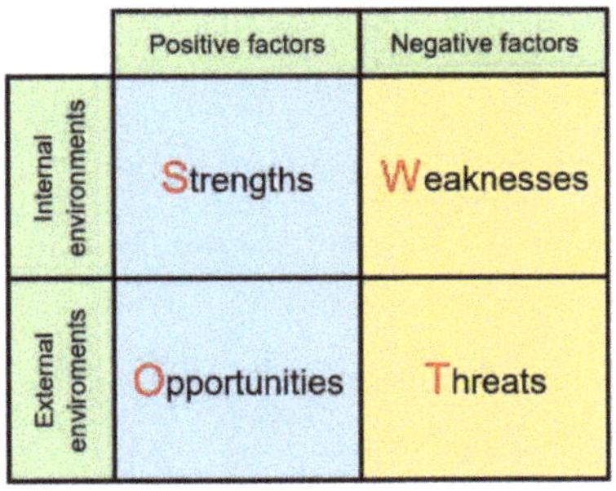

Figure 3. SWOT matrix.

3. Results and Discussion

3.1. LCA of Scenarios 1–3 and MCI of Scenarios 1–3

The results of the study by Pinho et al. [8], presented below, serve as the primary reference for preparing the SWOT matrix. Subsequently, the results from the Global Warming (GW) category were adopted as a reference to continue the analyses. Table 3 demonstrates that MDF and MDP exhibit the greatest environmental benefits per ton of waste and for the annual production of product, compared to Scenario 3. Considering the concept of the cascading use of wood, prioritizing use scenarios for manufacturing products is essential, with disposal for burning reserved as an end-of-life option [9,10,30,31].

Table 3. Comparative LCA results regarding GW (Scenarios 1–3).

Scenarios		Environmental Benefit (GW)	Annual Environmental Benefit (GW)
MDF	1	9.49×10^1	4.30×10^6
MDP	2	3.74×10^1	1.85×10^6
BRICK	3	2.90×10^1	1.70×10^4

The results of the sensitivity analysis regarding waste transport distances were also significant for this study. It demonstrates the feasibility of utilizing waste from the furniture industry throughout Grande Vitória. Only the truck modal from the southern region appears to be environmentally unviable. The modals that integrate trains and trucks show promise in all studied regions. Furthermore, considering that the amount of waste required to meet the monthly production of the projected scenarios and the estimated monthly generation of wood waste (6700 tons) in the furniture hubs of Colatina and Linhares, more than one scenario could be addressed simultaneously; however, the amount of wood waste is sensitive to market demands.

In addition to the Life Cycle Assessment study, this article also considers the calculations of the circularity index (MCI) and the product linearity index (LFI). Bertani et al. [23] highlighted the importance of joining MCI with LCA in search of circular and sustainable solutions. LCA is a consolidated methodology that follows ISO standards, allowing for the study of the entire product life cycle and obtaining information on different impact categories at different stages of the life cycle, whereas, for the MCI, information about the origin of the raw material, lifespan, function, and destination after use is necessary. Another feature of MCI is to present the result in a single value.

In Table 4, it is possible to see that the MDF and MDP indicated a strong tendency towards circularity, given the fact that they presented indices close to 1 for MCI. Regarding the linearity indicator, the value is 0.1346 for MDF and 0.1173 for MDP, indicating low linearity [2]. The brick presents moderate results for MCI and LFI. Appendix A, Appendix B, and Appendix C contains the tables prepared in Microsoft Excel 2016, with the data and equations used in calculating the LFI and MCI.

Table 4. Results of the circularity and linearity index of products (Scenarios 1–3).

Scenarios		Linearity Index (LFI)	Circularity Indicator (MCI)
MDF	1	0.1346	0.8789
MDP	2	0.1173	0.8945
BRICK	3	0.5910	0.4681

3.2. SWOT Matrix of Scenarios 1–3

Based on the literature review conducted by Pinho and Calmon [5], along with field research and the findings presented in Tables 3 and 4, the SWOT matrix was developed using the scenarios of the studied products. Upon completing the diagnosis, the scenarios were analyzed in light of the strengths, weaknesses, opportunities, and threats identified in the matrix quadrants.

In Figure 4, it is possible to observe that the strengths listed positively influence the panel industry in the redesign of the MDF produced, making it more eco-efficient product. Thus, the MDF was improved with the proposal to include 20% w/w of wood waste, replacing a percentage of virgin sources. It was possible to reduce the use of urea formaldehyde resin (UF) from 14% to 13.5% of the total weight of the board, influencing the reduction in environmental loads. In relation to MDP, corporative forces encourage the company to include it in its product portfolio, since the panel factory in the state of Espírito Santo, Brazil, has its production concentrated on the manufacture of MDF. Within the study analyzed, MDP incorporates 20% w/w of wood waste in its composition, making the panel more environmentally friendly, this being a practice that does not require a complex process for recycling. In addition to the region studied, the sensitivity analysis proved that all panel manufacturing can also receive waste from the entire Metropolitan Region of Grande Vitória. The proposed redesign aimed to enhance the eco-efficiency of the panels, resulting in environmental benefits compared to products manufactured solely with wood from virgin sources. The MCI showed a strong tendency towards circularity and the LFI showed a low tendency towards linearity in the analyses of the proposed panels that incorporate the WW with raw material.

Weaknesses are variables that can create unfavorable conditions for the adoption of the products proposed by the panel company. The factories in the state of Espírito Santo, Brazil, only have MDF in their portfolio; thus, the expansion of these factories would be required to enable MDP manufacturing. A thermo-hydrolytic treatment, in which wood fibers are recovered, requires an increase in water consumption, and, as a result, wastewater generated by the MDF manufacturing process will increase, which contributes to increased environmental impacts. The factory studied uses UF resin in the manufacture of panels, which was a relevant item in accounting for environmental impacts. Furthermore, a percentage of this adhesive is added to wood fibers in the recycling process. Therefore, more environmentally friendly solutions, as already reported in the literature, should be pursued for the manufacturing of wood panels in Brazil [32].

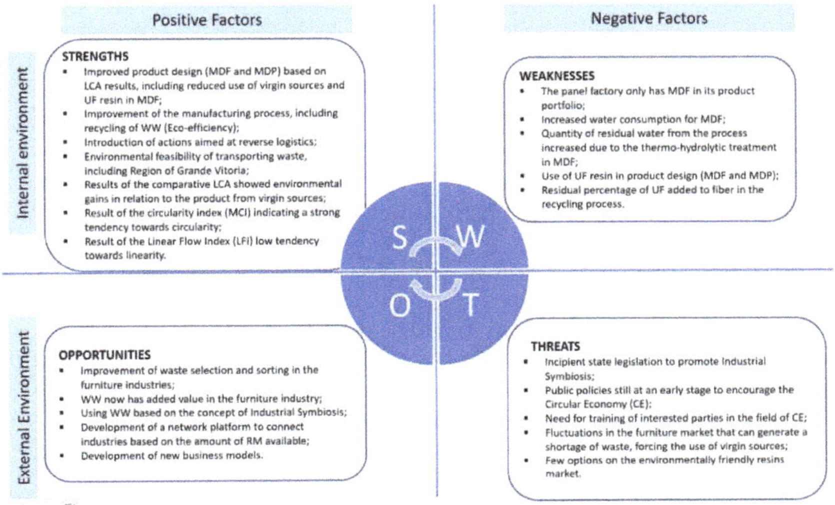

Figure 4. SWOT matrix (Scenarios 1 and 2) of the environmental performance of the production of MDF and MDP with wood waste.

Opportunities are external factors that, when worked on, positively stimulate the panel industry, in the readjustment of MDF and in the manufacture of MDP. Furthermore, they bring environmental and economic benefits to the furniture industries, also boosting CE in the state of Espírito Santo, Brazil. The selection and sorting of WW must be improved so that this material has added value and becomes raw material for the panel industry. This context is in line with the concept of industrial symbiosis, in which WW from the furniture industry is incorporated to produce new wooden panels. Therefore, it is appropriate to develop a network platform to monitor the amount of available WW. This platform will allow the panel industry to schedule its production with existing WW.

Threats are external variables that can create difficulties in implementing the proposed products by the panel company. To foster industrial symbiosis, it is necessary to create legislation that encourages connections between industries. In regards to the Circular Economy, there is a need for training interested parties and public policies are still incipient in this field. The availability of wood waste is sensitive to fluctuations in the furniture market, but the panel industry must be programmed to work with only virgin wood during these periods. In this context, it is important to highlight that there are still few large-scale produced options on the market to replace UF resin with more environmentally friendly products.

In relation to Scenario 3, in Figure 5, the highlighted strengths positively influence the red ceramic industry, improving the brick produced and making it a more eco-efficient product. The incorporation of 11% w/w of wood waste generates a reduction in the amount of clay, making the product lighter. The sensitivity analysis also proved that the entire manufacturing process of the proposed bricks can receive waste from the entire Metropolitan Region of Grande Vitória. Furthermore, the redesign proposal aimed at making the bricks more eco-efficient resulted in environmental gains compared to products manufactured solely with clay.

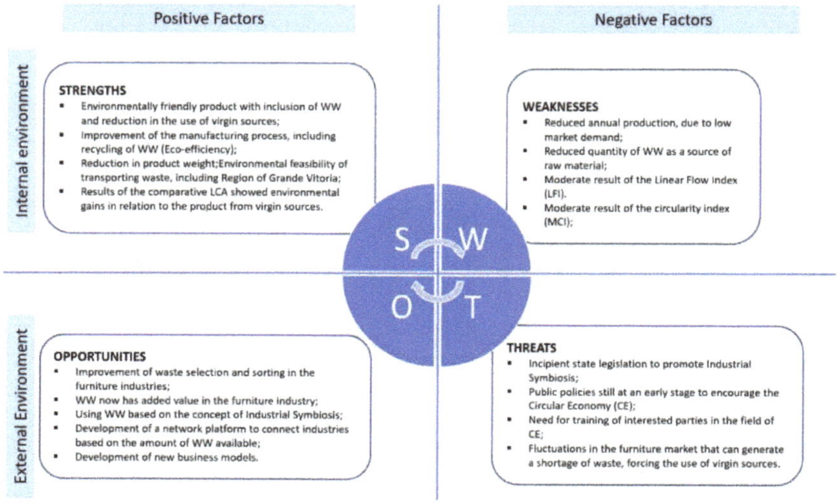

Figure 5. SWOT matrix (Scenario 3).

Weaknesses are variables that can create unfavorable conditions, particularly concerning the Material Circularity Indicator (MCI) and Linearity Flow Index (LFI), which exhibit moderate results. The production of solid brick is relatively low in the region under study, and an increase in demand could lead to greater environmental benefits compared to the current annual environmental benefit of manufacturing this product.

Opportunities are external factors that can positively stimulate the industry in the decision to improve the existing product in its portfolio. The selection and sorting of wood waste (WW) must be improved, generating added value and transforming waste into raw material for the ceramic industry, which will also lead to the qualification of labor. This proposed scenario is also in line with the concept of industrial symbiosis. Additionally, the network platform for monitoring the quantity of available WW will allow the ceramic industry to schedule its production with existing WW. In this context, there is a broad field for leveraging new sustainable business models.

Threats are external variables that can create difficulties in implementing improvements to be made to the product by the red ceramic industry, which are essentially similar to the challenges highlighted in the panel industries.

In the context of this analysis (Scenarios 1–3), it is also highlighted that this research includes an extensive phase of technical visits that took place over 4 years. It also adopts a regional scale, which enables the consideration of the positive and negative factors in the SWOT matrix. It all takes into consideration the characteristics and industrial synergies of the studied region. Furthermore, stakeholders involved in the generation and use of wood waste should collaborate to convert weaknesses and threats into strengths and opportunities.

3.3. Guidelines and Strategies

With the basis obtained through the proposed methodology, guidelines and strategies for the management of wood waste were developed, seeking to enable new cycles under the Circular Economy. This final stage was designed to support decisions for both external and internal stakeholders within the scope of industries implementing Scenarios 1, 2, and 3 as proposed.

As shown in Figure 6, the guidelines are connected to support decision-making within the scope of wood waste management in the furniture industry in Espírito Santo, Brazil. The three basic guidelines proposed are as follows:

- Guideline 1—Discourage disposal of natural wood waste (WW from virgin wood) and panels (WW from wood panels) to landfill;
- Guideline 2—Encourage the use of WW based on the concept of industrial symbiosis;
- Guideline 3—Promote the creation of a market for the use of WW as a raw material in new products, such as MDF, MDP, and ceramic brick.

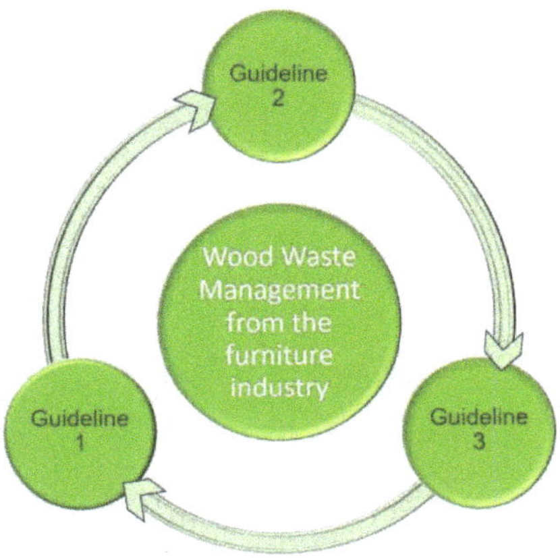

Figure 6. Guidelines for WW management from the furniture industry.

The first step expressed in Guideline 1 would be to inhibit the disposal of virgin wood waste and wood panel waste to landfill, as classified by Pinho et al. [8]. The strengths and opportunities detected for the manufacture of products in Scenarios 1, 2, and 3 show promising scenarios for the allocation of WW, which justifies the first guideline. Given its availability for disposal as a raw material, it is necessary to encourage the use of WW based on the concept of industrial symbiosis, as expressed in Guideline 2. Guideline 3 supports the reinsertion of wood waste into industries that can utilize it as a raw material for their products. It should also be noted that, once the market is established, efforts must be made to sustain it efficiently. In this regard, the study proposes the partial incorporation of wood waste, as the quantity of raw material fluctuates depending on market conditions. Therefore, in a scenario of scarcity, production can be adjusted according to the available quantities of waste and virgin wood, depending on the percentage available.

The three proposed guidelines form the fundamental framework, delineating a pathway for effective wood waste management. These guidelines were formulated with the proposition of strategies, representing necessary actions to achieve the proposed objectives. Moreover, it is imperative for industries in the sector, entities associated with the furniture industry, and policymakers to establish quantifiable time-bound goals that align with the strategies outlined in each guideline. These goals should be regularly monitored to assess compliance with the proposed strategies, ensuring continual improvement of the wood waste management system in the north–central region of the state of Espírito Santo, Brazil.

3.3.1. Guideline 1

The strategies outlined in the first guideline aim to discourage the disposal of raw material to landfills, as it can be repurposed as a raw material for other industries. The proposed strategies include the following: training stakeholders involved in the life cycle of renewable wood; establishing high tariffs for landfill disposal as a means of promoting industrial symbiosis; and monitoring the transport of small and large generators, as shown in Figure 7.

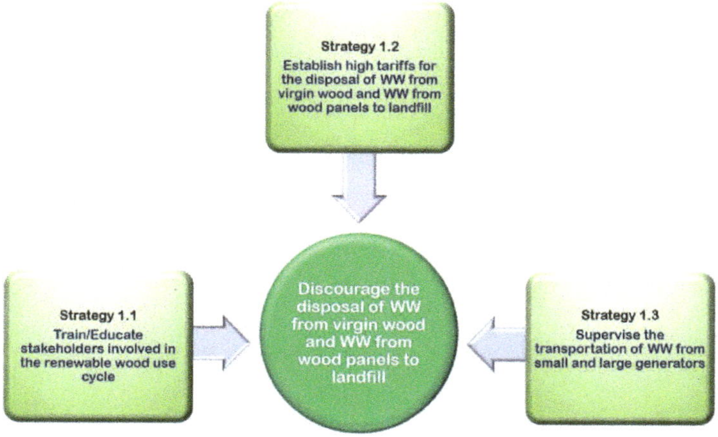

Figure 7. Guideline 1 for the management of wood waste from the furniture industry.

By discouraging the disposal of potentially usable wood waste in new cycles, a stock of raw material will be available for use in industry. In this way, Guideline 2 advances, encouraging connections between industries.

3.3.2. Guideline 2

The second guideline presented in Figure 8 concerns the need to encourage the use of WW based on the concept of industrial symbiosis. Relevant strategies include the following: the creation of legislation and standards that determine the sorting of waste based on type and quality; the creation of platforms to supply the quantity of WW available for use as a raw material in other industries; and the identification of existing synergies between generators and industries that can insert WW into the production cycle, such as the panel industry and the ceramic industry that were the subject of analysis in this study, and encourage the development of research into "eco-friendly" components for new products.

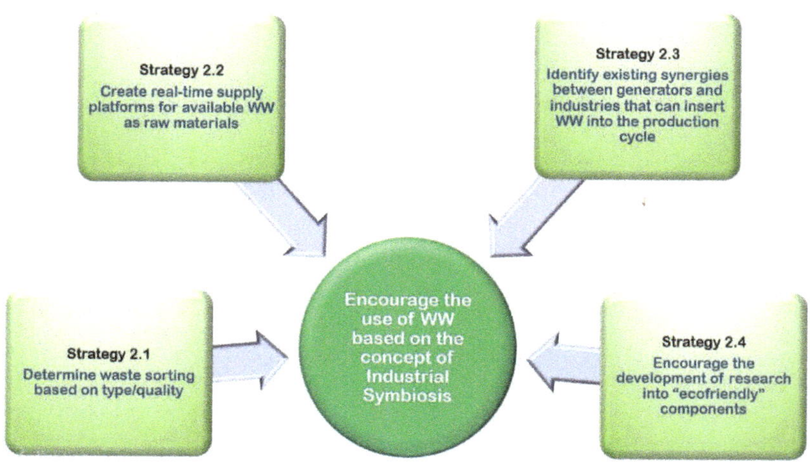

Figure 8. Guideline 2 for the management of wood waste from the furniture industry.

Based on quality, wood waste from the furniture industry can be destined for different scenarios, thus avoiding competition among them. In addition, in the study conducted by Bütün Buschalsky and Mai [16], MDF was recovered up to the third generation of panels using the thermo-hydrolytic disintegration process, and Kim and Song [33] stated that MDP can be recycled up to 16 times after the first use. This demonstrates the potential life cycle stages that these products present.

With these industries mapped and connected, with the available stock being supplied in real time, Guideline 3 deals with creating and maintaining the market so that WW management is sustainable.

3.3.3. Guideline 3

The third guideline aims to encourage the creation of a market for the use of WW as a raw material in new products. Within this proposal, the strategies include the following: the valorization of the railway modal as an alternative for WW transport, which proved to be promising in the analysis conducted for the products studied; the creation of legislation that encourages recycling as the first destination option for virgin wood waste and wood panel waste; the encouragement of new business models that use WW as a raw material in the manufacture of new products (open or closed cycles); and the drafting of legislation that encourages the use of WW in the product portfolio of companies that use virgin wood as a raw material, as shown in Figure 9.

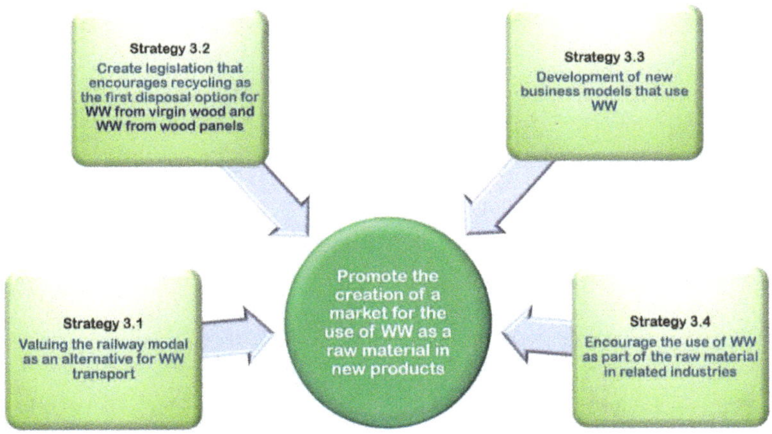

Figure 9. Guideline 3 for the management of wood waste from the furniture industry.

Using the proposed guidelines, it will be possible to reduce the amount of wood waste discarded in landfills and increase the recycling rate, all of this being covered by a market created and maintained for the adequate management of wood waste. It should also be noted that, after the possibility of recycling has been exhausted, the waste can be used for energy production, as long as the mixing proportion with panel waste is within the limits established by law, as presented in Pinho et al. [8] for the ceramic industry.

4. Conclusions and Prospects for Future Work

This article was based on studies published by Pinho and Calmon and Pinho et al. [5,8]. In the first article, an in-depth review of the literature of works related to LCA was conducted for the management of WW in the period from 2011 to 2021. Furthermore, a guiding proposal was presented, which highlights the main points to be considered for better LCA practice. In the second study, a comparative LCA was conducted to analyze the environmental aspects and impacts of various management scenarios of wood waste generated in the furniture industries of Espírito Santo, with a focus on the north–central region of the state. Thus, this work aimed to develop guidelines to support decision-making on furniture waste management, based on the association of three distinct methodologies applied from the perspective of the Circular Economy.

Based on the findings of Pinho et al. [8], which included the calculation of the circularity index and field studies, the achieved results were analyzed in relation to the strengths, weaknesses, opportunities, and threats proposed in the SWOT matrix methodology. This final stage was developed to support decision-making for both external stakeholders, concerning the creation of legislation and incentive/support programs, and internal stakeholders within the industries adhering to the proposed Scenarios 1, 2, and 3.

As the results found show the recycling scenarios as promising, prior to energy recovery, the guidelines studied indicate this direction. These basic pillars are connected to support decision-making in the management of wood waste from the furniture industry in Espírito Santo, Brazil. The three basic guidelines proposed are as follows:

- Guideline 1—Discourage the disposal of natural wood waste and wood panel waste to landfills;
- Guideline 2—Promote the utilization of wood waste under the principles of industrial symbiosis;
- Guideline 3—Foster the establishment of a market for the utilization of wood waste as a raw material in the manufacturing of new products, such as MDF, MDP, and ceramic bricks.

The guidelines were outlined with the proposition of strategies, which are essential for achieving the proposed objectives. Additionally, industries, furniture sector entities, and policymakers need to establish measurable time-bound goals that align with the strategies outlined in each guideline. These goals should be measured regularly to assess compliance with the proposed strategies, thereby ensuring continuous improvement of the wood waste management system designed for the north–central region of the state of Espírito Santo.

Finally, with the guidelines outlined in this research, it will be possible to reduce the amount of wood waste discarded in landfills and increase the recycling rate, all of this being covered by a market created and maintained for the adequate management of wood waste. It should also be noted that, after the possibility of recycling has been exhausted, the waste can be used for energy production. Additionally, an economic analysis of the projected scenarios must be incorporated into the decision-making process regarding the management of wood waste from the furniture industry, considering the region studied.

From the analysis performed, several perspectives have been identified for future studies. The most relevant ones are summarized and highlighted below:

- Faced with environmental emergencies, the demand for information regarding the life cycle is growing exponentially. Consequently, increasing scalability in conducting LCA studies is essential for companies aiming to grow sustainably. This is facilitated by the rising use of sensors, controllers, and digitization of processes in the context of Industry 4.0, which generate data that can be used in LCA. Therefore, new studies that address this objective should be encouraged.
- Future studies should also explore, through a consequential LCA, the impact of the increased use of wood waste across various production sectors.
- Studies that integrate methodologies with fuzzy membership could be beneficial and offer valuable insights. Therefore, it is a study field that should be further explored.
- Since the use and final destination phase are not included within the scope of this study, future research should investigate the use and final disposal phase of bricks and panels (cradle-to-grave or cradle-to-cradle).

- Wood panels with coatings were not discussed in this paper. Therefore, they require further study.
- Within this research field, there are few studies that explore how the efficiency of transportation is affected when the wood waste is shredded before being sent to the panel factory. Consequently, its effects should be investigated further.

Author Contributions: Conceptualization, G.C.d.S.P., J.L.C. and D.L.M.; methodology, G.C.d.S.P., J.L.C. and D.L.M.; software, G.C.d.S.P.; validation, G.C.d.S.P., J.L.C. and D.L.M.; formal analysis, G.C.d.S.P., J.L.C. and D.L.M.; investigation, G.C.d.S.P.; resources, J.L.C.; data curation, G.C.d.S.P.; writing—original draft preparation, G.C.d.S.P., J.L.C., D.L.M. and C.P.; writing—review and editing, G.C.d.S.P., J.L.C., D.L.M. and C.P.; visualization, G.C.d.S.P. and C.P.; supervision, J.L.C.; project administration, J.L.C. All authors have read and agreed to the published version of the manuscript.

Funding: This research received no external funding.

Institutional Review Board Statement: Not applicable.

Informed Consent Statement: Not applicable.

Data Availability Statement: The original contributions presented in this study are included in the article, and further inquiries can be directed to the corresponding author.

Acknowledgments: The authors of this study would like to thank the Espírito Santo Research and Innovation Support Foundation (FAPES) for its support.

Conflicts of Interest: The authors declare no conflicts of interest.

Appendix A

CIRCULAR ECONOMY FACTORS—MDF			
FACTOR	DEFINITION	VALUE	UNITY
M	Mass of a product.	690	kg
Fr	Fraction of mass of a product's feedstock from recicled sources.	0.1570	%
Fu	Fraction of mass of a product's feedstock from reused sources.	0.00	%
Fs	Fraction of a product's biological feedstock from Sustained Produc-tion. Biological material that is recycled or reused is captured as recycled or reused material, not biological feedstock.	0.6280	%
V	Material that is not from reuse, recicling or, for the purpose of this methodology, biological materials for sustained production.	**148.35**	kg
Cc	Fraction of mass of a product being collected to go into a composting process.	0.00	%
Ce	Fraction of mass of a product being collected for energy recovery where the materials satisfies the requiriments for inclusion.	0.00	%
Cr	Fraction of mass of a product being collected to go into a recycling process.	0.95	%
Cu	Fraction of mass of a product going into componet reuse.	0.00	%
Ec	Eficiency of the recycling process used for a portion of a product collected for recycling.	0.95	
EE	Eficiency of the energy recovery process for biological materials satisfing the requiriments for inclusion.	**0.0000**	
EF	Eficiency of the recycling process used to produce recycled feedstock for a product.	0.95	

Bc	The carbon content of a biological material, by default a value of 45% is used unless supported by evidence to the contrary.	0.45	%
ER	Recovered energy.	0.00	Mj
HHV	Higher heating value.	0.00	Mj
MB	Mass of biological material.	0.00	kg
W	Mass of unrecoverable waste associated with a product.	**38.0477**	kg
Wo	Mass of unrecoverable waste through a product's material going into landfill, waste to energy and other type of process where the materials are no longer recoverable.	**34.5000**	kg
Wc	Mass of unrecoverable waste generated in the process of recycling parts of a product.	**1.3939**	kg
Wf	Mass of unrecoverable waste generated when producing recycled feedstock for a product.	**5.7016**	kg
LFI	Linear flow index	**0.1346**	
F(X)	Utility factor built as a function of the utility **X** of a product.	0.90	
X	Utility of a product.	**1.00**	
L	Actual average lifetime of a product.	20	years
Lav	Average lifetime of an industry-average product of the same type.	20	years
U	Actual average number of functional units achieved during the use phase of a product.	0.80	qtt
Uav	Average numbers of functional units achieved during the use phase of a industry-average product of the same type.	0.80	qtt
	OUTPUT		
MCIp	**Material circularity Indicador of a product.**	**0.8789**	(0 to 1)
	☐ Input Data		

Appendix B

	CIRCULAR ECONOMY FACTORS—MDP		
FACTOR	DEFINITION	VALUE	UNITY
M	Mass of a product.	630	kg
Fr	Fraction of mass of a product's feedstock from recicled sources.	0.16	%
Fu	Fraction of mass of a product's feedstock from reused sources.	0.00	%
Fs	Fraction of a product's biological feedstock from sustained prodution. Biological material that is recicled or reused is captured as recicled or reused material, not biological feedstock.	0.66	%
V	Material that is not from reuse, recicling or, for the purpose of this methodology, biological materials for sustained production.	**113.40**	kg
Cc	Fraction of mass of a product being collected to go into a composting process.	0.00	%

Ce	Fraction of mass of a product being collected for energy recovery where the materials satisfies the requiriments for inclusion.	0.00	%
Cr	Fraction of mass of a product being collected to go into a recycling process.	0.95	%
Cu	Fraction of mass of a product going into componet reuse.	0.00	%
Ec	Eficiency of the recycling process used for a portion of a product collected for recycling.	0.95	
EE	Eficiency of the energy recovery process for biological materials satisfing the requiriments for inclusion.	**0.000**	
EF	Eficiency of the recycling process used to produce recycled feedstock for a product.	0.95	
Bc	The carbon content of a biological material, by default a value of 45% is used unless supported by evidence to the contrary.	0.00	%
ER	Recovered energy.	0.00	Mj
HHV	Higher heating value.	0.00	Mj
MB	Mass of biological material.	0.00	kg
W	Mass of unrecoverable waste associated with a product.	**34.8496**	kg
Wo	Mass of unrecoverable waste through a product's material going into landfill, waste to energy and other type of process where the materials are no longer recoverable.	**31.5000**	kg
Wc	Mass of unrecoverable waste generated in the process of recycling parts of a product.	**1.3939**	kg
Wf	Mass of unrecoverable waste generated when producing recycled feedstock for a product.	**5.3053**	kg
LFI	Linear flow index	**0.1172**	
F(X)	Utility factor built as a function of the utility **X** of a product.	0.9	
X	Utility of a product.	**1.0000**	
L	Actual average lifetime of a product.	20	years
Lav	Average lifetime of an industry-average product of the same type.	20	years
U	Actual average number of functional units achieved during the use phase of a product.	0.80	qtt
Uav	Average numbers of functional units achieved during the use phase of a industry-average product of the same type.	0.80	qtt
	OUTPUT		
MCIp	**Material circularity Indicador of a product.**	**0.8945**	(0 to 1)
	☐ Input Data		

Appendix C

	CIRCULAR ECONOMY FACTORS—BRICK		
FACTOR	DEFINITION	VALUE	UNITY
M	Mass of a product.	1	kg
Fr	Fraction of mass of a product's feedstock from recicled sources.	0.1110	%

Fu	Fraction of mass of a product's feedstock from reused sources.	0.00	%
Fs	Fraction of a product's biological feedstock from sustained prodution. Biological material that is recicled or reused is captured as recicled or reused material, not biological feedstock.	0.00	%
V	Material that is not from reuse, recicling or, for the purpose of this methodology, biological materials for sustained production.	**0.8890**	kg
Cc	Fraction of mass of a product being collected to go into a composting process.	0.00	%
Ce	Fraction of mass of a product being collected for energy recovery where the materials satisfies the requiriments for inclusion.	0.00	%
Cr	Fraction of mass of a product being collected to go into a recycling process.	0.80	%
Cu	Fraction of mass of a product going into componet reuse.	0.00	%
Ec	Eficiency of the recycling process used for a portion of a product collected for recycling.	0.80	
EE	Eficiency of the energy recovery process for biological materials satisfing the requiriments for inclusion.	**0.0000**	
EF	Eficiency of the recycling process used to produce recycled feedstock for a product.	0.80	
Bc	The carbon content of a biological material, by default a value of 45% is used unless supported by evidence to the contrary.	0.00	%
ER	Recovered energy.	0.0000	Mj
HHV	Higher heating value.	0.0000	Mj
MB	Mass of biological material.	0.00	kg
W	Mass of unrecoverable waste associated with a product.	**0.2739**	kg
Wo	Mass of unrecoverable waste through a product's material going into landfill, waste to energy and other type of process where the materials are no longer recoverable.	**0.20**	kg
Wc	Mass of unrecoverable waste generated in the process of recycling parts of a product.	**0.12**	kg
Wf	Mass of unrecoverable waste generated when producing recycled feedstock for a product.	**0.0278**	kg
LFI	Linear flow index	**0.5910**	
F(X)	Utility factor built as a function of the utility **X** of a product.	0.90	
X	Utility of a product.	**1.00**	
L	Actual average lifetime of a product.	40	years
Lav	Average lifetime of an industry-average product of the same type.	40	years
U	Actual average number of functional units achieved during the use phase of a product.	1	qtt
Uav	Average numbers of functional units achieved during the use phase of a industry-average product of the same type.	1	qtt
	OUTPUT		
MCIp	**Material circularity Indicador of a product.**	**0.4681**	(0 to 1)
	☐ Input Data		

References

1. Veiga, J.E. O Prelúdio do Desenvolvimento Sustentável. In *Economia Brasileira: Perspectivas do Desenvolvimento*; Centro Acadêmico Visconde de Cairu: São Paulo, Brasil, 2005; pp. 243–266.
2. Ellen Macarthur Foundation. *Circularity Indicators—Methodology*; Ellen MacArthur Foundation: Cowes, UK, 2019.
3. George, D.A.R.; Lin, B.C.; Chen, Y. A circular economy model of economic growth. *Environ. Model Softw.* **2015**, *73*, 60–63. [CrossRef]
4. UNEP-SETAC. Global guidance principles for life cycle assessment databases—A basis for greener processes and products. In *Shonan Guidance Principles*; UNEP-SETAC: Paris, France, 2011.
5. Pinho, G.C.D.S.; Calmon, J.L. LCA of Wood Waste Management Systems: Guiding Proposal for the Standardization of Studies Based on a Critical Review. *Sustainability.* **2023**, *15*, 1854. [CrossRef]
6. Faraca, G.; Tonini, D.; Astrup, T.F. Dynamic accounting of greenhouse gas emissions from cascading utilisation of wood waste. *Sci. Total Environ.* **2019**, *651*, 2689–2700. [CrossRef] [PubMed]
7. Höglmeier, K.; Weber-Blaschke, G.; Richter, K. Utilization of recovered wood in cascades versus utilization of primary wood—A comparison with life cycle assessment using system expansion. *Int. J. Life Cycle Assess.* **2014**, *19*, 1755–1766. [CrossRef]
8. De Souza Pinho, G.C.; Calmon, J.L.; Medeiros, D.L.; Vieira, D.; Bravo, A. Wood Waste Management from the Furniture Industry: The Environmental Performances of Recycling, Energy Recovery, and Landfill Treatments. *Sustainability* **2023**, *15*, 14944. [CrossRef]
9. Thonemann, N.; Schumann, M. Environmental impacts of wood-based products under consideration of cascade utilization: A systematic literature review. *J. Clean Prod.* **2018**, *172*, 4181–4188. [CrossRef]
10. Carus, M.; Dammer, L. The Circular Bioeconomy—Concepts, Opportunities, and Limitations. *Ind. Biotechnol.* **2018**, *14*, 83–91. [CrossRef]
11. Peirano, M.G.S. *A Favor da Etnografia*, 1st ed.; Relume-Dumará: Rio de Janeiro, Brazil, 1995; pp. 31–59.
12. Afonso, T.; da Costa Silva, S.S.; Pontes, F.A.R.; Koller, S.H. O uso do diário de campo na inserção ecológica em uma família de uma comunidade ribeirinha Amazônica = Using the field diary in ecological integration of a family from the Amazon's riverine community. *Psicol. Soc.* **2015**, *27*, 131–141. [CrossRef]
13. Marconi, M.A.; Lakatos, E.M. *Metodologia do Trabalho Científico: Projetos de Pesquisa/Pesquisa Bibliográfica/Teses de Doutorado, Dissertações de Mestrado, Trabalhos de Conclusão de Curso*, 8th ed.; Atlas: São Paulo, Brazil, 2017; p. 169.
14. *ISO 14040:2006*; Environmental Management—Life Cycle Assessment—Principles and Framework. International Organization for Standardization (ISO): Geneva, Switzerland, 2006.
15. *ISO 14044:2006*; Environmental Management—Life Cycle Assessment—Requirements and Guidelines. International Organization for Standardization (ISO): Geneva, Switzerland, 2006.
16. Buschalsky, F.Y.B.; Mai, C. Repeated thermo-hydrolytic disintegration of medium density fiberboards (MDF) for the production of new MDF. *Eur. J. Wood Wood Prod.* **2021**, *79*, 1451–1459. [CrossRef]
17. Hossain, M.U.; Poon, C.S. Comparative LCA of wood waste management strategies generated from building construction activities. *J. Clean. Prod.* **2018**, *177*, 387–397. [CrossRef]
18. Iritani, D.R.; Silva, D.A.L.; Saavedra, Y.M.B.; Grael, P.F.F.; Ometto, A.R. Sustainable Strategies Analysis through Life Cycle Assessment: A Case Study in a Furniture Industry. *J. Clean. Prod.* **2015**, *96*, 308–318. [CrossRef]
19. Ju, S.G.; Roh, J. Manufacture of dyed recycling wood fiber using waste MDF. *J. Korean Wood Sci. Technol.* **2017**, *45*, 297–307.
20. Roffael, E.; Behn, C.; Schneider, T.; Krug, D. Bonding of recycled fibres with urea-formaldehyde resins. *Int. Wood Prod. J.* **2016**, *7*, 36–45. [CrossRef]
21. Teixeira, M.F. Substituição de Matéria-Prima Virgem por Matéria-Prima Alternativa na Indústria de Madeira Reconstituída. Master's Thesis, PPGAD, Ambiente e Desenvolvimento, UNIVATES, Lageado, Brazil, 2012.
22. Caetano, M.D.D.E.; Depizzol, D.B.; Reis, A.d.O.P.d. Análise do gerenciamento de resíduos sólidos e proposição de melhorias: Estudo de caso em uma marcenaria de Cariacica, ES. *Gestão Produção* **2017**, *24*, 382–394. [CrossRef]
23. Bertani, I.F.; Martins, M.G.; Belli, M.C.A.S.; Rocha, T.B.; Moris, V.A.D.S. Análise da Aplicabilidade de Indicadores de Circularidade em uma Empresa de Reciclagem de Resíduos de Equipamentos Eletroeletrônicos (REEE). In *Encontro Nacional de Engenharia de produção—ENEGEP*; UFSCar: São Carlos, Brazil, 2022.
24. Análise Comparativa do Ciclo de Vida de Paredes Construídas com Blocos Cerâmicos, Blocos de Concreto e Concreto Armado Moldado Moldado in Loco, 2012. Available online: http://anicer.com.br/acv/ACV%20Blocos%20Cer%C3%A2micos.pdf (accessed on 3 January 2023).
25. Mantanis, G.; Athanassiadou, E.; Nakos, P.; Coutinho, A. A new recycling process for waste panels. In Proceedings of the European COST E31 Conference: Management of Recovered Wood, Thessaloniki, Greece, 22–24 April 2004; Gallis, C., Ed.; pp. 204–210.
26. Freitas, L.A. Gestão de Resíduos Sólidos na Construção e Montagem de Empreendimentos Industriais: Uma Contribuição da Ecologia Industrial. Ph.D. Thesis, Federal University of Rio de Janeiro, Rio de Janeiro, Brazil, 2018.

27. Mendes, H.C.; Bezerra, J.M.; Arlindo, V.J.C. Diagnóstico do gerenciamento de resíduos sólidos por meio da análise SWOT ao município de Tabuleiro do Norte–CE. *Rev. Bras. Geogr. Física* **2022**, *15*, 92–106. [CrossRef]
28. Mor, S.; Kaur, K.; Khaiwal, R. SWOT analysis of waste management practices in Chandigarh, India and prospects for sustainable cities. *J. Environ. Biol.* **2016**, *37*, 327.
29. Kotler, P.; Keller, K.L. *Administração de Marketing*, 12th ed.; Pearson Education do Brasil Ltda: São Paulo, Brazil, 2013; 796p.
30. Besserer, A.; Troilo, S.; Girods, P.; Rogaume, Y.; Brosse, N. Cascading Recycling of Wood Waste: A Review. *Polymers* **2021**, *13*, 1752 [CrossRef] [PubMed]
31. Lubke, H.; Ihnát, V.; Kuňa, V.; Balberčák, J. A multi-stage cascade use of wood composite boards. *Wood Res.* **2020**, *65*, 843–854 [CrossRef]
32. Silva, D.A.L.; Lahr, F.A.R.; Garcia, R.P.; Freire, F.M.C.S.; Ometto, A.R. Life cycle assessment of medium density particleboard (MDP) produced in Brazil. *Int. J. Life Cycle Assess.* **2013**, *18*, 1404–1411. [CrossRef]
33. Kim, M.H.; Song, H.B. Analysis of the global warming potential for wood waste recycling systems. *J. Clean Prod.* **2014**, *69*, 199–207 [CrossRef]

applied sciences

MDPI

Article

Tea Waste as a Sustainable Catalyst Support for Enhanced Removal of Contaminants of Emerging Concern via the Electro-Fenton Process: A Circular Economy Approach

Lucas Moreno-Bermedo [1], Jonathan Correa-Puerta [2], Claudio González-Fuentes [3], Néstor Escalona [4,5], Christian Onfray [1,*] and Abdoulaye Thiam [1,*]

1 Instituto Universitario de Investigación y Desarrollo Tecnológico (IDT), Universidad Tecnológica Metropolitana (UTEM), Ignacio Valdivieso 2409, San Joaquín 8940577, Región Metropolitana, Chile

2 Departamento de Física, Universidad Técnica Federico Santa María, Av. España 1680, Valparaíso 2340000, Región de Valparaíso, Chile

3 Instituto de Física, Pontificia Universidad Católica de Chile, Vicuña Mackenna 4860, Macul 7821093, Región Metropolitana, Chile

4 Departamento de Química Física, Facultad de Química y Farmacia, Pontificia Universidad Católica de Chile, Avenida Vicuña Mackenna 4860, Macul 7821093, Región Metropolitana, Chile

5 Departamento de Ingeniería Química y Bioprocesos, Escuela de Ingeniería, Pontificia Universidad Católica de Chile, Avenida Vicuña Mackenna 4860, Macul 7821093, Región Metropolitana, Chile

* Correspondence: confrayc@utem.cl (C.O.); athiam@utem.cl (A.T.)

Featured Application: This study aims to turn tea waste into catalytic solutions for water remediation.

Abstract: Contaminants of emerging concern (CECs) in water, including pharmaceuticals and personal care products, represent a significant threat to environmental and human health. In this context, the electro-Fenton (EF) process has emerged as a highly effective technique for the removal of such pollutants. This study investigates the innovative use of tea waste material (TWM) in combination with copper-iron nanoparticles (FeCuNPs) to degrade a mixture of CECs. A central aspect of this research is the sustainable reuse of organic waste material, such as TWM, to support catalytic nanoparticles. This approach not only utilizes a resource that would otherwise be discarded but also promotes sustainability in the treatment of contaminated water, aligning with the principles of the circular economy. The as-prepared FeCuNPs@TWM catalyst was fully characterized, and critical parameters influencing the pollutant removal were assessed, including adsorption capacity, catalyst load, and applied current. Under optimized conditions, the EF process, enhanced by FeCuNPs@TWM, achieved complete degradation of the contaminants within 15 min of the electrochemical process, and the activity remained after five catalytic cycles. Results demonstrate that using tea waste functionalized with FeCu nanoparticles as a catalyst not only improves the efficiency of the EF process but also offers an eco-friendly and cost-effective alternative.

Keywords: circular economy; biomass-derived catalyst; Fenton-based processes; tea waste; magnetic composite; contaminants of emerging concern

1. Introduction

The presence of contaminants of emerging concern (CECs) in aquatic environments has become a critical concern in recent decades due to their persistence and potential harm to human health and ecosystems [1]. These contaminants include pharmaceuticals,

https://doi.org/10.3390/app15031418

personal care products, endocrine-disrupting chemicals, and industrial compounds that are not completely removed by conventional water treatment processes. As these pollutants accumulate in water bodies, their adverse effects on aquatic life, biodiversity, and public health have sparked an urgent need for more effective water treatment technologies.

One of the most promising technologies for addressing this issue is the electro-Fenton (EF) process, a type of advanced oxidation process (AOP). In the EF process, the hydrogen peroxide (H_2O_2) is generated electrochemically at the cathode (Equation (1)), which reacts with ferrous ions (Fe^{2+}) in the solution, serving as the catalyst to produce hydroxyl radicals ($^{\bullet}OH$) (Equation (2)) [2]. These radicals are highly reactive and degrade a wide range of organic pollutants, including contaminants of emerging concern (Equation (3)) [3].

$$O_2 + 2H^+ + 2e^- \rightarrow H_2O_2 \quad (1)$$

$$Fe^{2+} + H_2O_2 \rightarrow {}^{\bullet}OH + Fe^{3+} + OH^- \quad (2)$$

$$CECs + {}^{\bullet}OH \rightarrow \text{By-products} + {}^{\bullet}OH \rightarrow CO_2 + \text{ions} + H_2O \quad (3)$$

Despite its effectiveness, the high cost of traditional catalysts and the need for sustainable resource management have prompted researchers to explore innovative approaches for making the EF process more sustainable and economically viable. The efficiency of the EF process can be significantly enhanced by using suitable catalysts that promote the generation of hydroxyl radicals [4]. Metal oxides and nanoparticles, such as iron and copper-based catalysts, have been extensively studied for this purpose. When solid catalysts are used, the Fenton reaction primarily occurs on the catalyst's surface, and the surface-based reactions can be summarized by the reactions outlined in Equations (4)–(7). In particular, copper-iron (FeCu) nanoparticles have gained attention due to their magnetic properties, high surface area, and catalytic activity [5–7]. These nanoparticles can be easily separated and reused, making them an attractive choice for sustainable water treatment. In addition, the presence of copper in FeCu nanoparticles further enhances the electron transfer processes, aiding the Fenton-like reactions (Equations (8) and (9)) and eases the recovery of Fe^{2+} (Equation (10)), boosting the overall catalytic activity and efficiency of the reaction [8,9].

$$\equiv Fe^{2+} + H_2O_2 \rightarrow \equiv Fe^{3+} + {}^{\bullet}OH + OH^- \quad (4)$$

$$\equiv Fe^{3+} + H_2O_2 \rightarrow \equiv Fe^{2+} + HO_2{}^{\bullet} + H^+ \quad (5)$$

$$\equiv Fe^{3+} + HO_2{}^{\bullet} \rightarrow \equiv Fe^{2+} + O_2 + H^+ \quad (6)$$

$$\equiv Cu^{+} + H_2O_2 \rightarrow \equiv Cu^{2+} + {}^{\bullet}OH + OH^- \quad (7)$$

$$\equiv Cu^{2+} + H_2O_2 \rightarrow \equiv Cu^{+} + HO_2{}^{\bullet} + H^+ \quad (8)$$

$$\equiv Cu^{2+} + HO_2{}^{\bullet} \rightarrow \equiv Cu^{+} + O_2 + H^+ \quad (9)$$

$$\equiv Cu^{+} + \equiv Fe^{3+} \rightarrow \equiv Cu^{2+} + \equiv Fe^{2+} \quad (10)$$

However, the practical use of iron-based nanoparticles is often limited by challenges such as nanoparticle aggregation and rapid oxidation [10]. To enhance their stability and performance, various support materials like activated carbon, biochar, zeolites, silica, and clays have been explored [11–14]. These supports offer high surface area, mechanical strength, and active sites for adsorption and catalysis [15], while their porous structures prevent nanoparticle agglomeration, improving the reactivity and longevity of iron-based systems. However, the widespread application of nanoparticle-based catalysts, especially in the EF process, faces limitations due to the high cost of synthetic materials and concerns over nanoparticle release into the environment. To address these issues, researchers are increasingly turning to natural and waste-derived materials as nanoparticle support structures [16]. These alternatives provide a cost-effective and environmentally friendly solution, combining the benefits of traditional support materials with the sustainability of using waste-derived resources for enhanced catalytic processes.

In the context of a circular economy, which seeks to minimize waste and make the best use of resources, the reuse of organic waste materials as catalytic supports has garnered significant attention [17–19]. In recent years, the use of agricultural and organic waste as a low-cost, renewable resource has been investigated for various applications, including water treatment. The incorporation of waste-derived materials into water purification processes not only aligns with the principles of resource recovery and waste minimization but also offers an eco-friendly alternative to traditional synthetic materials.

Tea waste is one of the most abundant organic waste materials generated globally, with an estimated 3.5 million tons consumed annually worldwide. The Food and Agriculture Organization of the United Nations has estimated that the global production of black tea will reach 4.4 million tons by 2027 [20]. Almost 30% of the tea produced is discharged as low-quality and waste tea resulting from tea processing and garden maintenance [21]. Tea waste, typically discarded after tea consumption, contains a variety of bioactive compounds, including polyphenols, tannins, and polysaccharides, which can be extracted, generating value-added products from tea waste [21–26]. Additionally, tea waste offers versatile properties and can be processed through various methods, including thermo-chemical and fermentation, for multiple applications, such as producing bio-oils, gases, and biochar. These technologies highlight tea waste's value as a resource for large-scale and specialized uses.

Over the last few years, research on tea waste has focused on converting it into value-added products such as packaging materials, nanocomposites, supercapacitors, biochar, aerogels, and hydrogels [27–29]. These applications improve waste management and support sustainable economic growth, offering environmental and economic benefits [30,31].

Tea residues are rich in cellulose, hemicelluloses, lignin, condensed tannins, and structural proteins, which possess a wide variety of functional groups, such as carboxyl, hydroxyl, and phenolic, as well as O–/heteroatoms-containing groups [32]. These functional groups, combined with the porous structure of tea waste, make it an ideal candidate for high adsorption capacity for several applications [33,34]. Numerous studies have demonstrated the potential of tea waste as an adsorbent for the removal of heavy metals, dyes, and organic pollutants from water [35–40]. Ali and coworkers [38] prepared a low-cost adsorbent from waste green tea leaves for phenol elimination through the sono-adsorption process. A phenol removal of 92% was obtained under 60 min of treatment under optimal conditions. Similarly, Kumbhar and colleagues [41] prepared a tea waste/Fe_3O_4 magnetic composite for the adsorption of crystal violet dye, achieving 98.7% removal using 1.0 g L^{-1} after 90 min. Utilizing tea waste not only provides an effective alternative to virgin materials but also addresses economic and environmental challenges, promoting sustainable waste management practices.

By functionalizing tea waste with nanoparticles, its adsorption properties can be further enhanced. This improvement is attributed to the increase in surface area and the reactivity of the catalyst, allowing for the removal of more complex and persistent contaminants, such as pharmaceuticals and industrial chemicals [42]. Iranparast Poshtiri and coworkers [43] developed a heterogeneous photo-Fenton catalyst for the removal of methyl orange through Cu^{2+} adsorption by using a magnetite/tea waste-derived catalyst. They obtained 96.9% elimination of the dye under optimized conditions. The catalyst also presented excellent stability, as evidenced by its negligible iron and copper leaching and reusability over three consecutive degradation cycles without significant loss of performance. The reusability and durability of the biochar-supported FeCu catalyst highlights its practical applicability for real-world wastewater treatment, promoting both sustainability and efficiency in contaminant removal.

This study explores the use of tea waste as a sustainable support material for nanoparticles in the EF process to remove a mixture of contaminants of emerging concern (CECs) from water: acetylsalicylic acid (ASA), ibuprofen (IBU), and naproxen (NAP). Specifically, the research focuses on the integration of copper-iron ferrite (FeCu) nanoparticles, a class of magnetic catalysts known for their excellent reactivity and reusability, onto a tea waste support. The use of tea waste as a catalyst support in the EF process exemplifies the circular economic approach by converting a widely available organic waste material into a functional component of a sustainable water treatment system. This not only addresses the issue of solid waste management but also contributes to reducing the environmental and economic costs associated with the use of virgin materials. By combining the natural adsorption capacity of tea waste with the catalytic power of FeCu nanoparticles, this approach aims to enhance the overall efficiency of the EF process while adhering to the principles of the circular economy and resource utilization. The elimination process follows a two-step approach: an initial adsorption-equilibrium phase, followed by the heterogeneous EF process. Key performance metrics such as degradation efficiency, reaction kinetics, and catalyst reusability were thoroughly assessed. This study introduces sustainable and underutilized biomass (tea waste) as a critical component in a hybrid adsorption-electrochemical system for wastewater treatment. It sets a new precedent for combining waste valorization, nanotechnology, and circular economy principles, providing an innovative framework for addressing water pollution challenges while promoting environmental sustainability.

2. Materials and Methods

2.1. Chemicals

Iron (II) sulfate heptahydrate ($FeSO_4{\cdot}7H_2O$), copper sulfate pentahydrate ($CuSO_4{\cdot}5H_2O$), sodium hydroxide (NaOH), and sodium sulfate ($NaSO_4$) were purchased by Sigma Aldrich (St. Louis, MO, USA). Acetylsalicylic acid (ASA, $C_9H_8O_4$), ibuprofen (IBU, $C_{13}H_{18}O_2$), and naproxen sodium (NAP, $C_{14}H_{13}NaO_3$) were of analytical standard ($\geq$98%, GC) and also provided by Sigma-Aldrich. Solutions were prepared using Mili-Q water (18.2 MΩ cm^2). The tea leaf waste (Ceylon black tea) used in this study was obtained from a local coffee shop.

2.2. Instruments and Equipment

Magnetic stirrers (Thermo Scientific, Waltham, MA, USA, model SP88857105) were used for all experiments. Samples were dried using a vacuum dryer (Merck, Darmstadt, Germany, model LVO-2050), and nanoparticle calcination was carried out in a muffle furnace at 800 °C (Selecta, Barcelona, España, model N13L). Nanoparticle dispersions were achieved in an ultrasonic bath (Branson, Brookfield, CT, USA, model SFX L102C). The binary catalyst was synthesized using an overhead mechanical stirrer (Dlab, Beijing, China, model, OS40-PRO). Samples underwent pyrolysis in a tubular furnace (Siomm, China) under an ultrapure nitrogen (N_2) atmosphere. UV-VIS spectroscopy analyses were performed using a Spectroquant Pharo 300 Merck spectrophotometer (Darmstadt, Germany).

2.3. Preparation of Catalyst

2.3.1. Synthesis of CuFeNP

CuFe nanoparticles were synthesized using a coprecipitation method, following the procedure outlined in the literature [8] with minor modifications. First, 40 mmol of $FeSO_4{\cdot}7H_2O$ was dissolved in 100 mL of water in a beaker under magnetic stirring.

Then, 20 mmol of $CuSO_4 \cdot 5H_2O$ was added to the solution, which was maintained at 60 °C. A 4 mol L^{-1} NaOH solution was then added dropwise until a homogeneous precipitate formed. The mixture was kept at 60 °C with vigorous stirring for 2 h. The resulting precipitate was washed with deionized water until the pH was neutral. The dark-brown solid product was vacuum-dried at 80 °C for 12 h, then ground in an agate mortar. Finally, the dried sample was calcined at 800 °C for 3 h.

2.3.2. Tea Waste Preparation

Tea waste was thoroughly washed with water heated to 90 °C until the solution became clear. The solid residue was then filtered and dried under vacuum at 80 °C for 12 h. After drying, the sample was finely ground using an agate mortar to achieve a uniform powder. This material was labeled tea waste (TW) and stored in an amber vial for future applications.

2.3.3. Preparation of CuFeNPs@TWM

The binary catalyst was prepared as described as follows: (i) The treated tea waste was added in 100 mL of water at 60 °C and stirred mechanically until fully dispersed; (ii) CuFeNPs were added into the solution, maintaining a weight ratio of 5:1 for TW and CuFeNPs, respectively; (iii) the mixture was stirred mechanically for 6 h, followed by filtration and drying under vacuum at 80 °C overnight. The obtained sample was ground manually in an agata mortar until a homogeneous powder was obtained. Then, the powder mixture was placed in a quartz vessel and pyrolyzed in a tubular furnace at 800 °C for 2 h under N_2 atmosphere, with a heating rate of 10° C min^{-1}. The resulting binary catalyst was ground in a mortar for 30 min and then kept in amber vials at room temperature for further applications. This material was labeled as CuFeNPs@TWM. For comparative purposes, TW was also thermally treated by pyrolysis at 800 °C in an N_2 atmosphere, resulting in a sample labeled TWM.

2.4. *CuFeNPs@TWM Physicochemical Characterization*

The binary catalyst and its components were analyzed using various analytical techniques to gather detailed information on their morphology, structure, chemical composition, and particle size. X-ray diffraction (XRD) was employed to study the crystallinity of the catalysts, using a Bragg–Brentano powder diffractometer (PANalytical, Almelo, The Netherlands, model X'Pert PRO MPD θ/θ) equipped with a Co/Kα radiation source ($\lambda K\alpha$ = 1.7903 Å; $\lambda K\alpha_1$ = 1.7890 Å, $\lambda K\alpha_2$ = 1.7929 Å) operating at 40 kV and 45 mA. The average crystalline size was determined using the Scherrer method [44] based on the data obtained by XRD analysis. The surface composition and chemical state of the elements in the material were investigated using X-ray photoelectron spectroscopy (XPS) with an ESFOSCAN instrument based on the PHI 5000 VersaProbe from Physical Electronics (ULVAC-PHI), Chanhassen, MN, USA. Measurements were conducted with a monochromatic, focused X-ray source (Al/Kα line: 1486.6 eV), calibrated using the Ag $3d_{5/2}$ line (full width at half maximum of 0.6 eV). The analyzed area was a 100 μm diameter circle. The spectral resolution was set to 224 eV Pass Energy and 0.8 eV/step for the survey spectra and 27 eV Pass Energy and 0.1 eV/step for high-resolution spectra. All measurements were performed in an ultrahigh vacuum (UHV) chamber, maintaining pressures between 5×10^{-10} and 5×10^{-9} Torr. Raman spectroscopy was performed using a Renishaw Raman spectrometer (Wotton-under-Edge, UK) with a 532 nm green laser. Surface composition was assessed through X-ray photoelectron spectroscopy (XPS) using a specific equipment model. Fourier-transform infrared spectroscopy (FT-IR) on a Nicolet iS5 Thermo Scientific (Waltham, MA, USA) system was used to identify functional groups present in the catalyst. The morphology of CuFe@TWM was examined using scanning electron microscopy

(SEM) with a GeminiSEM 360 Zeiss FESEM (Oberkochen, Germany), operated under high vacuum conditions and coupled with an Ultim Max 40 Oxford energy-dispersive spectroscopy (SEM-EDS) detector (High Wycombe, UK). The magnetization properties of the catalyst were investigated at room temperature using a homemade magnetometer (Santiago, Chile). Finally, the specific surface area was determined by the Brunauer–Emmett–Teller (BET) method, using the Rouquerol criteria. For this purpose, N_2 adsorption/desorption isotherms at 77 K were performed in a 3 Flex Micromeritics instrument (Norcross, GA, USA). The total pore volume was determined for all materials at $p/p_0 = 0.99$.

The point of zero charge (pH_{pzc}) of the catalyst was determined by the pH drift method [45]. For this purpose, 0.1 mol L^{-1} of Na_2SO_4 solution was used as an inert electrolyte. The pH was studied in the range of 1–9 and measured using a digital pH meter (Thermo Fisher Scientific, Waltham, MA, USA; model Stara 3215 Orion). The pH of the series of 20 mL of the electrolyte solution was adjusted using 0.1 mol L^{-1} of H_2SO_4/NaOH solutions. These pH values were labeled as initial pH (pH_i). Catalyst loads (4 mg) were added to the different solutions in stoppered glass tubes and equilibrated for 24 h. After this, the final pH (pH_f) was measured and plotted against the pH_i. The pHpzc was taken when the pH crossed the line $pH_i = pH_f$.

2.5. Electrochemical Setup

Electrochemical treatments were performed at 25 °C using a 150 mL undivided cell containing 100 mL solution with 50 m mol L^{-1} Na_2SO_4. The anode used was a metal mixed anode (MMO) made of a Titanium substrate covered with RuO_2, provided by NMT electrodes (Australia); and a carbon-polytetrafluoroethylene (PTFE) air diffusion purchased from E-TEK (Natick, MA, USA) was used as a cathode. Both were of a geometric area of 2.5 cm^{-2}, and they were kept at 1 cm distance from each other. Electrodes were previously activated by applying 100 mA cm^{-2} for 5 h in a 50 mmol L^{-1} Na_2SO_4 solution at pH 3.0. Electrochemical experiments were performed using a power supply MCP (model M10-QD305, Taichung, Taiwan) potentiostat-galvanostat.

2.6. Elimination of a Mixture of Contaminants of Emerging Concern

Solutions containing 5 mg/L of acetylsalicylic acid (ASA), ibuprofen (IBU), and sodium naproxen (NAP) were each treated in the presence of 50 mM Na_2SO_4, used as electrolyte. The reported concentration represents the average of three measurements, with the error expressed as the standard error of the mean (SEM). To assess the CECs' adsorption capacity of the surface of the material, the catalyst was dispersed in the solution before starting the treatment and stirred for 30 min. The optimization of the process was assessed by studying different currents (10, 20, and 50 mA) and different catalyst loads (0, 10, 20, and 50 mg L^{-1}). The concentration of each contaminant was followed by Ultra Performance Liquid Chromatography (UPLC), using a Perkin Elmer equipment, model Altus-30 (Waltham, MA, UA), coupled to a photodiode array (PDA) detector and controlled by an Empower software (v3). Each compound was studied at $\lambda = 225$ nm, and their retention times (t_R) were 2.6, 5.6, and 8.6 min for ASA, NAP, and IBU, respectively. A Phenomenex Kinetex® (C18, 100 × 2.1 mm, 1.7 µm, 100 Å) LC column (Torrence, CA, USA) was used at 30 °C. A flow of 0.3 mL min^{-1} and a gradient elution program of (A) Formic Acid (0.1% in water) and (B) Acetonitrile was used and is described as follows: t_{min}/A%; $t_{0.0}$/80; $t_{0.5}$/80; $t_{1.0}$/60; $t_{5.5}$/60; $t_{8.0}$/25; $t_{9.0}$/80; $t_{12.0}$/80.

The adsorption capacity of the material was studied following Equation (11) [46]:

$$q = \frac{(C_0 - C_t)\cdot V}{W} \quad (11)$$

where q is the amount of each catalyst uptake (in mg g^{-1}); C_0 is the initial concentration of each compound, and C_t is the concentration at time t (both in mg L^{-1}); V is the solution volume (in L), and W is the mass of the catalyst (in g).

The concentration of each contaminant in the function of time during the EF process was followed by Equation (12):

$$Normalized\ concentration = \frac{C_t}{C_0} \quad (12)$$

where C_0 and C_t are the initial concentration of each compound, and C_t is the concentration at time t (both in mg L^{-1}), respectively.

When organic compounds are attacked by $^{\bullet}OH$ radicals (Equation (3)), it could be described by the Langmuir–Hinshelwood model (Equation (13)) [47,48], which on integration can be written as either Equations (14) or (15).

$$-\frac{dC}{dt} = \mathrm{k}\,[\mathrm{OH}^{\bullet}]\,[\mathrm{CECs}] = \mathrm{k_{app}}\,[\mathrm{CECs}] \quad (13)$$

$$\ln\frac{C_0}{C_t} = \mathrm{k_{app}} \cdot \mathrm{t} \quad (14)$$

$$\ln\frac{C_t}{C_0} = -\mathrm{k_{app}} \cdot \mathrm{t} \quad (15)$$

Furthermore, the kinetic degradation of the mixture of CECs was determined by the pseudo-first order rate constants, also known as apparent rate constant, k_{app}.

2.7. Catalyst Reusability and Stability

The reusability of the prepared catalyst was studied under the optimized conditions. After the first cycle, the catalyst was magnetically recovered, thoroughly washed with abundant water and ethanol, dried, and then reused for subsequent analysis. Each cycle considered 30 min of equilibrium time and 30 min of the EF process. This procedure was repeated for a total of five catalytic cycles.

To gain some insights into the stability of the catalyst during the EF process, the concentration of Fe^{2+} leached into the solution was quantified after each cycle. For this purpose, the *o*-phenanthroline (phen) method was employed [49]. A 4 mL sample of the treated solution was mixed with 1 mL of a 1 g L^{-1} phen solution and 1 mL of acetate buffer (pH 4.6). Prior to analysis, the samples were filtered using a nylon syringe filter (13 mm diameter, 0.22 μm pore size). The resulting-colored solution was analyzed using a UV-Vis spectrophotometer (Merck, Spectroquant Prove 300) at a wavelength of 510 nm. Fe^{3+} quantification was determined by difference between the total released iron, determined by adding ascorbic acid into the solution, and the determined Fe^{2+}.

3. Results

3.1. Catalyst Characterization

The properties of the prepared CuFeNPs@TWM and its components were studied by analyzing their physicochemical properties. The crystalline structure of the materials was studied by XRD analysis (Figure 1). The diffractogram for TWM (Figure 1a) revealed broad peaks around 2θ = 26 and 47°, which are likely attributed to the (002) and (100) planes of graphitic carbon. This graphitic carbon originated from the biomass waste generated during the heat treatment. The diffractogram of the synthetized CuFeNPs (Figure 1b) indicated high crystallinity, with peaks corresponding to the lattice planes (111), (220), (331), (222), (400), (511), (440), and (533), suggesting the formation of $CuFe_2O_4$ nanoparticles with a cubic structure [COD card number: 00-901-2841] [50,51]. Additionally, other lattice planes, such as (11-1), (111), (20-2), (020), (202), (11-3), (022), (31-1), and (220), corresponding to monoclinic CuO [COD card number: 00-901-6057] [52], were also observed. These may have resulted from an incomplete reaction during the nanoparticle synthesis. The diffractogram for the binary material base (Figure 1c) revealed the presence of multiple characteristic crystallographic planes. Peaks observed at (111), (220), (331), (222), (400), (422), (511), (440), and (533) positions indicate the formation of the cubic $CuFe_2O_4$ structure observed before. Additionally, peaks at (110) and (200) positions were identified, corresponding to a cubic metallic Fe phase (COD card number: 00-900-0657) [53]. Moreover, the crystallographic planes at (111), (200), and (220) positions are attributable to the presence of cubic metallic Cu (COD card number: 00-500-0216) [54]. These findings confirm the coexistence of cubic phases of $CuFe_2O_4$, metallic Fe, and metallic Cu in the synthesized sample, demonstrating a complex multiphase structure. The average crystalline size determined was 36.0, 1.6, and 22.99 nm for CuFeNPs, TWM, and CuFeNPs@TWM, respectively. The relatively large crystalline size of CuFeNPs suggests well-developed crystal structures, while the small size of TWM evidences its amorphous nature due to pyrolysis. The intermediate size of CuFeNPs@TWM suggests that interactions between the nanoparticles and the TWM support induce lattice strain or limit crystal growth, enhancing surface area and potentially improving catalytic performance.

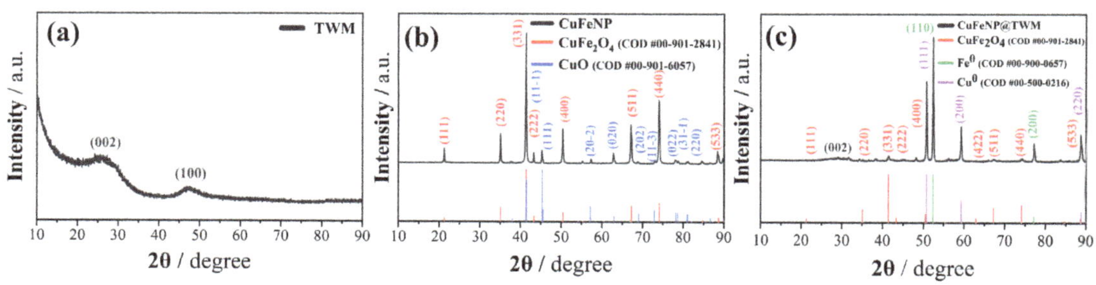

Figure 1. XRD diffractogram obtained for: (**a**) TWM, (**b**) CuFeNPs, and (**c**) CuFeNP@TWM.

Raman spectroscopy was used to study the vibrational and structural properties of the catalysts (Figure 2). For TWM, the typical D and G bands were found, which are associated with the A_{1g} symmetry phonon mode, induced from sp^3 hybridized carbon, and the E_{2g} phonon mode, associated with the in-plane vibration of sp^2-hybridization structure. In addition to a wide peak, around 2700–2800 cm^{-1} was observed, which is related to the 2D band, confirming the graphitic structure [55,56]. In the case of CuFeNPs, four peaks detected can be related to a $CuFe_2O_4$ structure. These peaks are around 164, 461, 536, and 681 cm^{-1} and have been associated with T_{2g}, F_{2g}, and A_{1g} Raman active nodes. The peak 461 cm^{-1} has been related to the O-site mode, representing the effect of the local lattice in the octahedral sublattice. The peak 680 cm^{-1} is related to the tetrahedral site mode that represents the local lattice effect in the tetrahedral sub-lattice [57–59]. Additionally, a peak at ~1300 cm^{-1} was found, which could be associated with α-Fe_2O_3 present in the structure [60,61]. In the case of the binary catalyst, only the D, G, and 2D bands were observed, which could be attributed to the higher intensity of these bands in comparison with that of the nanoparticles.

XPS was used to determine the elemental composition and the chemical states within the catalyst structure (Figure S1). The XPS analysis reveals significant differences in the elemental % at composition across the three samples. For the TWM sample (Figure S1a), the primary element detected is carbon (73.6%), accompanied by substantial amounts of oxygen (14.3%) and nitrogen (12.1%). In contrast, the CuFeNPs sample (Figure S1b) predominantly consists of oxygen (49.8%), iron (34.2%), and copper (11.1%). The composite sample CuFeNPs@TWM (Figure S1c) displays a mixed elemental composition, inheriting characteristics from both TWM and CuFeNPs. This sample contains significant amounts of carbon (47.2%) and oxygen (31.9%), with measurable quantities of iron (10.8%) and copper (8.6%), alongside a minor presence of nitrogen (1.6%). These values indicate that the synthesis process successfully integrates CuFe nanoparticles with the TWM matrix, resulting in a composite material with diverse elemental characteristics, potentially enhancing its functional properties for various applications.

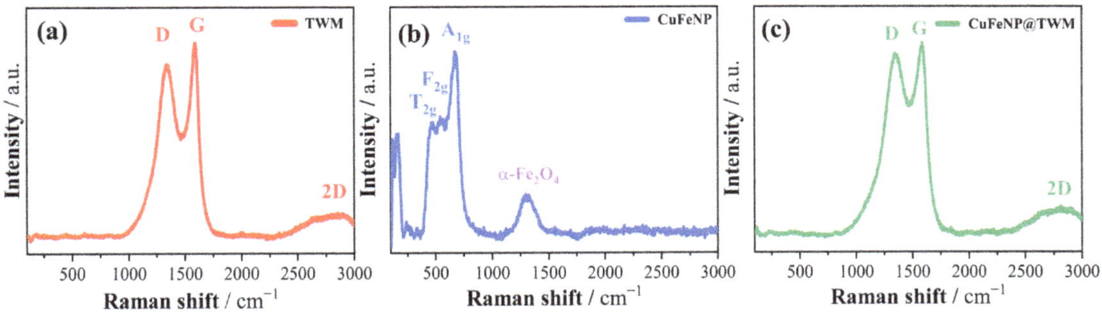

Figure 2. Raman spectra of (**a**) TWM, (**b**) CuFeNPs, and (**c**) CuFeNP@TWM.

In addition to the survey analysis, high-resolution spectra of the elements identified in CuFeNPs@TWM were examined (Figure 3), offering valuable insights into the chemical states and interactions within the material. The high-resolution C 1s spectrum (Figure 3a) revealed the presence of C=O, C–N, and C–C sites, which are characteristic of carbon-based materials derived from biomass. The C=O and C–N functionalities suggest the retention of oxygen- and nitrogen-containing groups after pyrolysis, which could enhance the material's hydrophilicity and possibly facilitate catalytic activity by providing reactive sites [62]. The C–C groups likely contribute to the overall stability and conductivity of the carbon matrix, supporting electron transfer processes essential in catalytic applications. The N 1s spectrum (Figure 3b) further confirmed the presence of nitrogen functionalities such as N–pyridinic, N–pyrrolic, and N–oxide. These nitrogen sites are particularly significant because they can enhance electron-donating properties, improve binding affinity with metal nanoparticles, and increase active site density, all of which can enhance catalytic performance [63,64]. N–pyridinic and N–pyrrolic groups, for example, are often associated with improved electron conductivity, while N–oxide groups could provide acidic sites or influence metal–support interactions, thereby stabilizing the nanoparticles [65]. For the O 1s spectrum (Figure 3c), three distinct sites were identified: C–O, C=O, and O–M (where M denotes Fe and/or Cu). The C–O and C=O sites are typically related to residual oxygen functionalities from the biomass and could play a role in modulating surface polarity and adsorption characteristics. M–O could be attributed to the metallic oxides of the prepared nanoparticles [66]. The Fe 2p spectrum (Figure 3d) exhibited characteristic peaks at binding energies corresponding to Fe $2p_{3/2}$ and $2p_{1/2}$, indicating the presence of both Fe^{2+} and Fe^{3+} oxidation states. This mixed-valence state of iron could be advantageous for redox reactions, as Fe^{2+}/Fe^{3+} cycling is often central to catalytic processes like the Fenton reaction [67,68]. The presence of both oxidation states also implies that the iron species are stabilized in a form conducive to electron exchange, which could enhance catalytic efficiency. Likewise, in the Cu 2p spectrum (Figure 3e), the detection of Cu^{+} and Cu^{2+} species highlights the coexistence of copper in different oxidation states, which could facilitate a wider range of redox reactions. The Cu^{+} species, often associated with high electron density, may provide active sites for specific catalytic processes, while the presence of Cu^{2+} could be beneficial for reactions requiring electron donation, such as the recuperation of the Fe^{2+} catalyst [6,69]. The coexistence of these oxidation states may enable a synergistic effect, potentially enhancing the overall catalytic performance and durability of the material.

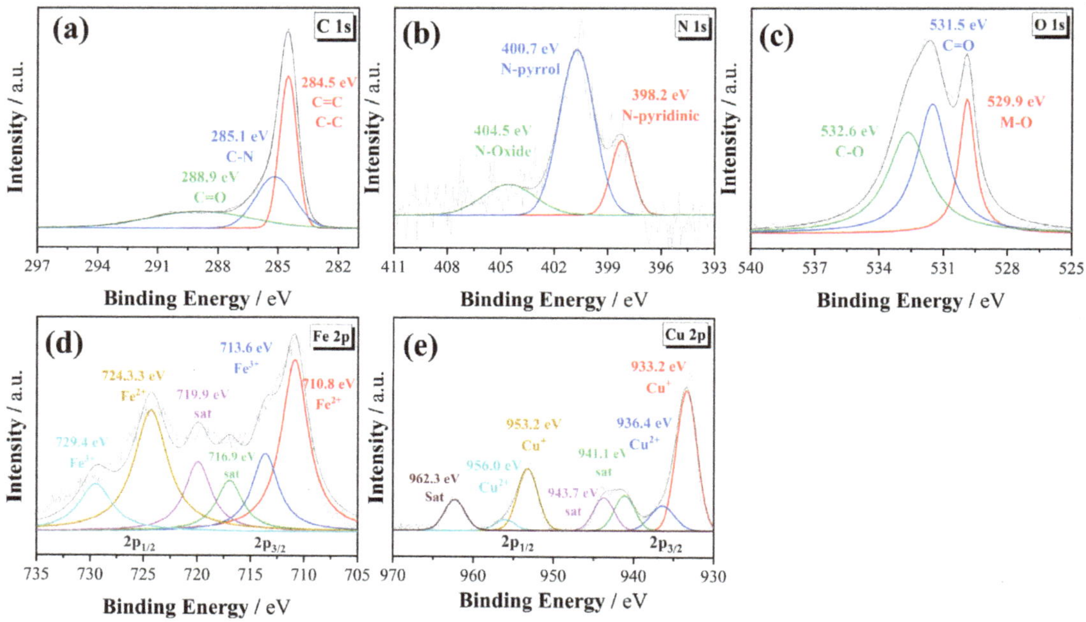

Figure 3. Elemental high-resolution spectra obtained by XPS for the CuFeNPs@TWM sample: (**a**) C 1s, (**b**) N 1s, (**c**) O 1s, (**d**) Fe 2p, and (**e**) Cu 2p.

Catalyst functionalities for different samples were analyzed by FT-IR (Figure 4). The FT-IR study shows typical peaks for biomass materials. For TWM, the peaks at 1010 and 1440 cm^{-1} are indicative of the stretching vibration of C–O and C–H bonds, the peaks at 1738 and 1617 cm^{-1} are likely associated with the stretching of C=O and C=C bonds, and peaks at 2849 and 2919 cm^{-1} correspond to the asymmetric stretching of C–H bonds within methyl ($-CH_3$) groups [16]. Additionally, a broad peak in the region of 3100–3600 cm^{-1} is attributed to the O–H stretching vibration of hydroxyl groups. For CuFeNPs, the stretch bonds associated with Fe–O and Cu–O were observed at 536 and 417 cm^{-1}, respectively [7]. In the case of the binary compound, CuFeNPs@TWM, peaks observed for its components were maintained, confirming the formation of the binary.

The morphology of the prepared materials was examined using SEM (Figure 5), providing insights into particle distribution, shape, and interactions within the catalyst structure. On the one hand, SEM analysis of TWM (Figure 5a) revealed a complex, irregular structure featuring a network of open pores and fibrous textures. This morphology likely arises from structural changes during the pyrolysis process, where high temperatures cause the breakdown and volatilization of organic compounds [32]. The observed opening of stomata, along with the development of interconnected pores and fibers, suggests that the biomass undergoes significant physical transformation, leaving behind a highly porous carbonaceous matrix [34]. The TWM support provides a carbonaceous matrix derived from biomass, potentially adding additional active sites or functionalities that could enhance metal–support interactions. On the other hand, SEM analysis revealed that in CuFeNPs (Figure 5b), the nanoparticles were highly agglomerated, forming irregular clusters with near-spherical morphologies. This clustering is likely driven by magnetic interactions between the CuFeNPs, a common phenomenon in materials with high magnetic susceptibility. Such agglomeration could impact the surface area and accessibility of active sites, which are critical factors in catalytic efficiency. The average particle size was measured at 133 ± 1 nm (Figure S2), suggesting a nanoscale distribution that can enhance reactivity by providing a high surface-to-volume ratio. Additionally, the analysis of the binary catalyst (Figure 4 confirmed that the CuFeNPs were well-dispersed on the TWM surface. The strong anchoring of the CuFeNPs on the TWM surface likely results from both chemical bonding and physical interactions, which may improve catalyst durability by minimizing nanoparticle detachment during use. The composition and elemental distribution of the materials were analyzed using EDS and elemental mapping, offering detailed insights into the elemental composition of the catalyst and support interactions. The EDS spectrum for TWM (Figure S3) indicated the presence of mostly C (82.7 at.%) and O (21.8 at.%) and traces of N, Na, K, Ca, and Mg, which were not quantified. For CuFeNPs, EDS (Figure S4) revealed that only Cu, Fe, and O were present, with atomic percentages of 12.2%, 29.7%, and 58.1%, respectively. In the CuFeNPs@TWM sample, elemental analysis revealed percentages of 90.4 at.% for C, 6.7 at.% for Fe, 0.9 at.% for Cu, and 2.2 at.% for O. The high carbon content indicates a strong presence of the TWM support, which serves as a carbon-rich matrix for metallic nanoparticles. The above is consistent with the high intensity of the D, G, and 2D bands of the binary observed via Raman spectroscopy. This carbon matrix not only provides a large surface area and electron-conductive network but also enhances the dispersion of the CuFe nanoparticles [70]. Interestingly, nitrogen was not detected in the EDS analysis, which could be due to the limitations of EDS in detecting low concentrations of light elements. However, XPS analysis confirmed the presence of nitrogen in low amounts, possibly from nitrogen doping or residual biomass nitrogen incorporated during pyrolysis. Although undetectable by EDS, these nitrogen sites may still play an important role in modifying the electronic properties of the carbon matrix, potentially enhancing its binding affinity and catalytic performance.

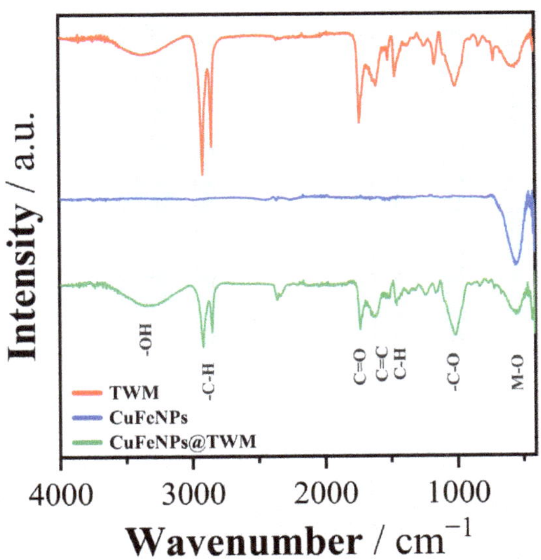

Figure 4. FT-IR spectra acquired from the samples.

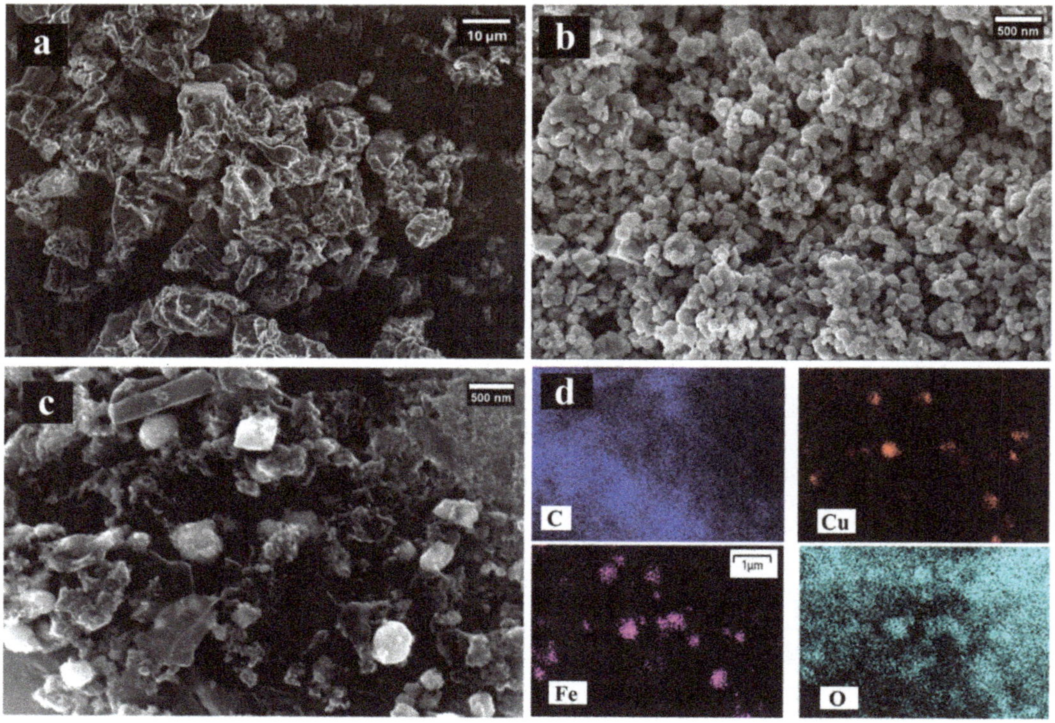

Figure 5. SEM images for (**a**) TWM, (**b**) CuFeNPs, and (**c**) CuFeNPs@TWM; and (**d**) EDS elemental mapping for (**c**).

The magnetization behavior of the samples was examined via hysteresis curves at room temperature, offering valuable insights into their magnetic properties and their implications for catalytic applications (Figure 6). As anticipated, CWM showed no magnetic properties, consistent with its carbonaceous and non-magnetic nature. In contrast, the CuFeNPs sample displayed a classic magnetic hysteresis loop, achieving a saturation magnetization of 31.5 emu g^{-1}. The presence of this loop indicates ferromagnetic behavior with notable magnetic retention. This ferromagnetism is attributed to the magnetic interactions between Fe and Cu ions within the spinel structure of the nanoparticle [71]. However, this retention can lead to nanoparticle agglomeration, a common drawback for ferromagnetic materials, as the particles retain magnetization even after removing the magnetic field. Such agglomeration can limit particle dispersion, reducing the accessibility of catalytic sites. Upon analyzing the binary composite CuFeNPs@TWM, an increase in magnetization was observed, reaching a saturation value of 40.1 emu g^{-1} at 1 T applied field. Interestingly, no remanent magnetization was detected, indicating a shift to superparamagnetic behavior. This transformation suggests that the synthesis conditions of the binary composite, particularly the high-temperature process, altered the magnetic interactions within the material [72]. High-temperature synthesis has been shown to influence crystallinity, particle size, and atomic arrangements, which can enhance magnetization by aligning magnetic domains more effectively without inducing remanence. The superparamagnetic behavior observed in CuFeNPs@TWM, characterized by the S-shaped magnetization curve and zero remanence, is particularly advantageous. Superparamagnetic materials exhibit high magnetic susceptibility, allowing them to become strongly magnetized in the presence of an external magnetic field but immediately return to zero magnetization once the field is removed. This is ideal for applications where reversible magnetic manipulation is required, such as catalyst separation in water treatment. The immediate loss of magnetization upon removing the field prevents agglomeration, which improves particle stability and maintains a high surface area accessible for catalytic activity. Additionally, high saturation magnetization enhances the ease of separation, as a modest magnetic field can effectively collect the particles without risking residual magnetization that could lead to re-agglomeration.

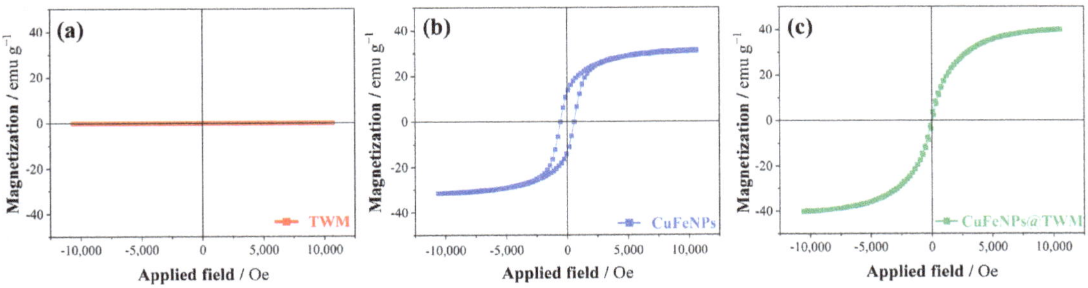

Figure 6. Magnetization curves obtained for: (**a**) TWM, (**b**) CuFeNPs, and (**c**) CuFeNPs@TWM.

The textural properties of the catalysts were studied using sortometry. Nitrogen adsorption/desorption is shown in Figure S5. The analysis showed IV-type isotherm for TWM and CuFeNPs@TWM, associated with mesoporous structures [73,74], and III-type isotherm for CuFeNPs, which are characteristics of agglomerates of nanoparticles, with low interaction with the adsorbate [75]. In addition, H3-type hysteresis was found for TWM and CuFeNPs@TWM, suggesting slit-like pores, characteristic of graphitic materials. Calculated BET surface area (S_{BET}) was 489, 6, and 295 $m^2\ g^{-1}$ for TWM, CuFeNPs, and CuFeNPs@TWM, respectively; and total pore volume was 0.221, 0.029, and 0.157 $cm^3\ g^{-1}$, for TWM, CuFeNPs, and CuFeNPs@TWM, respectively (Table 1). Both S_{BET} and V_{Total} suggest that CuFeNPs occupy part of the pore volume of the carbon material, diminishing the S_{BET} and V_{total} of the binary catalyst, such as was observed in SEM images (Figure 5c).

Table 1. BET-surface area, total volume, and average diameter pore for different samples.

Catalyst	S_{BET} ($m^2\ g^{-1}$)	V_{Total} ($cm^3\ g^{-1}$)	D Pore * (nm)
TWM	489	0.221	2
CuFeNPs	6	0.029	18
CuFeNPs@TWM	295	0.157	2

* Average pore diameter determined by 4V/BET.

In order to understand the effect of the reaction solution pH on the surface charge of the catalyst, the point of zero charge was studied. As is shown in Figure S6, the pH_{pzc} was found to be 2.3. At solutions with a pH below pH_{pzc}, the positive of the catalyst is positively charged, attracting anions from the solution. While, at solutions with a pH greater than pH_{pzc}, the surface of the catalyst is negatively charged, attracting cations [5].

3.2. Catalytic Performance of CuFeNPs@TWM for the Elimination of CECs

The catalytic performance of the binary material and its components were studied towards the degradation of a mixture of CECs, including ASA, NAP, and IBU, by a heterogeneous EF process. Figure 7 shows the normalized concentration decay for each contaminant during the treatment, including two steps: (i) 30 min of equilibrium/adsorption (considered as negative time, colored zone), and (ii) 30 min of the heterogeneous EF process (considered as positive time). Results indicated a low adsorption capacity using only TWM (Figure 7a) and CuFeNPs (Figure 7b), achieving a remotion of the mixture of contaminants that ranges between 2–3% and 1–2%, respectively, after the 30 min of adsorption. TWM exhibited modest adsorption capacities for all three contaminants (Table 2). The adsorption values for ASA, NAP, and IBU were 0.64 ± 0.01, 1.64 ± 0.05, and 1.39 ± 0.04 mg g^{-1}, respectively. This suggests that while TWM provides some level of porosity and adsorption sites, its performance is limited due to the absence of metallic active sites or surface modifications that allow its interaction with the contaminants. CuFeNPs showed slightly enhanced adsorption for ASA (0.81 ± 0.03 mg g^{-1}) compared to TWM, but reduced performance for NAP (1.30 ± 0.04 mg g^{-1}) and for IBU (0.08 ± 0.05 mg g^{-1}). This may be attributed to the specific affinity of CuFeNPs for ASA, potentially due to its interaction with active sites provided by the metallic oxides, while steric or chemical incompatibility with NAP and IBU hindered their adsorption. When the current was applied, low degradation was achieved using TWM, reaching removals that range from 3 to 17% within 30 min of the electrochemical process. This removal could be attributed to the •OH radicals generated on the anode surface and the lower-oxidant H_2O_2 electrogenerated on the cathode [76]. Likewise, for CuFeNPs, low oxidation percentages (4–14%) were achieved, which is explained by the agglomeration of the magnetic material that reduces the interaction between the Fenton and Fenton-like sites present in the catalyst with the H_2O_2 [11].

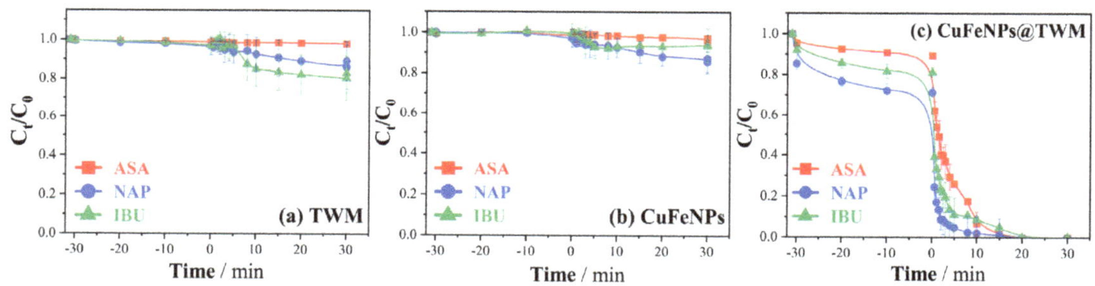

Figure 7. Concentration decay of each CEC using different materials as catalysts for heterogeneous EF process: (**a**) TWM, (**b**) CuFeNPs, and (**c**) CuFeNPs@TWM. Experimental conditions: Catalyst load, 20 mg L^{-1}; applied current, 20 mA; electrolyte, Na_2SO_4 50 mmol L^{-1}; pH, 3.0.

Table 2. Adsorption capacity of different catalysts, and for different catalysts dosage using CuFeNPs@TWM *.

Catalyst **	**q (ASA, mg g^{-1})**	**q (NAP, mg g^{-1})**	**q (IBU, mg g^{-1})**
TWM	0.64 ± 0.01	1.64 ± 0.05	1.39 ± 0.04
CuFeNPs	0.81 ± 0.03	1.30 ± 0.04	0.08 ± 0.05
CuFeNPs@TWM	5.32 ± 0.01	14.26 ± 0.02	9.39 ± 0.02
Catalyst dosage *** (mg L^{-1})	q (ASA, mg g^{-1})	q (NAP, mg g^{-1})	q (IBU, mg g^{-1})
10	4.01 ± 0.01	12.23 ± 0.01	9.97 ± 0.04
20	5.32 ± 0.01	14.26 ± 0.02	9.39 ± 0.02
50	3.26 ± 0.01	7.69 ± 0.04	4.09 ± 0.04

* Adsorption capacity reached after 30 min of equilibrium. ** Catalyst dosage used was 20 mg L^{-1}. *** CuFeNPs@TWM was used as catalyst.

Nevertheless, a remotion that ranges from 11 to 29% was achieved using the binary catalyst, CuFeNPs@TWM (Figure 7c), within 30 min of the equilibrium step. As can be seen in Table 2, using the binary compound, higher adsorption capacities for all contaminants were found, reaching 5.32 ± 0.01, 14.26 ± 0.02, and 9.39 ± 0.02 mg g^{-1} for ASA, NAP, and IBU, respectively. The enhancement in adsorption could be explained by a synergistic effect of the CuFe nanoparticles and the TWM support. While the first enhances the chemical interaction with the CECs, the latter increases the surface area available for adsorption, as was previously reported by Chen and coworkers [77]. Additionally, when CuFeNPs@TWM was used, complete degradation of three contaminants within 15 min of the electrochemical process was reached. This enhancement of the effectiveness could be attributed to the deagglomeration of the nanoparticles on the TWM surface, which increases the available area of the catalyst to react with the H_2O_2 electrogenerated to produce $^{\bullet}OH$ efficiently by Fe and Cu sites (Fenton-like reactions). In addition, some functional groups on the carbon surface of TWM, including carbonyl, carboxyl, and quinone groups, may activate the H_2O_2 generated and help with the catalyst recovery, along with Cu sites [78,79].

Key parameters of the heterogeneous EF process were studied using the binary catalyst, such as applied current and catalyst concentration. Figure 8 shows the elimination of the CECs by applying different currents. When no current was applied (Figure 8a), the remotion ranges from 11 to 29% after 30 min of adsorption. However, after 60 min, compared with the electrochemical process, the adsorption process reached 29.4%, 40.2%, and 42.5% of ASA, NAP, and IBU, respectively. This elimination was only attributed to an adsorption process due to the absence of electrogeneration of oxidants. By applying 10 mA (Figure 8b), an increase in the elimination rate was observed, achieving remotion of 79.1%, 91%, and 98.7% for ASA, NAP, and IBU, respectively, within 30 min of the electrochemical process. This improvement of the remotion is explained by the electrogeneration of strong oxidants, such as H_2O_2 and $^{\bullet}OH$, by applying a current [80,81]. An increase in the elimination rate was obtained when applied 20 mA (Figure 8c), achieving eliminations of 99.6%, 98.9%, and 100% for ASA, NAP, and IBU, respectively, within 15 min of the process. However, when 50 mA was applied (Figure 8c), a slight increase in the removal rate was found for NAP and IBU, achieving a complete elimination within 10 min, but for ASA, the total elimination was reached at 20 min of the EF treatment. As can be seen, there was not a significative increase in the elimination rate compared to that obtained between 20 and 50 mA, which could be attributed to the favoring of secondary reactions such as O_2 evolution, reduction in H^+, and recombination of two adjacent $^{\bullet}OH$ to produce H_2O_2 [82]. By comparing the apparent rate constants (k_{app}), shown in Table 3, an increase in the initial kinetics of the electrochemical process were obtained when increasing the current, which is attributed to the higher generation of oxidants. After the first minutes of treatment, the electrogenerated species react with the oxidized forms of the CECs, decreasing the kinetics of the process. This effect is confirmed at the end of the EF process in Figure 8d. Therefore, 20 mA was selected as the optimal current for further applications.

Table 3. Apparent rate constants of the CECs by applying different currents.

Current (mA) *	CEC	k_{app} (min^{-1})	R_2
10	ASA	0.12	0.99
	NAP	0.25	0.99
	IBU	0.22	0.99
20	ASA	0.19	0.97
	NAP	0.66	0.99
	IBU	0.28	0.99
50	ASA	0.39	0.99
	NAP	1.17	0.99
	IBU	0.49	0.99

* The used catalyst was CuFeNPs@TWM and its dosage was 20 mg L^{-1}.

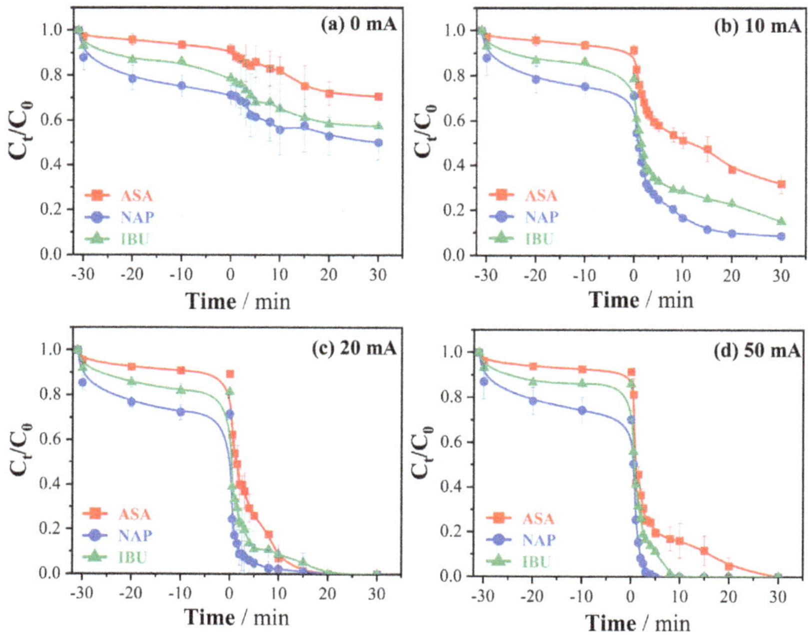

Figure 8. Normalized concentration decay for each CEC in the function of different applied currents by heterogeneous EF process: (**a**) Adsorption (0 mA), (**b**) 10 mA, (**c**) 20 mA, and (**d**) 50 mA. Experimental conditions: Catalyst, CuFeNPs@TWM; catalyst load, 20 mg L^{-1}; electrolyte, Na_2SO_4 50 mmol L^{-1}; pH, 3.0.

Once the applied current was optimized, the effect of catalyst load was studied in the range of 10–50 mg L^{-1}. As seen in Figure 9, there is a dependence between the amount of the catalyst used and the elimination of the contaminants, which involves both the adsorption and electrochemical processes. When 10 mg L^{-1} was used (Figure 9a), 4.1% (ASA), 12.3% (NAP), and 11.9% (IBU) were removed from the solution by adsorption after 30 min, achieving adsorption capacities of 4.01 ± 0.01, 12.23 ± 0.01, and 9.97 ± 0.04 mg g^{-1} for ASA, NAP, and IBU, respectively (Table 2). This dosage yielded high adsorption efficiency, particularly for NAP and IBU, suggesting that at lower dosages, the active sites are more concentrated and efficiently utilized, potentially reducing site competition between adsorbates [78,83]. For the same dosage, 29.1% (ASA), 29.9% (NAP), and 49.5% (IBU) were degraded after 30 min by the EF process.

By using 20 mg L^{-1} (Figure 9b), a significative increase in the elimination of the contaminants was observed. In this case, adsorption of 10.7% (ASA), 29.4% (NAP), and 19.8% (IBU) was reached, showing a higher retention capacity on the catalyst surface, achieving 5.32 ± 0.01, 14.26 ± 0.02, and 9.39 ± 0.02 mg g^{-1}, for ASA, NAP, and IBU, respectively. These values show higher levels of adsorption capacity than that obtained using 10 mg L^{-1} for ASA and NAP, while showing no significant variation for IBU [41,84]. After the equilibrium, a complete oxidation of the CECs was found at 15 min of electrochemical treatment, indicating that the increase in the amount of catalyst increases the number of active sites, not only for the adsorption but also for the oxidation process during the EF process.

A different effect was observed for 50 mg L^{-1} (Figure 9c) when an increase in remotion was only observed during the adsorption and not during the electrochemical process. After the first step, an elimination of 16.3% (ASA), 38.5% (NAP), and 20.5% (IBU) was found, favoring the remotion compared with that obtained using lower catalyst loads. However, if we compare by adsorption capacity, at 50 mg L^{-1}, the amount adsorbed decreased to

$3.26 \pm 0.01, 7.69 \pm 0.042$, and 4.09 ± 0.04 mg g^{-1} for ASA, NAP, and IBU, respectively. This decrease could be explained by the inefficient interaction of the catalyst with the contaminant, produced by different phenomena such as active site saturation, agglomeration, or competition between molecules for the same adsorption sites [32,85]. Alike, during the EF process, the oxidation rate was slower, achieving a complete elimination of ASA and NAP at 20 min of the treatment, but 8.7% of IBU remained after 30 min of the process. This effect could be attributed to the non-desired reactions, such as the interaction of Fe^{2+} with $^{\bullet}OH$ to generate Fe^{3+} and OH^-. This process reduces Fe^{2+} and $^{\bullet}OH$ available, thereby decreasing the oxidation efficiency of the CECs [86,87].

The pH of the solution is an important parameter in the EF process. To investigate its influence, the degradation efficiency at varying pH levels was studied using CuFeNPs@TWM (Figure S7). Results show that the contaminant removal by adsorption and degradation is strongly dependent on the initial solution pH. At pH 5 (Figure S7a), adsorption removal was below 8% for all contaminants, while degradation efficiencies after 30 min of the EF process were 14.0% for ASA, 33.6% for NAP, and 18.0% for IBU. At a pH of 7 (Figure S7b), no removal by adsorption was observed, and degradation efficiencies dropped to 8.8% for ASA, 14.6% for NAP, and 4.3% for IBU. As can be seen, the degradation efficiency at pH 3 >>> pH 5 > pH 7. This trend could be explained by the pH_{pzc} of the catalyst (pH_{pzc} 2.3). When the solution pH is 5 and 7 (i.e., pH >> pH_{pzc}), there was a negative charge due to the deprotonation. This negatively charged surface reduces interactions with the contaminants, which are also negatively charged at their respective pH higher than their pK_a values (ASA: 3.0, NAP: 4.2, and IBU: 4.9) [88–90]. Furthermore, at higher pH values, the generation of $^{\bullet}OH$ is reduced, contributing to the reduced degradation efficiency.

It is noteworthy that the catalytic activity of the prepared material significantly surpasses that of previously reported biochar-based catalysts used in Fenton-based processes for the removal of CECs. A detailed comparison is provided in Table 4. The prepared biochar composite demonstrates significantly higher catalytic efficiency compared to similar biochar composites synthesized using analogous procedures but with different biomass and nanoparticles. This superior performance in terms of the applied current and required treatment time highlights the potential of our biochar composite to set a new benchmark for developing highly efficient Fenton-based catalysts for CECs removal.

Table 4. Comparison of the contaminant removal process using catalysts used in this work and other related catalysts.

Catalyst	Contaminant	Process	Initial pH	Current	Remotion/Time	Reference
Waste green tea leaves	Phenol	Sono-adsorption	3.5	-	92% (60 min)	[91]
$CuFe_2O_4$/penicillin residue biochar	Tetracycline	Fenton-like	3-9	-	>95% (90 min)	[92]
Cu-adsorbed on magnetite/tea waste	Methyl orange	Photo-Fenton catalyst	3	-	96.9% (200 min)	[43]
$ZnFe_2O_4$ adsorbed on eucalyptus sawdust	Rhodamine B (RB) and reactive brilliant red X-3B (RBR)	Adsorption+ photo-Fenton	3	-	>99% (60 min)	[77]
Cu-Fe-FeC_3 on nitrogen-doped biochar	Amoxicillin	Electro-Fenton	4	120 mA	100% (35 min)	[93]
Nano-Fe_3O_4	Amoxicillin	Electro-Fenton	3	300 mA	100% (60 min)	[94]
CuFeNPs@TWM	Acetylsalicylic acid, Ibuprofen, Naproxen	Electro-Fenton	3	20 mA	100% (15 min)	This work

3.3. Catalyst Reusability and Stability

Exploring the reusability of heterogeneous catalysts is a key aspect of advanced research aimed at developing sustainable processes. To evaluate reusability of CuFeNPs@TWM, its performance in degrading a mixture of CECs was tested over multiple cycles under optimized conditions (Figure 10). As can be seen, the removal efficiency of CuFeNPs@TWM remained consistent for up to five catalytic cycles, achieving a complete elimination of the contaminants within each treatment cycle. This suggests good stability of the material under the studied conditions. For a more detailed assessment of the stability of the catalyst, the amount of Fe^{2+} and Fe^{3+} leached during each cycle were quantified (Figure S8). Stability tests showed that after the first cycle, the amount of Fe^{2+} reached almost 0.02 mmol L^{-1}. This value decreased during the second cycle but progressively increased with the subsequent catalytic cycle. This behavior suggests that a portion of the catalyst is being affected by the oxidized species generated during the treatment. However, as demonstrated in Figure 10, this phenomenon does not significantly impact catalytic activity. Additionally, the concentration of Fe^{3+} was found to be approximately one-third of the Fe^{2+} concentration, indicating that the catalyst undergoes effective recovery during the treatment.

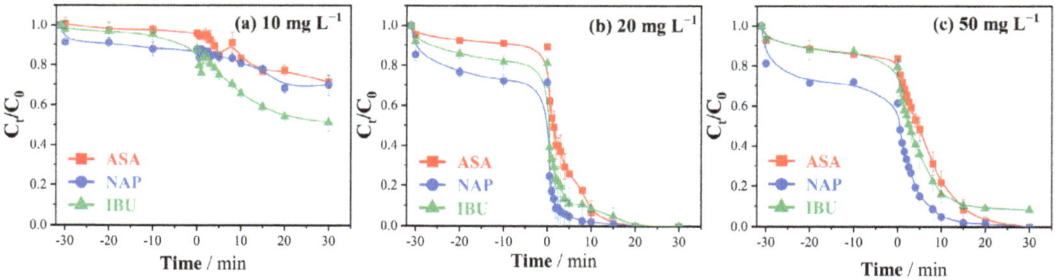

Figure 9. Normalized concentration decay for each CEC in the function of different catalyst loads by heterogeneous EF process: (**a**) 10 mg L^{-1} (**b**) 20 mg L^{-1}, and (**c**) 50 mg L^{-1}. Experimental conditions: Catalyst, CuFeNPs@TWM; applied current, 20 mA; electrolyte, Na_2SO_4 50 mmol L^{-1}; pH, 3.0.

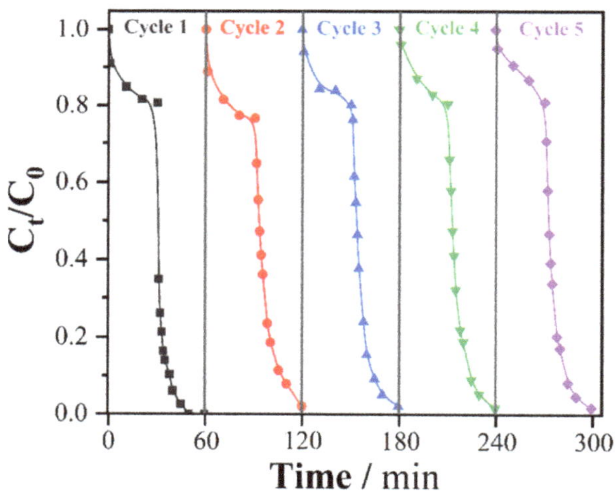

Figure 10. Normalized concentration of the mixture of CECs in the function of time of treatment at different catalytic cycles. Each cycle considered 30 min of adsorption and 30 min of EF process. Experimental conditions: Catalyst, CuFeNPs@TWM; catalyst load, 20 mg L^{-1}; applied current, 20 mA; electrolyte, Na_2SO_4 50 mmol L^{-1}; pH, 3.

As demonstrated in this research, tea waste has proven to be an excellent support material for magnetic nanoparticles, enhancing their stability, reactivity, and catalytic performance, particularly in the EF process. The porosity and functional groups of the material facilitate the adsorption of contaminants and ease the generation of strong oxidants, for effective degradation of contaminants of emerging concern.

The reuse of tea waste offers significant environmental and economic advantages by reducing organic waste and transforming it into a high-value material, aligning with circular economy principles. This approach addresses critical challenges such as waste management and water pollution, providing a low-cost, scalable solution for CECs removal while reducing the dependence on non-renewable catalyst materials.

4. Conclusions

This study demonstrates the potential of using nanostructured catalysts derived from biomass waste, specifically CuFe nanoparticles supported in tea waste material (CuNPs@TWM), for the efficient removal of pharmaceutical contaminants from water. By analyzing the adsorption behavior of Acetylsalicylic acid (ASA), naproxen (NAP), and ibuprofen (IBU), key insights were obtained regarding the material's performance and the influence of catalyst dosage. Results revealed that CuNPs@TWM exhibited superior adsorption capacities compared to TWM and CuFeNPs, with maximum adsorption values of 5.32 ± 0.01, 14.26 ± 0.02, and 9.39 ± 0.02 mg g^{-1} for ASA, NAP, and IBU, respectively. The enhanced performance of CuNPs@TWM can be attributed to the synergistic effect between the high surface area of TWM and the catalytic activity of CuFe nanoparticles, which provide abundant active sites for adsorption and chemical interactions. The investigation into catalyst dosage revealed that the adsorption efficiency was highly dosage-dependent, with an optimal dosage of 20 mg L^{-1} maximizing contaminant removal. At this dosage, a balance was achieved between the availability of active sites and contaminant concentrations. Complete degradation of the contaminants was achieved within 15 min by the EF process under optimal conditions (20 mA and 20 mg L^{-1}). The applied current facilitated the efficient generation of oxidants, while the catalyst supported their activation. At higher currents (50 mA), secondary reactions like oxygen evolution and hydroxyl radical recombination slightly reduced efficiency, confirming 20 mA as the optimal current.

Overall, the findings underscore the potential of combining biomass-derived supports with metal nanoparticles to create cost-effective, sustainable, and efficient adsorbents for water treatment applications. CuNPs@TWM offers a promising solution for removing CECs from contaminated water, aligning with global efforts to mitigate pharmaceutical pollution in aquatic environments. While the approach is promising at the laboratory scale, scaling up for industrial application may require further optimization and cost analysis. Future studies should explore the long-term stability and reusability of these materials, as well as their performance in complex water matrices containing numerous contaminants at pre-pilot plants. Additionally, the integration of these adsorbents into scalable water treatment systems could pave the way for practical applications in industrial and municipal settings.

Supplementary Materials: The following are available online at https://www.mdpi.com/article/10.3390/app15031418/s1: Figure S1. XPS spectra for samples: (a) TWM, (b) CuFeNPs, and (c) CuFeNPs@TWM; Figure S2. Size distribution of the synthesized CuFeNPs; Figure S3. EDS elemental mapping for TWM; Figure S4. EDS elemental mapping for CuFeNPs; Figure S5. N_2 adsorption and desorption isotherms obtained for (a) TWM, (b) CuFeNPs, and (c) CuFeNPs@TWM; Figure S6. The pH_{pzc} of the CuFeNPs@TWM; Figure S7. Normalized concentration decay for each CEC in the function of different solution initial pH: (a) pH 5, and (b) 7. Experimental conditions: Catalyst, CuFeNPs@TWM; catalyst load, 20 mg L^{-1}; applied current, 20 mA; electrolyte, Na_2SO_4 50 mmol L^{-1}; Figure S8. Concentration of Fe^{2+} and Fe^{3+} in solution after different catalytic cycles. Each cycle considered 30 min of adsorption and 30 min of EF process. Experimental conditions: Catalyst, CuFeNPs@TWM; catalyst load, 20 mg L^{-1}; applied current, 20 mA; electrolyte, Na_2SO_4 50 mmol L^{-1}; pH, 3.

Author Contributions: L.M.-B.: Methodology and investigation; J.C.-P.: formal analysis, writing—review and editing; C.G.-F.: formal analysis, writing—review and editing; N.E.: formal analysis, validation, writing—review and editing; C.O.: conceptualization, methodology, supervision, validation, visualization, investigation, funding acquisition, project administration and writing—original draft; A.T.: conceptualization, methodology, supervision, validation, visualization, funding acquisition, project administration and writing—original draft. All authors have read and agreed to the published version of the manuscript.

Funding: This research was funded by ANID/FONDECYT Chile, by Postdoctoral grant N° 3220510 and Regular grant N° 1210343; FONDEQUIP EQM 160070; and Millennium Science Initiative Program, Nucleus in NanoBioPhysics—NNBP # NCN2021_021.

Institutional Review Board Statement: Not applicable.

Informed Consent Statement: Not applicable.

Data Availability Statement: The original contributions presented in the study are included in the article; further inquiries can be directed to corresponding authors.

Acknowledgments: Authors thank Carla Herrera Cantera, from Instituto Universitario de Investigación y Desarrollo Tecnológico (IDT) from Universidad Tecnológica Metropolitana for her invaluable technical support throughout this work.

Conflicts of Interest: The authors declare no conflicts of interest.

References

1. Wu, X.; Ren, J.; Xu, Q.; Xiao, Y.; Li, X.; Peng, Y. Priority Screening of Contaminant of Emerging Concern (CECs) in Surface Water from Drinking Water Sources in the Lower Reaches of the Yangtze River Based on Exposure-Activity Ratios (EARs). *Sci. Total Environ.* **2023**, *856*, 159016. [CrossRef] [PubMed]
2. Almeida, L.C.; Garcia-Segura, S.; Arias, C.; Bocchi, N.; Brillas, E. Electrochemical Mineralization of the Azo Dye Acid Red 29 (Chromotrope 2R) by Photoelectro-Fenton Process. *Chemosphere* **2012**, *89*, 751–758. [CrossRef] [PubMed]
3. He, K.; Guo, X.; Wang, G.; Xiong, J.; Peng, H.; Lai, F.; He, J. Excellent Treatment Effect of Actual Wastewater from a Pesticide Plant Using Electro-Fenton Process Catalyzed by Natural Pyrite. *J. Environ. Chem. Eng.* **2023**, *11*, 111211. [CrossRef]
4. Nidheesh, P.V.; Trellu, C.; Vargas, H.O.; Mousset, E.; Ganiyu, S.O.; Oturan, M.A. Electro-Fenton Process in Combination with Other Advanced Oxidation Processes: Challenges and Opportunities. *Curr. Opin. Electrochem.* **2023**, *37*, 101171. [CrossRef]
5. Soufi, A.; Hajjaoui, H.; Elmoubarki, R.; Abdennouri, M.; Qourzal, S.; Barka, N. Heterogeneous Fenton-like Degradation of Tartrazine Using $CuFe_2O_4$ Nanoparticles Synthesized by Sol-Gel Combustion. *Appl. Surf. Sci. Adv.* **2022**, *9*, 100251. [CrossRef]
6. Campos, S.; Salazar, R.; Arancibia-Miranda, N.; Rubio, M.A.; Aranda, M.; García, A.; Sepúlveda, P.; Espinoza, L.C. Nafcillin Degradation by Heterogeneous Electro-Fenton Process Using Fe, Cu and Fe/Cu Nanoparticles. *Chemosphere* **2020**, *247*, 125813. [CrossRef]
7. Gamboa-Savoy, F.; Onfray, C.; Hassan, N.; Salazar, C.; Thiam, A. Enhanced Catalytic Reduction of Emerging Contaminant by Using Magnetic $CuFe_2O_4$@MIL-100(Fe) in Fenton-Based Electrochemical Processes. *Chemosphere* **2023**, *337*, 139231. [CrossRef]
8. Wu, J.; Wang, X.; Kang, H.; Zhang, J.; Yang, C. $CuFe_2O_4$ as Heterogeneous Catalyst in Degradation of P-Nitrophenol with Photoelectron-Fenton-like Process. *Int. J. Environ. Stud.* **2014**, *71*, 534–545. [CrossRef]
9. Thomas, N.; Dionysiou, D.D.; Pillai, S.C. Heterogeneous Fenton Catalysts: A Review of Recent Advances. *J. Hazard. Mater.* **2021**, *404*, 124082. [CrossRef]
10. Poza-Nogueiras, V.; Rosales, E.; Pazos, M.; Sanromán, M.Á. Current Advances and Trends in Electro-Fenton Process Using Heterogeneous Catalysts—A Review. *Chemosphere* **2018**, *201*, 399–416. [CrossRef]
11. Muzenda, C.; Nkwachukwu, O.V.; Jayeola, K.D.; Zinyemba, O.; Zhou, M.; Arotiba, O.A. Heterogenous Electro-Fenton Degradation of Sulfamethoxazole on a Polyethylene Glycol-Coated Magnetite Nanoparticles Catalyst. *Chemosphere* **2023**, *339*, 139698. [CrossRef] [PubMed]
12. Huang, B.; Yang, C.; Zeng, H.; Zhou, L. Multivalent Iron-Based Magnetic Porous Biochar from Peach Gum Polysaccharide as a Heterogeneous Fenton Catalyst for Degradation of Dye Pollutants. *Int. J. Biol. Macromol.* **2023**, *253*, 126753. [CrossRef] [PubMed]
13. Khalil, K.M.S.; Elhamdy, W.A.; Said, A.E.A.A. Nanostructured Sustainable Carbon Derived from Biomass as Catalyst Support for Alumina in Catalytic Methanol Conversion to DME as Hydrogen Carrier. *Int. J. Hydrog. Energy* **2024**, *51*, 819–833. [CrossRef]
14. Xuan, S.; Wang, Y.X.J.; Yu, J.C.; Leung, K.C.F. Preparation, Characterization, and Catalytic Activity of Core/Shell Fe_3O_4@polyaniline@Au Nanocomposites. *Langmuir* **2009**, *25*, 11835–11843. [CrossRef] [PubMed]

15. Huo, J.; Tessonnier, J.-P.; Shanks, B.H. Improving Hydrothermal Stability of Supported Metal Catalysts for Biomass Conversions: A Review. *ACS Catal.* **2021**, *11*, 5248–5270. [CrossRef]
16. Chai, Y.; Bai, M.; Chen, A.; Xu, X.; Tong, Z.; Yuan, J.; Peng, L.; Shao, J.; Xiong, J.; Peng, C. Upcycling Contaminated Biomass into Metal-Supported Heterogeneous Catalyst for Electro-Fenton Degradation of Thiamethoxam: Preparation, Mechanisms, and Implications. *Chem. Eng. J.* **2023**, *453*, 139814. [CrossRef]
17. Sherwood, J. The Significance of Biomass in a Circular Economy. *Bioresour. Technol.* **2020**, *300*, 122755. [CrossRef]
18. Hsiao, C.-J.; Hu, J.-L. Biomass and Circular Economy: Now and the Future. *Biomass* **2024**, *4*, 720–739. [CrossRef]
19. Chew, K.W.; Chia, S.R.; Chia, W.Y.; Cheah, W.Y.; Munawaroh, H.S.H.; Ong, W.J. Abatement of Hazardous Materials and Biomass Waste via Pyrolysis and Co-Pyrolysis for Environmental Sustainability and Circular Economy. *Environ. Pollut.* **2021**, *278*, 116836. [CrossRef]
20. Guo, S.; Kumar Awasthi, M.; Wang, Y.; Xu, P. Current Understanding in Conversion and Application of Tea Waste Biomass: A Review. *Bioresour. Technol.* **2021**, *338*, 125530. [CrossRef]
21. Sui, W.; Xiao, Y.; Liu, R.; Wu, T.; Zhang, M. Steam Explosion Modification on Tea Waste to Enhance Bioactive Compounds' Extractability and Antioxidant Capacity of Extracts. *J. Food Eng.* **2019**, *261*, 51–59. [CrossRef]
22. Abdeltaif, S.A.; Sirelkhatim, K.A.; Hassan, A.B. Estimation of Phenolic and Flavonoid Compounds and Antioxidant Activity of Spent Coffee and Black Tea (Processing) Waste for Potential Recovery and Reuse in Sudan. *Recycling* **2018**, *3*, 27. [CrossRef]
23. Abraham, A.M.; Alnemari, R.M.; Brüßler, J.; Keck, C.M. Improved Antioxidant Capacity of Black Tea Waste Utilizing Plantcrystals. *Molecules* **2021**, *26*, 592. [CrossRef] [PubMed]
24. Xue, Z.; Wang, J.; Chen, Z.; Ma, Q.; Guo, Q.; Gao, X.; Chen, H. Antioxidant, Antihypertensive, and Anticancer Activities of the Flavonoid Fractions from Green, Oolong, and Black Tea Infusion Waste. *J. Food Biochem.* **2018**, *42*, e12690. [CrossRef]
25. Sökmen, M.; Demir, E.; Alomar, S.Y. Optimization of Sequential Supercritical Fluid Extraction (SFE) of Caffeine and Catechins from Green Tea. *J. Supercrit. Fluids* **2018**, *133*, 171–176. [CrossRef]
26. Muniandy, Y.; Mohamad, S.; Raoov, M. Green and Efficient Magnetic Micro-Solid Phase Extraction Utilizing Tea Waste Impregnated with Magnetic Nanoparticles for the Analysis of Ibuprofen in Water Samples by Using UV-Vis Spectrophotometry. *RSC Adv.* **2024**, *14*, 11977–11985. [CrossRef]
27. Akbayrak, S.; Özçifçi, Z.; Tabak, A. Activated Carbon Derived from Tea Waste: A Promising Supporting Material for Metal Nanoparticles Used as Catalysts in Hydrolysis of Ammonia Borane. *Biomass Bioenergy* **2020**, *138*, 105589. [CrossRef]
28. Malhotra, M.; Suresh, S.; Garg, A. Tea Waste Derived Activated Carbon for the Adsorption of Sodium Diclofenac from Wastewater: Adsorbent Characteristics, Adsorption Isotherms, Kinetics, and Thermodynamics. *Environ. Sci. Pollut. Res.* **2018**, *25*, 32210–32220. [CrossRef]
29. Peng, C.; Yan, X.B.; Wang, R.T.; Lang, J.W.; Ou, Y.J.; Xue, Q.J. Promising Activated Carbons Derived from Waste Tea-Leaves and Their Application in High Performance Supercapacitors Electrodes. *Electrochim. Acta* **2013**, *87*, 401–408. [CrossRef]
30. Wang, Z.; Ahmad, W.; Zhu, A.; Zhao, S.; Ouyang, Q.; Chen, Q. Recent Advances Review in Tea Waste: High-Value Applications, Processing Technology, and Value-Added Products. *Sci. Total Environ.* **2024**, *946*, 174225. [CrossRef]
31. Mandal, S.; Yadav, A.; Kumar, S.; Murasingh, S. An Extensive Review Study on Bioresources Recovery from Tea Waste and Its Emerging Applications. *Waste Manag. Bull.* **2024**, *2*, 155–166. [CrossRef]
32. Yang, X.; Cui, X. Adsorption Characteristics of Pb (II) on Alkali Treated Tea Residue. *Water Resour. Ind.* **2013**, *3*, 1–10. [CrossRef]
33. Wu, D.; Shi, Y.; Jing, H.; Wang, X.; Song, X.; Si, D.; Liang, S.; Hao, C. Tea-Leaf-Residual Derived Electrocatalyst: Hierarchical Pore Structure and Self Nitrogen and Fluorine Co-Doping for Efficient Oxygen Reduction Reaction. *Int. J. Hydrog. Energy* **2018**, *43*, 19492–19499. [CrossRef]
34. Gundogdu, A.; Duran, C.; Senturk, H.B.; Soylak, M.; Imamoglu, M.; Onal, Y. Physicochemical Characteristics of a Novel Activated Carbon Produced from Tea Industry Waste. *J. Anal. Appl. Pyrolysis* **2013**, *104*, 249–259. [CrossRef]
35. Nordin, A.H.; Ngadi, N.; Nordin, M.L.; Noralidin, N.A.; Nabgan, W.; Osman, A.Y.; Shaari, R. Spent Tea Waste Extract as a Green Modifying Agent of Chitosan for Aspirin Adsorption: Fixed-Bed Column, Modeling and Toxicity Studies. *Int. J. Biol. Macromol.* **2023**, *253*, 126501. [CrossRef]
36. Tuli, F.J.; Hossain, A.; Kibria, A.K.M.F.; Tareq, A.R.M.; Mamun, S.M.M.A.; Ullah, A.K.M.A. Removal of Methylene Blue from Water by Low-Cost Activated Carbon Prepared from Tea Waste: A Study of Adsorption Isotherm and Kinetics. *Environ. Nanotechnol. Monit. Manag.* **2020**, *14*, 100354. [CrossRef]
37. Çelebi, H.; Gök, G.; Gök, O. Adsorption Capability of Brewed Tea Waste in Waters Containing Toxic Lead(II), Cadmium (II), Nickel (II), and Zinc(II) Heavy Metal Ions. *Sci. Rep.* **2020**, *10*, 17570. [CrossRef]
38. Al-Qahtani, S.D.; Alhasani, M.; Alkhathami, N.; Abu Al-Ola, K.A.; Alkhamis, K.; El-Desouky, M.G.; El-Bindary, A.A. Effective Levofloxacin Adsorption and Removal from Aqueous Solution onto Tea Waste Biochar; Synthesis, Characterization, Adsorption Studies, and Optimization by Box–Behnken Design and Its Antibacterial Activity. *Environ. Technol.* **2024**, *45*, 4928–4950. [CrossRef]
39. Khalil, U.; Bilal Shakoor, M.; Ali, S.; Rizwan, M.; Nasser Alyemeni, M.; Wijaya, L. Adsorption-Reduction Performance of Tea Waste and Rice Husk Biochars for Cr(VI) Elimination from Wastewater. *J. Saudi Chem. Soc.* **2020**, *24*, 799–810. [CrossRef]

40. Kabir, M.M.; Mouna, S.S.P.; Akter, S.; Khandaker, S.; Didar-ul-Alam, M.; Bahadur, N.M.; Mohinuzzaman, M.; Islam, M.A. Shenashen, M.A. Tea Waste Based Natural Adsorbent for Toxic Pollutant Removal from Waste Samples. *J. Mol. Liq.* **2021** *322*, 115012. [CrossRef]
41. Kumbhar, P.; Narale, D.; Bhosale, R.; Jambhale, C.; Kim, J.H.; Kolekar, S. Synthesis of Tea Waste/Fe_3O_4 magnetic Composite (TWMC) for Efficient Adsorption of Crystal Violet Dye: Isotherm, Kinetic and Thermodynamic Studies. *J. Environ. Chem. Eng* **2022**, *10*, 107893. [CrossRef]
42. Raj, R.; Tripathi, A.; Das, S.; Ghangrekar, M.M. Is Waste-Derived Catalyst Mediated Electro-Fenton a Sustainable Option for Mitigating Emerging Contaminants from Wastewater? *Curr. Opin. Environ. Sci. Health* **2024**, *37*, 100523. [CrossRef]
43. Iranparast Poshtiri, N.; Dadvand Koohi, A.; Ebrahimian Pirbazari, A.; Esmaeili Khalil Saraei, F. Catalytic Activity of Magnetite Fish-Scales and Tea-Waste as Heterogeneous Photo-Fenton Catalysts for Removal of Methyl Orange through Cu(II) Adsorption *Mater. Chem. Phys.* **2022**, *290*, 126525. [CrossRef]
44. Uvarov, V.; Popov, I. Metrological Characterization of X-Ray Diffraction Methods for Determination of Crystallite Size in Nano-Scale Materials. *Mater. Charact.* **2007**, *58*, 883–891. [CrossRef]
45. Velinov, N.; Najdanović, S.; Radović Vučić, M.; Mitrović, J.; Kostić, M.; Bojić, D.; Bojić, A. Biosorption of Loperamide by Lignocellulosic-Al_2O_3 Hybrid: Optimization, Kinetics, Isothermal and Thermodynamic Studies. *Cellul. Chem. Technol.* **2019** *53*, 175–189. [CrossRef]
46. Iglesias, O.; Fernández de Dios, M.A.; Pazos, M.; Sanromán, M.A. Using Iron-Loaded Sepiolite Obtained by Adsorption as a Catalyst in the Electro-Fenton Oxidation of Reactive Black 5. *Environ. Sci. Pollut. Res.* **2013**, *20*, 5983–5993. [CrossRef]
47. Najdanović, S.M.; Petrović, M.M.; Slipper, I.J.; Kostić, M.M.; Prekajski, M.D.; Mitrović, J.Z.; Bojić, A.L. A New Photocatalyst Bismuth Oxo Citrate: Synthesis, Characterization, and Photocatalytic Performance. *Water Environ. Res.* **2018**, *90*, 719–728 [CrossRef]
48. Kusvuran, E.; Gulnaz, O.; Irmak, S.; Atanur, O.M.; Yavuz, H.I.; Erbatur, O. Comparison of Several Advanced Oxidation Processes for the Decolorization of Reactive Red 120 Azo Dye in Aqueous Solution. *J. Hazard. Mater.* **2004**, *109*, 85–93. [CrossRef]
49. Droguett, C.; Salazar, R.; Brillas, E.; Sirés, I.; Carlesi, C.; Marco, J.F.; Thiam, A. Treatment of Antibiotic Cephalexin by Heterogeneous Electrochemical Fenton-Based Processes Using Chalcopyrite as Sustainable Catalyst. *Sci. Total Environ.* **2020**, *740*, 140154 [CrossRef]
50. Mexmain, J. Contribution a l'etude Du Ferrite Cuivreux et de Ses Solutions Solides Avec Le Ferrite Cuivrique. *Annu. Chim.* **1971**, *1971*, 297–308.
51. Laokul, P.; Amornkitbamrung, V.; Seraphin, S.; Maensiri, S. Characterization and Magnetic Properties of Nanocrystalline $CuFe_2O_4$, $NiFe_2O_4$, $ZnFe_2O_4$ Powders Prepared by the Aloe Vera Extract Solution. *Curr. Appl. Phys.* **2011**, *11*, 101–108. [CrossRef]
52. Asbrink, S.; Waskowska, A. CuO: X-Ray Single-Crystal Structure Determination at 196 K and Room Temperature. *J. Phys. Condens. Matter* **1991**, *3*, 8173–8180. [CrossRef]
53. Wilburn, D.R.; Bassett, W.A. Hydrostatic Compression of Iron and Related Compounds; an Overview. *Am. Mineral.* **1978**, *63*, 591–596.
54. Swanson, H.E.; Tatge, E. Standard X-Ray Diffraction Powder Patterns. National Bureau of Standards (US). *Circular* **1953**, *359*, 1.
55. Sadezky, A.; Muckenhuber, H.; Grothe, H.; Niessner, R.; Pöschl, U. Raman Microspectroscopy of Soot and Related Carbonaceous Materials: Spectral Analysis and Structural Information. *Carbon N. Y* **2005**, *43*, 1731–1742. [CrossRef]
56. Reich, S.; Thomsen, C. Raman Spectroscopy of Graphite. *Philos. Trans. R. Soc. London. Ser. A Math. Phys. Eng. Sci.* **2004**, *362*, 2271–2288. [CrossRef]
57. Li, T.; Wu, Q.; Wang, W.; Xiao, Y.P.; Liu, C.; Yang, F. Solid-Solid Reaction of $CuFe_2O_4$ with C in Chemical Looping System: A Comprehensive Study. *Fuel* **2020**, *267*, 117163. [CrossRef]
58. Verma, K.; Kumar, A.; Varshney, D. Effect of Zn and Mg Doping on Structural, Dielectric and Magnetic Properties of Tetragonal $CuFe_2O_4$. *Curr. Appl. Phys.* **2013**, *13*, 467–473. [CrossRef]
59. Wu, L.K.; Wu, H.; Zhang, H.B.; Cao, H.Z.; Hou, G.Y.; Tang, Y.P.; Zheng, G.Q. Graphene Oxide/$CuFe_2O_4$ Foam as an Efficient Absorbent for Arsenic Removal from Water. *Chem. Eng. J.* **2018**, *334*, 1808–1819. [CrossRef]
60. Costa, A.F.; Pimentel, P.M.; Aquino, F.M.; Melo, D.M.A.; Melo, M.A.F.; Santos, I.M.G. Gelatin Synthesis of $CuFe_2O_4$ and $CuFeCrO_4$ Ceramic Pigments. *Mater. Lett.* **2013**, *112*, 58–61. [CrossRef]
61. Genchev, G.; Erbe, A. Raman Spectroscopy of Mackinawite FeS in Anodic Iron Sulfide Corrosion Products. *J. Electrochem. Soc.* **2016**, *163*, C333–C338. [CrossRef]
62. Li, O.L.; Qin, L.; Takeuchi, N.; Kim, K.; Ishizaki, T. Effect of Hydrophilic/Hydrophobic Properties of Carbon Materials on Plasma-Sulfonation Process and Their Catalytic Activities in Cellulose Conversion. *Catal. Today* **2019**, *337*, 155–161. [CrossRef]
63. Wang, X.; Zuo, L.; Wang, Y.; Zhen, M.; Xu, L.; Kong, W.; Shen, B. Electrochemical Performance of Nitrogen Self-Doping Carbon Materials Prepared by Pyrolysis and Activation of Defatted Microalgae. *Molecules* **2023**, *28*, 7280. [CrossRef]
64. Nidheesh, P.V.; Gopinath, A.; Ranjith, N.; Praveen Akre, A.; Sreedharan, V.; Suresh Kumar, M. Potential Role of Biochar in Advanced Oxidation Processes: A Sustainable Approach. *Chem. Eng. J.* **2021**, *405*, 126582. [CrossRef]

65. Zhang, T.; Sun, L.; Sun, X.; Dong, H.; Yu, H.; Yu, H. Radical and Non-Radical Cooperative Degradation in Metal-Free Electro-Fenton Based on Nitrogen Self-Doped Biochar. *J. Hazard. Mater.* **2022**, *435*, 129063. [CrossRef]
66. Wang, X.; Zhang, C.; Song, G.; Jing, J.; Li, S.; Zhou, M.; Dewil, R. Highly Efficient FeS/Fe_3O_4 @ Biomass Carbon Bifunctional Catalyst with Enriched Oxygen Vacancies for Heterogeneous Electro-Fenton Catalysis. *J. Hazard. Mater.* **2024**, *479*, 135692. [CrossRef]
67. Sun, K.; Sun, C.; Wang, X.; Hou, J.; Shi, W.; Li, C.; Guo, F. Construction of Cost-Effective Biomass Iron-Based Photocatalysis-Self-Fenton Degradation System for Antibiotic Removal. *J. Environ. Chem. Eng.* **2024**, *12*, 113568. [CrossRef]
68. Xue, Q.; Lin, H.; Feng, Q.; Yang, Y.; Dong, M.; Hu, K.; Song, B.; Sean Goh, P.; Shen, X. Synergistic Photocatalysis and Fenton-like Process Driven by a Biochar-Supported Biochar/Iron Hydroxide Oxide/Bismuth Molybdate S-Type Heterojunction for Tetracycline Degradation: Mechanistic Insights and Degradation Pathways. *Appl. Surf. Sci.* **2025**, *679*, 161277. [CrossRef]
69. Xu, A.; Chen, R.; Khan, A.; Zhao, W.; Lin, J.; Hu, R.; Li, W.; Liu, X.; Li, X.; Zhao, S. Modulating Electronic Structure of Fe Site by Adjacent Cu Site through Orbital Coupling in Cu-Fe_2O_3 Catalyst Derived from Fenton Sludge Quenching. *Chem. Eng. J.* **2024**, *493*, 152489. [CrossRef]
70. Chen, T.; Zhu, Z.; Wang, Y.; Zhang, H.; Qiu, Y.; Yin, D. Efficient Organics Heterogeneous Degradation by Spinel $CuFe_2O_4$ Supported Porous Carbon Nitride Catalyst: Multiple Electron Transfer Pathways for Reactive Oxygen Species Generation. *Chemosphere* **2022**, *300*, 134511. [CrossRef]
71. Liu, Z.; Gao, B.; Zhao, P.; Fu, H.; Kamali, A.R. Magnetic Luffa/Graphene/$CuFe_2O_4$ Sponge for Efficient Oil/Water Separation. *Sep. Purif. Technol.* **2024**, *337*, 126347. [CrossRef]
72. Priyadharsini, R.; Das, S.; Venkateshwarlu, M.; Deenadayalan, K.; Manoharan, C. The Influence of Reaction and Annealing Temperature on Physical and Magnetic Properties of $CuFe_2O_4$ Nanoparticles: Hydrothermal Method. *Inorg. Chem. Commun.* **2022**, *140*, 109406. [CrossRef]
73. Deivayanai, V.C.; Karishma, S.; Thamarai, P.; Saravanan, A.; Yaashikaa, P.R. Efficient Red Azo Dye Removal from Wastewater Using Magnetic Nanoparticle Impregnated Prosopis Juliflora Biomass: ANN Modeling Approach. *Desalination Water Treat.* **2024**, *320*, 100746. [CrossRef]
74. Shi, J.; Ye, R.; Yue, Y.; Liu, J. Insights into Interfacial Catalysis for Aqueous Reduction of 4-Nitrophenol Catalyzed by Ni Metal Nanoparticles Encapsulated within Biomass-Derived N-Doped Carbon Matrix. *Chem. Eng. Sci.* **2024**, *294*, 120122. [CrossRef]
75. Thommes, M.; Kaneko, K.; Neimark, A.V.; Olivier, J.P.; Rodriguez-Reinoso, F.; Rouquerol, J.; Sing, K.S.W. Physisorption of Gases, with Special Reference to the Evaluation of Surface Area and Pore Size Distribution (IUPAC Technical Report). *Pure Appl. Chem.* **2015**, *87*, 1051–1069. [CrossRef]
76. Daniel, G.; Zhang, Y.; Lanzalaco, S.; Brombin, F.; Kosmala, T.; Granozzi, G.; Wang, A.; Brillas, E.; Sirés, I.; Durante, C. Chitosan-Derived Nitrogen-Doped Carbon Electrocatalyst for a Sustainable Upgrade of Oxygen Reduction to Hydrogen Peroxide in UV-Assisted Electro-Fenton Water Treatment. *ACS Sustain. Chem. Eng.* **2020**, *8*, 14425–14440. [CrossRef]
77. Chen, C.; Gao, W.; Fan, H.; Huang, X.; Tong, Z. Facile Single-Step Synthesis of $ZnFe_2O_4$@biochar for Synergistic Adsorption and Photo-Fenton Degradation of RB and RBR Binary Dyes. *J. Taiwan Inst. Chem. Eng.* **2024**, *163*, 105641. [CrossRef]
78. Deng, F.; Olvera-Vargas, H.; Zhou, M.; Qiu, S.; Sirés, I.; Brillas, E. Critical Review on the Mechanisms of Fe^{2+} Regeneration in the Electro-Fenton Process: Fundamentals and Boosting Strategies. *Chem. Rev.* **2023**, *123*, 4635–4662. [CrossRef]
79. Garcia-Segura, S.; Brillas, E.; Cornejo-Ponce, L.; Salazar, R. Effect of the Fe3+/Cu2+ Ratio on the Removal of the Recalcitrant Oxalic and Oxamic Acids by Electro-Fenton and Solar Photoelectro-Fenton. *Sol. Energy* **2016**, *124*, 242–253. [CrossRef]
80. Sirés, I.; Brillas, E.; Oturan, M.A.; Rodrigo, M.A.; Panizza, M. Electrochemical Advanced Oxidation Processes: Today and Tomorrow. A Review. *Environ. Sci. Pollut. Res.* **2014**, *21*, 8336–8367. [CrossRef]
81. Brillas, E.; Sirés, I.; Oturan, M.A. Electro-Fenton Process and Related Electrochemical Technologies Based on Fenton's Reaction Chemistry. *Chem. Rev.* **2009**, *109*, 6570–6631. [CrossRef]
82. Oturan, N.; Bo, J.; Trellu, C.; Oturan, M.A. Comparative Performance of Ten Electrodes in Electro-Fenton Process for Removal of Organic Pollutants from Water. *ChemElectroChem* **2021**, *8*, 3294–3303. [CrossRef]
83. Liu, T.; Gao, C.; Luan, P.; Xu, D.; Wang, J.; Cheng, Z.; Yan, B.; Chen, G. Performance Evaluation of Different Nitrogenous Biomass Pyrolysis Chars: Physicochemical Properties and Ibuprofen Adsorption. *Biomass Bioenergy* **2024**, *189*, 107346. [CrossRef]
84. Al-Yousef, H.A.; Alotaibi, B.M.; Aouaini, F.; Sellaoui, L.; Bonilla-Petriciolet, A. Adsorption of Ibuprofen on Cocoa Shell Biomass-Based Adsorbents: Interpretation of the Adsorption Equilibrium via Statistical Physics Theory. *J. Mol. Liq.* **2021**, *331*, 115697. [CrossRef]
85. Tang, J.; Wang, Y.; Ma, J.; Chen, Y.; Chen, M. High-Density Co-Nx Sites Anchored on Biomass-Based Carbon Nanotubes for Efficient Removal of Plant Viruses in Water. *J. Environ. Chem. Eng.* **2024**, *12*, 113022. [CrossRef]
86. Temur Ergan, B.; Aydin, E.S.; Gengec, E. Improving Electro-Fenton Degradation Performance Using Waste Biomass-Derived-Modified Biochar Electrodes: A Real Environment Textile Water Treatment. *J. Environ. Chem. Eng.* **2023**, *11*, 111439. [CrossRef]

87. Chen, C.; Yan, Z.; Ma, Z.; Ma, D.; Xing, S.; Li, W.; Yang, J.; Han, Q. Magnetic Fe_3O_4 Nanoparticles Supported on Carbonized Corncob as Heterogeneous Fenton Catalyst for Efficient Degradation of Methyl Orange. *Chin. J. Chem. Eng.* **2024**, *77*, 144–155. [CrossRef]
88. Chatton, J.Y.; Besseghir, K.; Roch-Ramel, F. Salicylic Acid Permeability Properties of the Rabbit Cortical Collecting Duct. *Am. J Physiol. -Ren. Physiol.* **1990**, *259*, F613–F618. [CrossRef]
89. Li, X.; Cooper, M.A. Measurement of Drug Lipophilicity and PK a Using Acoustics. *Anal. Chem.* **2012**, *84*, 2609–2613. [CrossRef]
90. Kostić, M.; Najdanović, S.; Radović Vučić, M.; Velinov, N.; Bojić, D.; Nikolić, G.; Bojić, A. A New Catalyst with the Superior Performance for Treatment of Water Polluted by Anthraquinone Compounds. *Bull. Mater. Sci.* **2021**, *44*, 219. [CrossRef]
91. Ali, A.; Siddique, M.; Chen, W.; Han, Z.; Khan, R.; Bilal, M.; Waheed, U.; Shahzadi, I. Promising Low-Cost Adsorbent from Waste Green Tea Leaves for Phenol Removal in Aqueous Solution. *Int. J. Environ. Res. Public Health* **2022**, *19*, 6396. [CrossRef] [PubMed]
92. Li, Z.; Liu, R.; Gao, W.; Zhang, W.; Li, C.; Liu, X.; Wang, N. $CuFe_2O_4$/Penicillin Residue Biochar Heterogeneous Fenton-like Catalyst Used in the Treatment of Antibiotic Wastewater: Synthesis, Performance and Working Mechanism. *J. Water Process Eng.* **2023**, *55*, 104124. [CrossRef]
93. Qi, H.; Pan, G.; Shi, X.; Sun, Z. Cu–Fe–FeC3@nitrogen-Doped Biochar Microsphere Catalyst Derived from $CuFe_2O_4$@chitosan for the Efficient Removal of Amoxicillin through the Heterogeneous Electro-Fenton Process. *Chem. Eng. J.* **2022**, *434*, 134675. [CrossRef]
94. Kalantary, R.R.; Farzadkia, M.; Kermani, M.; Rahmatinia, M. Heterogeneous Electro-Fenton Process by Nano-Fe_3O_4 for Catalytic Degradation of Amoxicillin: Process Optimization Using Response Surface Methodology. *J. Environ. Chem. Eng.* **2018**, *6*, 4644–4652. [CrossRef]

applied sciences

Article

Design of a Waste Classification System Using a Low Experimental Cost Capacitive Sensor and Machine Learning Algorithms

Juan Carlos Vesga Ferreira [1,*], Harold Esneider Perez Waltero [1] and Jose Antonio Vesga Barrera [2]

1 Telecommunications Engineering Program, School of Basic Sciences, Technology and Engineering (ECBTI), Universidad Nacional Abierta y a Distancia, Bogotá 110231, Colombia; harold.perez@unad.edu.co
2 Faculty of Engineering, Corporación Universitaria de Ciencia y Desarrollo, Bogotá 110111, Colombia; jose.vesga@unicienciabga.edu.co
* Correspondence: juan.vesga@unad.edu.co; Tel.: +57-3212759948

Abstract: The management and classification of solid waste is one of the most important challenges worldwide. The objective is to design a basic waste classification system at the source using a low-cost experimental capacitive sensor and machine learning algorithms. For this, two types of sensor models were established (Traditional Model (MT) and Non-Traditional Model (MNT)), which were built with recyclable material and tested with different types of materials, in order to evaluate their behavior and sensitivity level. The results obtained demonstrated that the two sensors responded with acceptable sensitivity levels for each of the materials used as a test; however, the MNT was the one that generated the values with the greatest variability, an aspect that is deemed highly significant, because, thanks to this type of response to various types of materials, it facilitates the classification processes through the use of machine learning algorithms. Finally, the two prototypes of sensors manufactured can be considered of significant relevance for the development of more complex solutions, related to the classification and possible characterization of materials, when compared to the capacitive sensors found on the market, which only then allow us to identify if there is a presence or not of some object through adjustment by potentiometer, generating as a result a digital output. This aspect largely limits the use of commercial capacitive sensors to applications exclusively related to presence or level detection.

Keywords: sensor; capacitive; solid waste; machine learning; artificial intelligence

1. Introduction

Solid waste management stands among the most pressing global challenges today. The generation of waste has steadily increased, driven by factors such as rapid population growth, accelerated industrialization, and economic development [1]. This trend has given rise to a series of environmental and social problems, particularly regarding public health and the conservation of the natural environment [2]. Proper waste management is essential to mitigate these negative effects, requiring the implementation of efficient processes that encompass waste collection at the source, transportation, treatment, segregation, and final disposal [3]. Developing countries, in particular, face the challenge of optimizing these processes, whose effectiveness largely depends on the knowledge and resources available to the institutions responsible for their implementation [4].

Within the framework of environmental policies, solid waste management holds a fundamental position. Its proper implementation aims not only to reduce environmental

 https://doi.org/10.3390/app15031565

impact but also to improve living conditions, prevent public health issues, and promote sustainability [5]. Solid waste classification has gained prominence as a key measure to achieve sustainable economic development goals, reduce poverty, and preserve ecosystems. The adoption of efficient solid waste classification systems is crucial to creating jobs, stimulating the recycling sector, and fostering social well-being [6]. Therefore, efforts must focus on improving waste separation practices at the source, benefiting both the environment and the communities that rely on these economic activities.

Given this scenario, there is a need to develop innovative technologies that optimize waste classification at its origin, enabling a more effective resolution of this issue. With the advancement of emerging technologies such as the Internet of Things (IoT) and Artificial Intelligence, governments worldwide are working toward the creation of "smart cities," where efficient waste management plays a fundamental role [7]. These technologies have great potential to automate and optimize waste classification processes. However, most solutions developed for waste classification focus on image recognition using techniques such as convolutional neural networks, optical methods, and spectrophotometry, among others. Below are some of the works related to this field:

Machine learning-based systems for specific materials:

- In [8], researchers designed recycling machines equipped with ML to classify glass and plastic bottles. Although effective, these solutions are limited to transparent materials due to the optical properties required for their detection.

Hybrid solutions with spectrophotometry and ML:

- In [9], a study integrated spectrophotometry with ML, achieving high accuracy in waste classification. However, the high cost of implementation makes it difficult to adopt in resource-limited environments.

Classification using convolutional neural networks (CNN):

- In [10], researchers proposed YOLO-Green, a model optimized for real-time detection that classifies waste from images. Although efficient, its performance decreases considerably when processing contaminated waste, a common problem in recycling centers. In [11], researchers developed a system to classify recyclable materials using CNN, achieving an accuracy of 94.5%. However, its implementation depends on expensive machine vision equipment and controlled conditions, which limits its applicability in contaminated environments.

Deep Learning Techniques:

- In [12], researchers propose ConvoWaste, which uses deep neural networks to classify waste into categories such as plastics and metals, achieving an accuracy of 98%.
- In [13], the research proposes an intelligent model to categorize waste using deep learning techniques. AlexNet, DenseNet121, and SqueezeNet have been implemented to perform the classification tasks. The obtained results showed great success in the classification process. DenseNet121 achieved the best performance with a value of 0.9415 in terms of accuracy. However, its reliance on specific databases restricts its generalization to other contexts.

While these methods offer high levels of effectiveness, they often require expensive equipment and are, in many cases, sensitive to the presence of impurities. This makes them unsuitable for environments with high levels of contamination, such as waste collection points, significantly increasing implementation costs and limiting their large-scale adoption.

In this context, the present article proposes an innovative solution through the development of a low-cost capacitive sensor with the capability and sensitivity to facilitate the classification of various types of solid waste at the source, including glass, plastic, metal, and organic waste. This solution utilizes machine learning algorithms, offering an efficient and sustainable alternative to this pressing issue.

2. Materials and Methods

2.1. Solid Waste Classification

The classification or segregation of solid waste is considered by numerous studies as one of the fundamental pillars of the Solid Waste Management System [14]. These wastes are composed of materials, many of which are recyclable, such as paper, cardboard, glass, plastic, rubber, and ferrous and non-ferrous metals, which can be reused in various industrial processes. On the other hand, organic waste has the potential to be transformed for the production of biogas and biofertilizers. However, due to inadequate waste management and incorrect collection practices, common in many cities, serious social, health, and environmental problems have been generated, in addition to economic, resource, and energy losses [15]. As a consequence of this poor management, recycling rates are extremely low and the amount of waste deposited in landfills exceeds their capacity, which negatively impacts society and the natural environment.

When talking about the correct way to separate waste, the method best known and implemented by institutions and at the domestic level is the manual classification method, which consists of depositing waste in different containers. Taking into account their material, this method is supported by pedagogy programs that instruct people on the subject [16]. However, the technology of garbage sorting and environmental monitoring has developed rapidly in recent years, where the combination of the Internet of Things, artificial intelligence and environmental technology has become a trend for solving environmental and social problems. [17].

2.2. Sensors

A sensor is a device that is used to measure or detect a change in the environment or in a physical variable under study, which will deliver an output value under a specific format [18]. An analog sensor produces a voltage or continuous output signal, which is generally proportional to the quantity being measured. Digital sensors, on the other hand, produce discrete digital output signals or voltages that are a digital representation of the quantity being measured. Digital sensors produce a binary output signal in the form of a logic "1" or a logic "0", ("ON" or "OFF") or can be represented as a "byte" (8 bits).

2.3. Capacitive Sensor

Capacitive proximity sensors are devices that, without the need for contact, allow the detection of the presence or absence of practically any object regardless of the material (glass, paper, metal, plastic, biological, etc.). They use the electrical property called "capacitance", where by means of a change in the electric field produced by the object under observation ("dielectric"), it generates a change in the value of the resulting capacitance, thereby facilitating the detection of an object and the possibility of identifying the type of material that composes it [8]. Figure 1 shows the appearance of the capacitive sensor.

When an object is located within the sensitive zone, an alteration of the electric field is produced, varying the resulting capacitance and therefore the frequency of the oscillator. There are two categories of targets that capacitive sensors can detect: the first is conductive and the second is non-conductive. Conductive targets include metal, water, blood, acids, bases, and salt water, among other similar materials. These targets have a higher capacitance and the dielectric strength of the targets is irrelevant [19]. When it comes to metallic elements, the electric field decreases since a phenomenon equivalent to series capacitances occurs.

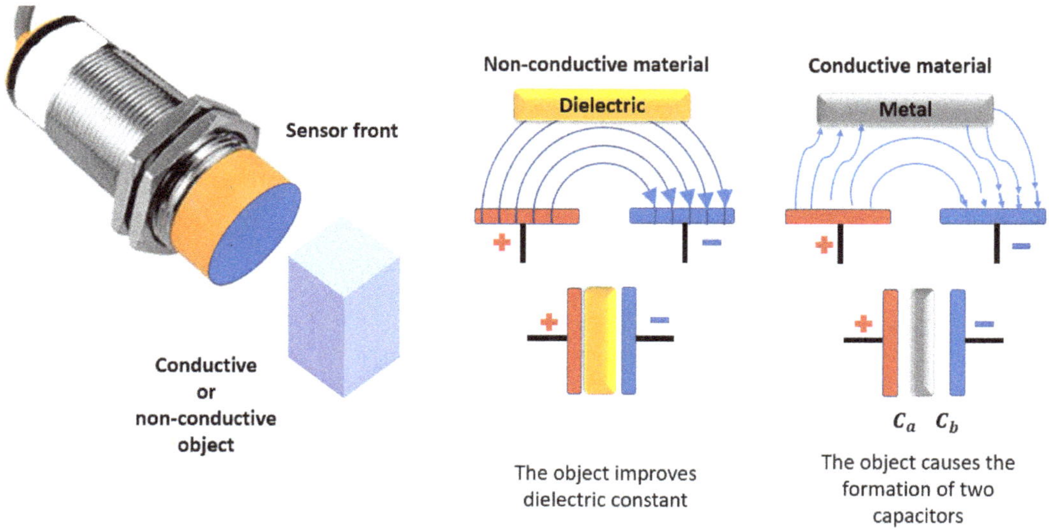

Figure 1. Appearance of the capacitive sensor.

On the other hand, when dealing with a non-conductive target, it acts as an insulator for the electrode of the sensors. A target dielectric constant, also sometimes called a dielectric constant, is the measure of insulation properties used to determine the sensing distance reduction factor. For non-metallic objects, there is an increase in the electric field due to the increase in the dielectric constant. Capacitive sensors are normally used for the detection of non-metallic materials such as glass, oil, water, cardboard, ceramics, plastic, wood, and paper, among others [20].

Although there are several very high-precision methods for measuring parameters related to magnetic permeability, electrical permittivity, and conductivity of a material, among which microwave measurement methods stand out, these are usually very expensive to implement in materials and equipment, which is why they are not considered suitable for the solution required for the problems associated with the identification of materials for the classification of waste at the source. Given this situation, it is considered feasible to carry out the classification process alternately, using a low-cost sensor, articulated with classification techniques-based machine learning algorithms.

2.4. Design of the Proposed Capacitive Sensors

For the fabrication of the two proposed sensors, recycled aluminum sheets from beer cans were utilized. This material was selected due to its accessibility, reusability, and alignment with the overarching objectives of the project. It is important to note that beer or soft drink cans can be made of either steel or aluminum, depending on the application and the production region, where the availability of these materials may vary. However, in most cases, cans are made of aluminum, as in this study, due to its effectiveness in preserving beverages by protecting them from air, light, and potential contamination risks, including bacteria, viruses, hazardous chemicals, and physical agents.

Moreover, aluminum is an economical and sustainable material, thanks to its high recyclability [21]. Cans are typically coated with a thin layer of paint and plastic. The plastic serves to separate the beverage from the aluminum, preventing chemical reactions between the liquid and the metal or contamination from direct contact. The most common plastics used for this type of coating include polyethylene (PE), polypropylene (PP), polystyrene (PS), polyethylene terephthalate (PET), and polyvinyl chloride (PVC). As an insulating material, plastic possesses a relative electrical permittivity, which directly influences the capacitance value of each sensor. Table 1 provides the relative electrical permittivity values for the aforementioned types of plastic [22].

Table 1. Electrical permittivity values for some types of plastic.

Plastic Type	Electrical Permittivity (ε_r)
Polyethylene	2.3
Polystyrene	2.6
Polypropylene	2.2 to 2.6
PVC	2.9
Polyethylene Terephthalate PET	2.8

During the construction process of the MT and MNT sensors, a plastic layer (polypropylene) was used as protection against the environment in which they will be used, establishing a value of $\varepsilon_r = 2.6$. In addition, this layer allowed maintaining a constant distance between the plates, avoiding direct contact between them.

2.4.1. Sensor Design MT

This sensor is made up of two concentric plates, where one of the plates is circular with radius a and the other plate has the shape of a circular crown with inner and outer radius b and c, respectively. Figure 2 shows the proposed diagram for the MT sensor, accompanied by the distribution of electrical charges on each of the plates and electric field lines, in order to establish the conditions and key elements for estimating the capacitance value.

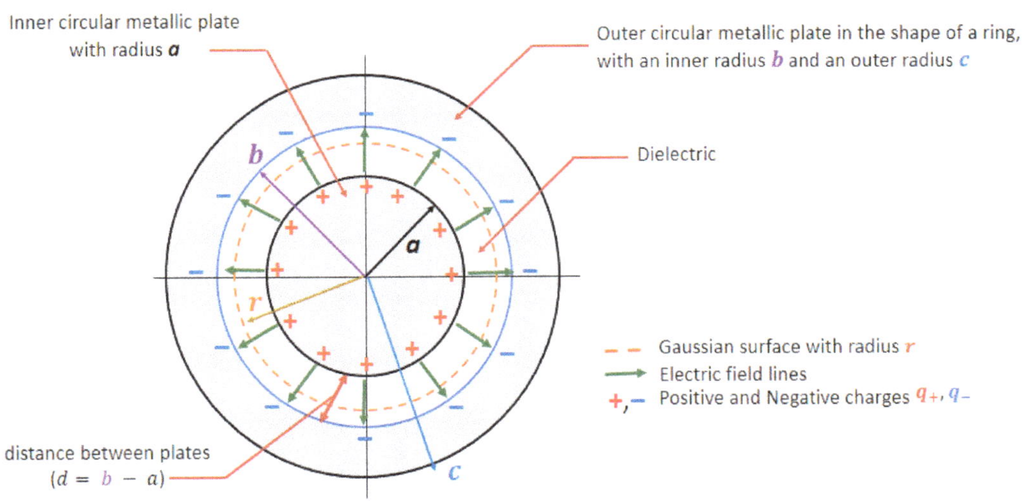

Figure 2. Diagram of the MT sensor, illustrating the concentric plates and electric field lines. The dielectric distance (d) and charge distribution are fundamental parameters for calculating the capacitance.

For the process to estimate the capacitance value for the MT sensor model, it begins with Equation (1), which corresponds to Gauss' Law [23], which defines the mathematical expression that allows calculating the electric field (E), by defining a Gaussian surface radius (r) containing the total charge of the capacitor (q) uniformly distributed over the surface area (A) and electrical permittivity $\varepsilon(\varepsilon = \varepsilon_0\varepsilon_r)$.

$$EA = \frac{q}{\varepsilon} = \frac{q}{\varepsilon_0\varepsilon_r} \quad (1)$$

where

- ε_o : Electrical permittivity in a vacuum ($8.85 \times 10^{-12}[C^2/Nm^2]$)
- ε_r : Relative permittivity of the dielectric between the plates of the capacitor.

Solving for the electric field (E) and replacing the area of the circumference $(A = \pi r^2)$ in expression (1) results in Equation (2):

$$E = \frac{q}{\varepsilon A} = \frac{q}{\varepsilon(\pi r^2)} \tag{2}$$

Subsequently, we proceed to define the mathematical expression that allows us to calculate the value of the voltage (V) existing between points a and b. The result of this process is described in Equation (3):

$$V = -\int_b^a Ed_r = -\int_b^a \frac{qd_r}{\varepsilon(\pi r^2)} = -\frac{q}{\pi\varepsilon}\int_b^a \frac{d_r}{r^2} V = \frac{q}{\pi\varepsilon}\left[\frac{1}{a} - \frac{1}{b}\right] = \frac{q}{\pi\varepsilon}\left[\frac{b-a}{ab}\right] \tag{3}$$

Finally, the value of the capacitance for the MT sensor (C_{MT}) is given by the relationship between the total charge (q) and the voltage (V), as presented in Equation (4):

$$C_{MT} = \frac{q}{V} = \pi\varepsilon\left[\frac{ab}{b-a}\right] = \pi\varepsilon_o\varepsilon_r\left[\frac{ab}{b-a}\right] \tag{4}$$

2.4.2. MNT Sensor Design

This sensor consists of two "E" shaped plates, a design inspired by that of a resistive humidity sensor. In this case, both the area of the plates ($A = A_1 = A_2$) and the number of slots interspersed between them are equal, and are separated by a distance (d). Figure 3 presents the proposed diagram for the MNT sensor, whose parameters are defined by the dimensions a, b, c, and d. Furthermore, illustrates how the positive and negative electric charges are distributed on the plates of the capacitive sensor, showing the electric field lines and their similarity to the model of a parallel plate capacitor, characterized by an area (A) and separation (d).

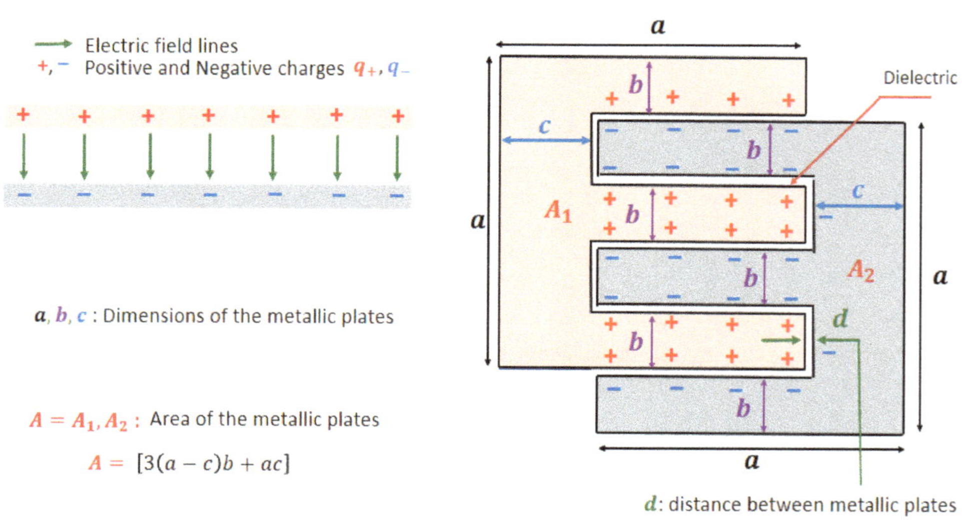

Figure 3. Diagram of the MNT sensor, illustrating two flat plates in the shape of an 'E', separated by the dielectric distance (d). The dimensions (a, b, c), the area (A) and charge distribution are key parameters for calculating the capacitance.

Therefore, the capacitance value for the MNT capacitive sensor (C_{MNT}) can be calculated from Equation (5):

$$C_{MNT} = \frac{\varepsilon A}{d} = \frac{\varepsilon_o \varepsilon_r A}{d} = \frac{\varepsilon_o \varepsilon_r [3(a-c)b + ac]}{d} \tag{5}$$

3. Results

3.1. Comparative Analysis of Theoretical and Experimental Capacitance for MT and MNT Sensors

In the experiment, a UNI-T UT603 LCR digital capacitance and inductance meter (UNI-T, Shenzhen, China) with a tolerance of 1% was used to determine the capacitance in pF of each of the proposed sensors, considering the conditions and type of waste material.

Figure 4 shows a capacitance value $C'_{MT} = 21\ pF$ delivered by the LCR for the implemented prototype of the MT sensor, when it is free of residues, accompanied by the proposed diagram. From Equation (4), we proceed to calculate the capacitance value (C_{MT}) according to the sensor dimensions that were considered at the discretion of the researchers for its construction (a = 2.1 cm, b = 2.3 cm, c = 3 cm, and d = 0.2 cm).

$$C_{MT} = \pi \varepsilon_o \varepsilon_r \left[\frac{ab}{b-a} \right] = 17.458\ pF \tag{6}$$

$$e_{abs} = C'_{MT} - C_{MT} = 3.542 \times 10^{-12} \tag{7}$$

$$e_r = \frac{C'_{MT} - C_{MT}}{C_{MT}} = 0.203 \tag{8}$$

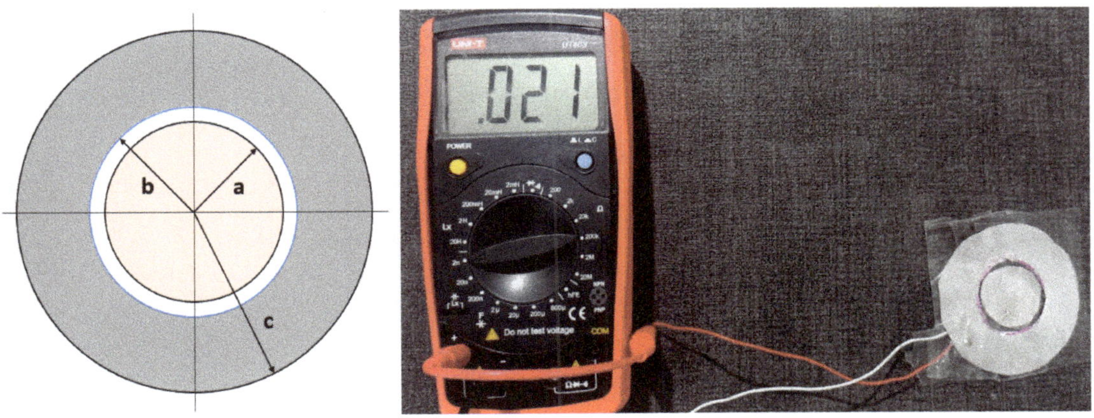

Figure 4. Capacitance value with prototype sensor traditional model (MT) without residuals for reading.

Equations (6)–(8) present the result of this process accompanied by the absolute error (e_{abs}) and the relative error (e_r) [24] in contrast to the result obtained experimentally, where it can be seen that the theoretical value and the one obtained experimentally are very similar to each other, with reduced error levels taking into account that the capacitance values are of the order of $10^{-12}(pF)$, an aspect that is very common in commercial capacitors, which establish a tolerance value given by the manufacturer and considering that the relative permittivity values of the materials used for the construction of the sensors may vary a little compared to those that were theoretically adopted.

The proposed diagram and the capacitance value of $C'_{MNT} = 28\ pF$ for the implemented MNT sensor prototype, when free of residues, are presented in Figure 5. Using Equation (5), the capacitance value (C_{MNT}) is determined based on the sensor's dimensions, which were selected by the researchers during its construction (a = 5 cm, b = 0.9 cm, c = 1.5 cm and d = 0.1 cm).

$$C_{MNT} = \frac{\varepsilon_0 \varepsilon_r A}{d} = 24.5\ pF \tag{9}$$

$$e_{abs} = C'_{MNT} - C_{MNT} = 3.49 \times 10^{-12} \tag{10}$$

$$e_r = \frac{C'_{MNT} - C_{MNT}}{C_{MNT}} = 0.143 \tag{11}$$

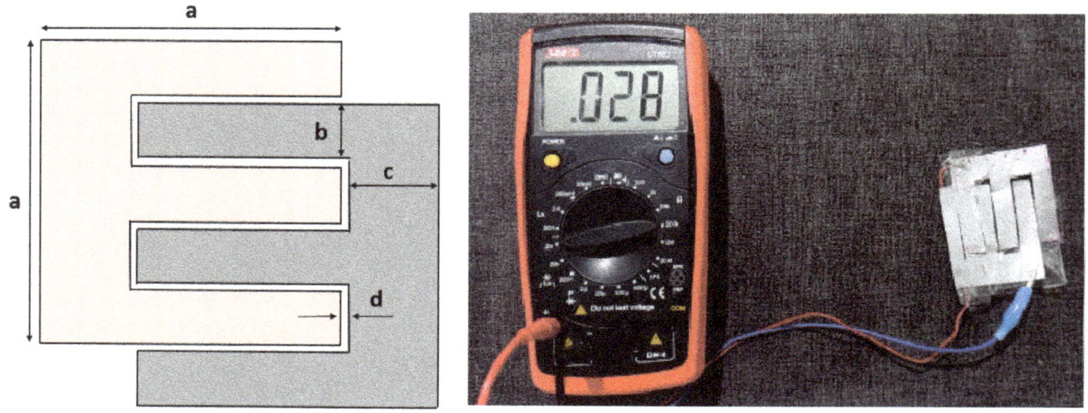

Figure 5. Capacitance value with non-traditional model sensor prototype (MNT) without residuals for reading.

The results of this process are presented in Equations (9)–(11), together with the absolute error (e_{abs}) and the relative error (e_r) when compared to the results obtained experimentally. It can be observed that the theoretical and experimental values are very similar, as in the previous case, showing low levels of absolute and relative error. From this, it can be concluded that the mathematical expressions proposed for each type of sensor are consistent with the experimental results, which allows establishing a valuable reference point for future research in which it is planned to use any of the proposed models or make modifications regarding their design or dimensions.

3.2. Estimation of the Number of Samples and Analysis of the Information

To carry out the sampling campaigns, the estimation of the sample size required for two independent samples was made, since they are two different sensors (MT and MNT), establishing a power of the test ($\beta = 0.9$), a confidence level of 95% ($\alpha = 0.05$) and an effect size of 0.5 [25]. By using the XLSTAT tool, a minimum size of 85 was obtained as a result for each sample. However, in order to improve accuracy and ensure statistical representativeness, a final size of 100 samples per sensor (MT and MNT) was established.

Additionally, the distribution of the samples was considered uniform for the four types of waste (plastic, glass, metal, and organic), with 25 records per category, establishing an adequate balance in the classification, avoiding biases during the training and testing phase of the machine learning model.

Tables 2 and 3 show the mean, standard deviation, and confidence intervals for each characteristic, according to the type of material or waste used during the sampling campaign depending on the type of sensor.

Table 2. Statistical results of capacitance measurements with sensor MT.

Material	Capacitance [pF]		
	Mean	Standard Deviation	Confidence Interval (95%)
Plastic	24.680	1.519	(24.084, 25.275)
Glass	25.080	1.394	(24.538, 25.622)
Metal	25.120	1.394	(24.573, 25.666)
Organic	251.120	120.823	(203.757, 298.482)

Table 3. Statistical results of capacitance measurements with sensor MNT.

Material	Capacitance [pF]		
	Mean	Standard Deviation	Confidence Interval (95%)
Plastic	34.680	1.574	(34.063, 35.297)
Glass	34.880	1.666	(34.227, 35.533)
Metal	31.960	0.934	(31.593, 32.326)
Organic	680.960	324.396	(553.799, 808.120)

Figures 4–7 present some of the results obtained during the measurement campaigns, both in the absence of material or waste (Figures 4 and 5), and in the presence of reusable waste such as plastic, glass, and metal, as well as in various types of organic waste. Figures 4 and 6 show some of the results recorded in terms of capacitance using the MT capacitive sensor, registering a value of 21 pF in the absence of material or residue. In turn, values between 22 and 27 pF for reusable materials and values between 80 and 450 pF when dealing with organic waste. On the other hand, Figures 5 and 7 present some of the results in terms of capacitance for the MNT capacitive sensor, registering a value of 28 pF in the absence of material or residue. In turn, values between 31 and 37 pF for reusable materials and values between 150 and 1200 pF when dealing with organic waste.

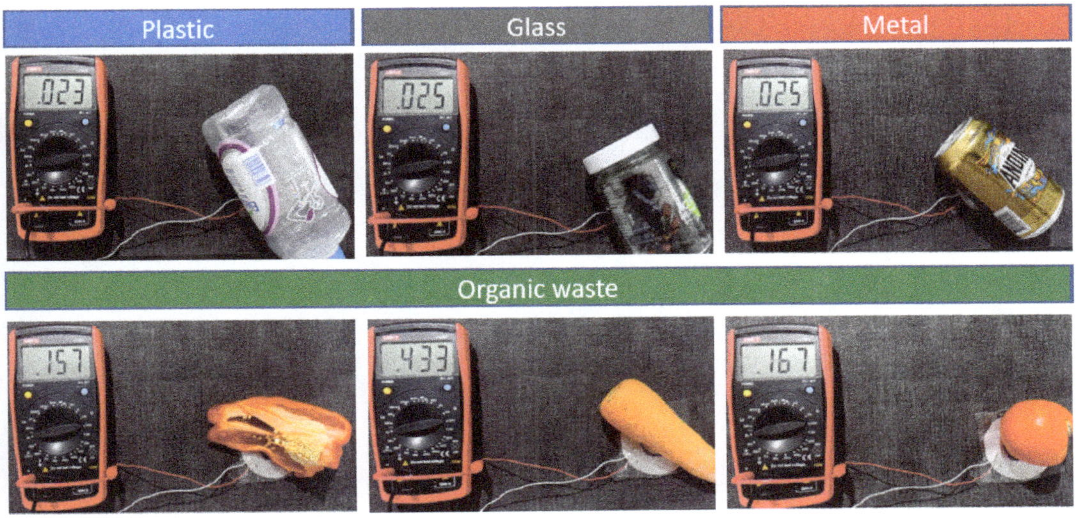

Figure 6. Records with Sensor—Traditional Model (MT) for various types of waste.

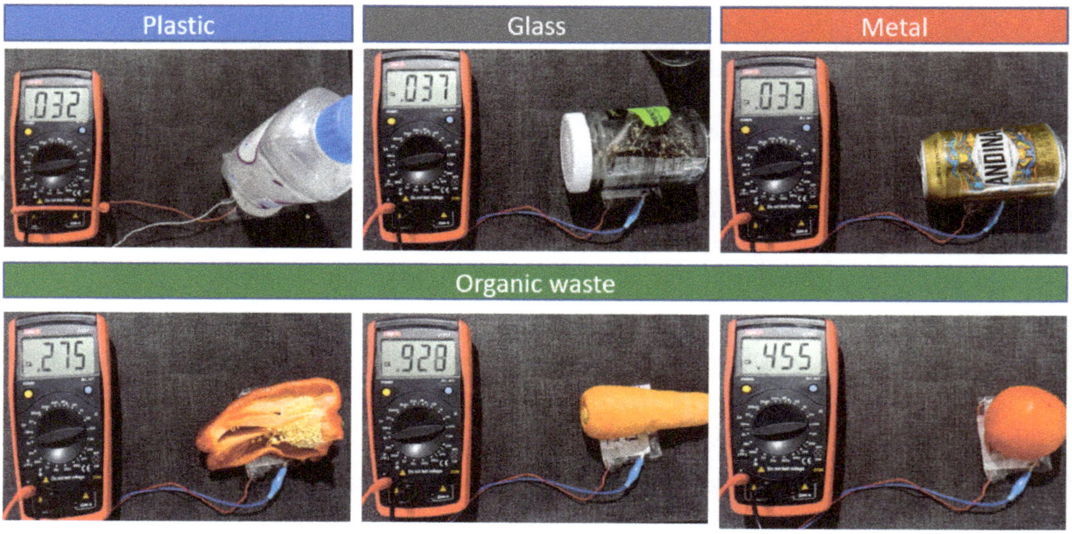

Figure 7. Records with sensor—non-traditional model (MNT) for various types of waste.

To visually represent the distribution of data recorded during the sampling campaigns and estimate the variability levels for each type of residue considered as a reference, box plots were generated for each case. These diagrams also aided in evaluating the behavior of the proposed sensors and identifying outliers in the dataset. The results are presented in Figures 8 and 9.

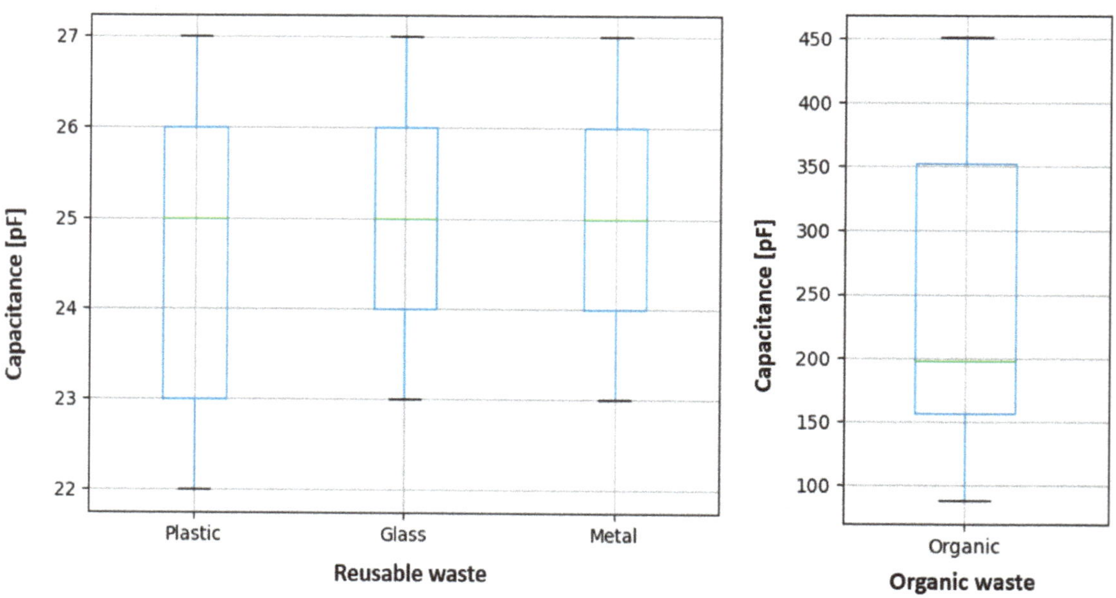

Figure 8. Plot of boxes related to capacitance [pF] for the MT sensor.

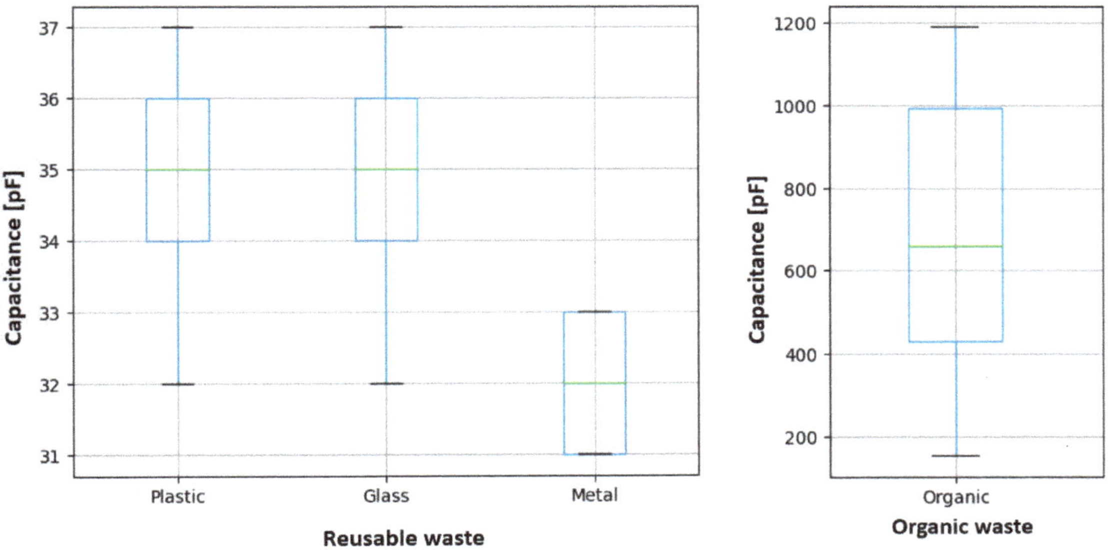

Figure 9. Plot of boxes related to capacitance [pF] for the MNT sensor.

Additionally, the possibility was considered that the size of the sample that was established for the development of the project could have affected in some way the appearance of each of the resulting figures. However, from a practical point of view, according to [26], it is possible to consider that a box plot can be reliable as long as the sample size established for each item is equal to or greater than 20. Taking into account that 25 samples were established for each type of residue, it can be considered that this recommendation is fulfilled and with this the result of each figure can be considered reliable.

Figures 9 and 10 show the box diagrams related to the capacitance [pF] recorded by the MT and MNT sensors. In them it can be seen that the behavior of the capacitance for the readings recorded by the MNT sensor describes a normal distribution for the case of the three types of reusable waste (plastic, glass, and metal), a situation that occurs in a similar way with waste corresponding to glass and metal in the case of the MT sensor. However, for the particular case of plastic in the MT sensor, it describes a positive asymmetry, describing a higher concentration of the values recorded between 50 and 75%.

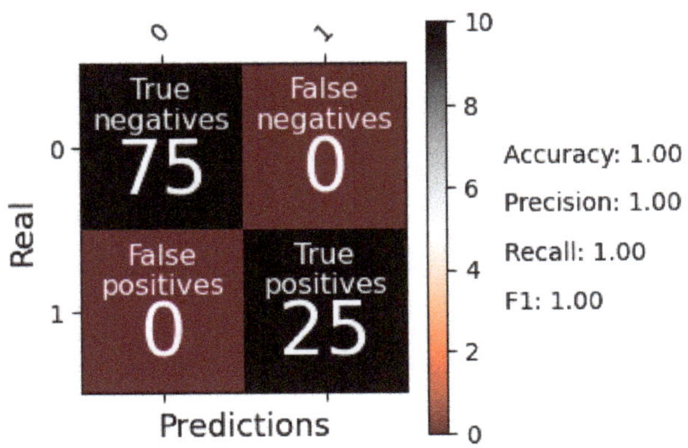

Figure 10. Confusion matrix obtained for the Random Forests model.

This is not the case for the organic residues for the two sensors, where a negative asymmetry is observed, thus establishing an indication of a higher concentration of readings for this type of residue between the Q1 and Q2 quartiles, a phenomenon that describes a higher incidence in the MT sensor. In turn, the readings related to plastic for the MT sensor slightly show a higher level of dispersion compared to the readings of the other reusable waste for both types of sensors. Additionally, it can be seen that the values corresponding to the median for the readings recorded by the MT sensor for reusable waste present a common value of approximately 25 [pF] in the three cases, an aspect that occurs similarly for the MNT sensor, where the waste related to plastic and glass describe a median value or Q2 of 35 and for metal, a value of approximately 32. For the particular case of organic waste, approximate values are obtained for the MT and MNT sensors of 200 and 660, respectively, with a higher level of dispersion for the case of the MNT sensors.

The comparative statistical analysis of the MT and MNT sensors reveals significant differences in capacitance measurement for different materials, with the MNT sensor's performance standing out due to its greater dynamic range. For reusable materials (plastic, glass, and metal), both sensors show relatively homogeneous capacitance values, with similar means and low standard deviations. However, the 95% confidence intervals (CI) of the MNT sensors are slightly wider, which could be attributed to greater sensitivity in the measurement. For example, for glass, the MT sensor presents a CI of (24.538, 25.622), while the MNT sensor widens it to (34.227, 35.533), indicating a shift towards higher capacitance values and a more differentiated measurement between materials.

For organic waste, the superiority of the MNT sensor is evident due to its ability to capture higher capacitance values and greater relative variability, thereby reducing the overlap with reusable materials. The MT sensor records a mean of 251.120 pF with a standard deviation of 120.823 and a wide CI of (203.757, 298.482), while the MNT sensor reaches a mean of 680,960 pF with a standard deviation of 324.396 and a CI of (553.799, 808.120). Not only does this demonstrate that the MNT sensor can more effectively differentiate organic waste from the rest, but its greater dynamic range, as quantified by the ratio of means to standard deviations (coefficient of variation), strengthens its classification capability. In organic waste, the coefficient of variation for the MT sensor is 48.1%, while for the MNT sensor it rises to 47.6%, indicating that, although both present high dispersions, the MNT sensor is more effective in separation thanks to its greater width in the distribution of values.

Capacitance was selected as the sole characteristic for the sorting process due to its high sensitivity to the dielectric properties of materials. This sensitivity allows us to differentiate waste according to its composition in a quantitative way, since organic waste presents significantly higher capacitance values, attributed to its higher dielectric constant and inherent moisture content. In contrast, reusable materials such as plastic, glass, and metal show lower and more homogeneous capacitance values, given their reduced dielectric character. These behavior patterns are clearly evident in the results presented in Figures 8 and 9.

3.3. Comparison of MT Vs. MNT Treatments

In this study, a paired-sample *t*-test was used due to the characteristics of the experimental design where the capacitance measurements for the MT and MNT sensors were obtained using the same waste samples, which establishes a direct dependency relationship between the observations. This implies that the observations in both groups are not independent, but are directly related, since each pair of data corresponds to the same waste sample evaluated by both sensors.

The paired-sample t-test is appropriate in this case because it evaluates whether the mean of the differences between the measurements of the MT and MNT sensors is significantly different from zero. This approach allows controlling the variability between the samples, focusing only on the differences attributable to the performance of the sensors.

Additionally, although the t-test requires that the differences between pairs follow an approximately normal distribution, this criterion was met after verifying the normality of the data using graphical and statistical methods, ensuring that the results obtained are valid and reliable to establish significant differences between the sensors evaluated.

Since the differences between measurements presented a verifiable normal distribution, the use of the paired t-test was the most appropriate choice, allowing reliable results to be obtained with greater sensitivity to detect significant differences between the sensors evaluated.

Let X_i and Y_i represent the recorded capacitance values (the most relevant characteristic) for MT and MNT sensors, respectively, with $d_i = X_i - Y_i$ denoting the difference between the samples. To assess whether the MNT sensor shows superior performance compared to the MT sensor, the following hypotheses are posed:

$$H_o: \mu_x \leq \mu_y$$

$$H_a: \mu_x > \mu_y$$

where μ_x and μ_y are the means corresponding to the capacitance values obtained during the sampling campaigns for the two types of MT and MNT sensors, respectively. The H_o hypothesis states that MNT is better than the MT and the H_a establishes the opposite condition. To accept or reject the proposed hypotheses, it is necessary to perform a hypothesis contrast on the difference in means with paired sampling, using the so-called paired t-test. For this, the following steps are established [27]:

Step 1: A new random variable $Z = X - Y$ is defined and the mean value and standard deviation for the variable Z are calculated. The result of this process yielded the values of -111.28 and 217.816 for $\overline{Z}$ and S_z, respectively. On the other hand, when defining a new variable Z, it is necessary to make an adjustment in the hypotheses as follows:

$$H_o: \mu_x \leq \mu_y \rightarrow \mu_x - \mu_y \leq 0 \rightarrow \mu_z \leq 0$$

$$H_a: \mu_x > \mu_y \rightarrow \mu_x - \mu_y > 0 \rightarrow \mu_z > 0$$

Step 2: We proceed to calculate the value of the statistic established for the test by using the following expression:

$$d = \frac{\overline{Z}}{S_z}\sqrt{n} = \frac{-111.28}{214.816}\sqrt{100} = -5.180$$

where d is the value of the statistic and n obeys the number of samples for the two proposed scenarios.

Step 3: Establish the acceptance range of the H_o for $\left\{t : t < T_{(\alpha;n-1)}\right\}$ at 5% significance ($\alpha = 0.05$) and $n-1$ degrees of freedom. For the particular case, the value of $T(0.05;99) = 1.6603$, defining the range of acceptance of the H_o between $(-\infty, 1.6603)$. When evaluating the value of the d statistic, it is observed that it is within the acceptance interval, which is why H_o is not rejected. In view of the above, it is concluded that the sensor supported in the MNT is better than the MT sensor for the proposed scenario, with

95% confidence. Additionally, the MNT sensor describes a higher variance (greater response sensitivity, in the presence of different types of materials and even of the same type) compared to the MT sensor, which is very favorable for the identification and classification of waste or materials.

3.4. Classification of Solid Waste with MNT Sensor Using Machine Learning Algorithms

Among the various machine learning algorithms available, Random Forest was selected for the classification process using the MNT sensor due to its robustness and ability to handle highly variable data. Its ensemble learning approach, combining multiple decision trees, minimizes the risk of overfitting, ensuring stable and accurate predictions even in noisy environments. Furthermore, Random Forest allows the interpretation of feature importance, such as capacitance, facilitating the identification of patterns relevant to waste classification.

While notable differences in capacitance exist between organic and recyclable waste, Random Forest ensures robust and automated classification in cases with smaller dielectric variations or in complex environments with significant data noise. Compared to more advanced algorithms like Support Vector Machines (SVM) or neural networks, Random Forest achieves an optimal balance between accuracy and computational efficiency, making it particularly suitable for resource-constrained deployments.

The training, testing, and validation phases were carried out using the dataset collected during sampling campaigns with the MNT sensor, consisting of 100 records with capacitance (pF) as a feature distributed into two classes: reusable waste (Class 0) and organic waste (Class 1). The data analysis processes were implemented using Python 3.6.3 libraries such as sklearn, matplotlib, pandas, numpy, and scipy, through the Google Colab platform.

For the Random Forest algorithm configuration process, the following parameters were established:

- Number of trees (n_estimators): 100 trees. This value is commonly used as a standard in practical scenarios, and as a recommendation in default configurations of the Scikit-learn library, establishing a balance point between performance and computing time.
- Split criterion: Gini. Used to evaluate the purity of the partitions in each node of the tree.
- Maximum depth (max_depth): No limit. Allowing full growth of the trees and ensuring the capture of complex patterns in the data.
- Minimum leaf sample size (min_samples_leaf): In order to avoid overfitting by reducing variability in the final predictions.

To ensure transparency and minimize experimental errors, the following additional techniques were applied:

- Data normalization: The RobustScaler technique from the sklearn library was used to adjust the capacitance values and mitigate the influence of outliers on the model.
- Data splitting: The dataset was split into 80% for training and 20% for testing using the train_test_split function, using a random seed to ensure reproducibility.

Figure 10 presents the confusion matrix obtained for the Random Forests model, in which there were no false positive or false negative records, correctly classifying the 75 records belonging to Class 0 (reusable waste) and the 25 records of Class 1 (organic waste), demonstrating adequate levels of precision and reliability during the classification process according to the reading made by the sensor. In turn, performance metrics such as accuracy, precision, recall, and F1-score were calculated for the Random Forests model, obtaining a perfect value of 1.0 in all cases. This result reflects optimal performance. Additionally, the absence of overfitting was validated by the consistency of all metrics, thus avoiding relying solely on the accuracy metric, which can be misleading under bias conditions.

Figure 11 presents the graphic validation of the proposed classification model, in which the decision limits used to separate the waste classes can be clearly seen. In turn, the reference line observed horizontally reflects consistent thresholds for the capacitance values corresponding to each type of waste. Finally, for both reusable and organic waste, the capacitance thresholds are in an intermediate tolerance zone, which significantly reduces the risk of incorrect classification when the new data deviate slightly from the values established during the training process.

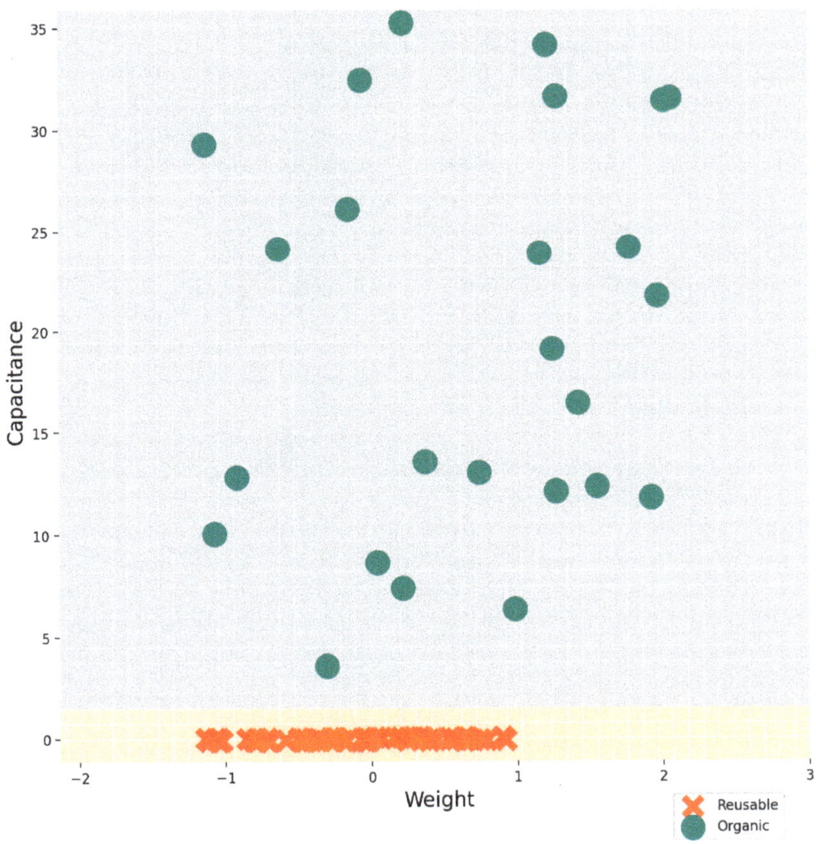

Figure 11. Validation of the Random Forest.

The MT and MNT sensors, made from recycled materials such as aluminum, represent an economical and sustainable solution compared to commercial sensors. This approach not only significantly reduces production costs but also promotes sustainable practices by incorporating recycled components into their design.

Unlike commercial capacitive sensors, which only provide binary detection (presence or absence), the MT and MNT sensors offer continuous capacitance values, enabling more precise and effective material differentiation. This outstanding capability makes them suitable for applications requiring high classification accuracy, such as the separation of organic and recyclable waste. Additionally, they surpass commercial alternatives based on optical or spectrometry technologies in terms of simplicity and cost.

Another distinguishing feature of the MT and MNT sensors is their sensitivity to dielectric materials, which offers a significant advantage in environments where environmental conditions vary considerably. Unlike commercial optical solutions, which can be affected by external factors such as lighting or dust, the MT and MNT sensors maintain consistent and reliable performance.

These sensors are an innovative and efficient alternative that combine precision, low cost, and sustainability, making them ideal for real-world applications. Their accessible design enhances their feasibility in communities with limited resources compared to advanced technologies that, although precise, require expensive equipment and specialized maintenance. This accessibility also makes them a valuable tool for waste classification at the source, particularly in residential settings, marking a significant difference in waste management and contributing to environmental protection.

Among the limitations identified in the present study is the possibility that the sensitivity of the sensors may be influenced by environmental factors, such as humidity and temperature, which could alter the recorded capacitance values. However, it is reasonable to assume that reusable waste, such as plastics, metals and glass, present significantly lower humidity levels than organic waste, where the humidity concentration could have a more representative impact on the measurements.

Additionally, the long-term durability and scalability of sensors made from recycled materials constitute challenges inherent to any emerging technology. Although these aspects and hypotheses were not analyzed in detail in this manuscript, they do not compromise the validity of the results obtained. This is because the main objective of the study was to demonstrate the technical feasibility of low-cost capacitive sensors and their integration with machine learning algorithms for waste classification.

Although the capacitance values recorded by the sensors are sufficiently differentiable to classify organic and reusable waste without the need to use machine learning algorithms, the inclusion of a classification algorithm such as Random Forest provides significant benefits to the design of the proposed system.

Firstly, it offers robustness against possible fluctuations in capacitance values derived from external factors, such as humidity, variations in temperature, interference from electromagnetic fields generated by nearby devices and other environmental elements. This allows for more accurate classification in real scenarios where operating conditions are often less controlled and data may be affected by noise.

In addition, the implementation of machine learning not only increases the scalability of the system, but also allows integration with more complex datasets, incorporating additional features such as temperature and chemical composition of the waste or historical values. This capability increases the accuracy and adaptability of the model, providing a basis for optimizing both sensor design and classification methodologies in future developments.

From a practical perspective, the use of Random Forest simplifies system automation by reducing the need for human intervention. The model can dynamically adapt to new operating conditions without requiring manual reconfiguration, which is especially valuable in decentralized applications. For example, in communities with limited resources, having a robust and autonomous system is crucial to ensure reliable and continuous performance. In turn, the inclusion of the algorithm contributes significantly to the quantitative and qualitative analysis of the data, improving the reproducibility and validation of the results both in experimental scenarios and in real applications, reinforcing the reliability of the system in line with current trends in the design of smart, sustainable, and scalable technologies.

Finally, these challenges represent clear opportunities for future research, in which controlled tests can be carried out to evaluate the stability of the sensors under variable environmental conditions and to carry out lifespan studies in real application scenarios.

In this context, the results presented in this article constitute a fundamental starting point for the advancement of knowledge in this area, providing a solid foundation for future developments that address these limitations with greater depth and rigor.

4. Conclusions

Within the broad spectrum of problems related to environmental protection, solid waste management occupies one of the most important places within environmental management policies, where the main objective is to establish mechanisms that allow this process to be carried out efficiently and in compatibility with the environment and public health. For this reason, the use of technologies such as capacitive sensors and Artificial Intelligence could help to implement economic and efficient solutions that allow the classification of solid waste at the source, minimizing the time necessary for its classification compared to separation process manuals, reduce the health risks that may arise when working with contaminated waste and increase the percentage of waste that can be reused and reprocessed later thanks to the proper separation process. According to the results obtained, the results demonstrated that the two models of capacitive sensors proposed describe an adequate behavior in terms of sensitivity and coherence with the readings registered for each type of material or test residue that was used for its evaluation. However, the MNT sensor model was the one that showed optimal performance in relation to sensitivity and variability for each type of waste, with 95% confidence, allowing waste classification processes through the use of machine learning algorithms. Establishing an economic and reliable alternative for the implementation of this type of solutions at the source and in turn promoting the development of future projects related to the classification of materials in a more selective way, could be oriented to the separation of reusable waste such as glass, paper, plastic, and metal, among others independently or for the specific characterization of materials, with possible industrial applications.

Author Contributions: J.C.V.F.: Conceptualization, methodology, software, validation, formal analysis, investigation, data curation, supervision, project administration and funding acquisition; H.E.P.W.: validation, investigation, data curation, and supervision; J.A.V.B.: resources, visualization, supervision, project administration, and funding acquisition. All authors have read and agreed to the published version of the manuscript.

Funding: The authors appreciate the financial support provided by UNAD (PG0501ECBTI2022) and the company Smarttic during the development of this study.

Institutional Review Board Statement: Not applicable.

Informed Consent Statement: Not applicable.

Data Availability Statement: The raw data supporting the conclusions of this article will be made available by the authors on request.

Conflicts of Interest: The authors declare no conflicts of interest.

References

1. Pan, Z.; Sun, X.; Huang, Y.; Liang, T.; Lu, J.; Zhang, L.; Qi, C. Anaerobic Co-Digestion of Food Waste and Microalgae at Variable Mixing Ratios: Enhanced Performance, Kinetic Analysis, and Microbial Community Dynamics Investigation. *Appl. Sci.* **2024**, *14*, 4387. [CrossRef]
2. Gan, B.; Zhang, C. Research on the algorithm of urban waste classification and recycling based on deep learning technology. In Proceedings of the 2020 International Conference on Computer Vision, Image and Deep Learning (CVIDL), Chongqing, China, 10–12 July 2020; Institute of Electrical and Electronics Engineers, Inc.: Piscataway, NJ, USA, 2020; pp. 232–236. [CrossRef]
3. Kokoulin, A.N.; Uzhakov, A.A.; Tur, A.I. The Automated Sorting Methods Modernization of Municipal Solid Waste Processing System. In Proceedings of the 2020 International Russian Automation Conference (RusAutoCon), Sochi, Russia, 6–12 September 2020; Institute of Electrical and Electronics Engineers, Inc.: Piscataway, NJ, USA, 2020; pp. 1074–1078. [CrossRef]

4. Dadic, B.; Ivankovic, T.; Spelic, K.; Hrenovic, J.; Jurisic, V. Natural Materials as Carriers of Microbial Consortium for Bioaugmentation of Anaerobic Digesters. *Appl. Sci.* **2024**, *14*, 6883. [CrossRef]
5. Colasante, A.; D'Adamo, I. The circular economy and bioeconomy in the fashion sector: Emergence of a "sustainability bias" *J. Clean Prod.* **2021**, *329*, 129774. [CrossRef]
6. Blumberga, D.; Ozarska, A.; Indzere, Z.; Chen, B.; Lauka, D. Energy, Bioeconomy, Climate Changes and Environment Nexus *Environ. Clim. Technol.* **2019**, *23*, 370–392. [CrossRef]
7. Saranya, M. A Survey on Health Monitoring System by using IOT. *Int. J. Res. Appl. Sci. Eng. Technol.* **2018**, *6*, 778–782. [CrossRef]
8. Dhulekar, P.; Gandhe, S.T.; Mahajan, U.P. Development of Bottle Recycling Machine Using Machine Learning Algorithm. In Proceedings of the 2018 International Conference On Advances in Communication and Computing Technology (ICACCT), Sangamner, India, 8–9 February 2018; Institute of Electrical and Electronics Engineers, Inc.: Piscataway, NJ, USA, 2018; pp. 515–519. [CrossRef]
9. Saranya, A.; Bhambri, M.; Ganesan, V. A Cost-Effective Smart E-Bin System for Garbage Management Using Convolutional Neural Network. In Proceedings of the 2021 International Conference on System, Computation, Automation and Networking (ICSCAN), Puducherry, India, 30–31 July 2021; Institute of Electrical and Electronics Engineers, Inc.: Piscataway, NJ, USA, 2021; pp. 1–6. [CrossRef]
10. Lin, W. YOLO-Green: A Real-Time Classification and Object Detection Model Optimized for Waste Management. In Proceedings of the 2021 IEEE International Conference on Big Data (Big Data), Orlando, FL, USA, 15–18 December 2021; Institute of Electrical and Electronics Engineers, Inc.: Piscataway, NJ, USA, 2021; pp. 51–57. [CrossRef]
11. Liu, K.; Liu, X. Recycling Material Classification using Convolutional Neural Networks. In Proceedings of the 2022 21st IEEE International Conference on Machine Learning and Applications (ICMLA), Nassau, Bahamas, 12–14 December 2022; Institute of Electrical and Electronics Engineers, Inc.: Piscataway, NJ, USA, 2022; pp. 83–88. [CrossRef]
12. Nafiz, M.S.; Das, S.S.; Morol, M.K.; Al Juabir, A.; Nandi, D. ConvoWaste: An Automatic Waste Segregation Machine Using Deep Learning. In Proceedings of the 2023 3rd International Conference on Robotics, Electrical and Signal Processing Techniques (ICREST), Dhaka, Bangladesh, 7–8 January 2023; Institute of Electrical and Electronics Engineers, Inc.: Piscataway, NJ, USA, 2023; pp. 181–186. [CrossRef]
13. Ba Alawi, A.E.; Saeed, A.Y.A.; Almashhor, F.; Al-Shathely, R.; Hassan, A.N. Solid Waste Classification Using Deep Learning Techniques. In Proceedings of the 2021 International Congress of Advanced Technology and Engineering (ICOTEN), Taiz, Yemen, 4–5 July 2021; Institute of Electrical and Electronics Engineers, Inc.: Piscataway, NJ, USA, 2021; pp. 1–5. [CrossRef]
14. Wang, B.; Zhou, W.; Shen, S. Garbage classification and environmental monitoring based on internet of things. In Proceedings of the 2018 IEEE 4th Information Technology and Mechatronics Engineering Conference (ITOEC), Chongqing, China, 14–16 December 2018; Institute of Electrical and Electronics Engineers, Inc.: Piscataway, NJ, USA, 2018; pp. 1762–1766. [CrossRef]
15. Ravi, S.; Jawahar, T. Smart city solid waste management leveraging semantic based collaboration. In Proceedings of the 2017 International Conference on Computational Intelligence in Data Science(ICCIDS), Chennai, India, 2–3 June 2017; Institute of Electrical and Electronics Engineers, Inc.: Piscataway, NJ, USA, 2018; pp. 1–4. [CrossRef]
16. Rismiyati; Endah, S.N.; Khadijah; Shiddiq, I.N. Xception Architecture Transfer Learning for Garbage Classification. In Proceedings of the 2020 4th International Conference on Informatics and Computational Sciences (ICICoS), Semarang, Indonesia, 10–11 November 2020; Institute of Electrical and Electronics Engineers, Inc.: Piscataway, NJ, USA, 2020; pp. 1–4. [CrossRef]
17. Vesga Ferreira, J.C.; Sepulveda, F.A.A.; Perez Waltero, H.E. Smart Ecological Points, a Strategy to Face the New Challenges in Solid Waste Management in Colombia. *Sustainability* **2024**, *16*, 5300. [CrossRef]
18. Vesga, J.C.; Barrera, J.A.; Sierra, J.E. Design of a prototype remote medical monitoring system for measuring blood pressure and glucose measurement. *Indian J. Sci. Technol.* **2018**, *11*, 1–8. [CrossRef]
19. Harlacher, B.L.; Stewart, R.W. Common mode voltage measurements comparison for CISPR 22 conducted emissions measurements. In Proceedings of the 2001 IEEE EMC International Symposium. Symposium Record. International Symposium on Electromagnetic Compatibility (Cat. No.01CH37161), Montreal, QC, Canada, 13–17 August 2001; Institute of Electrical and Electronics Engineers, Inc.: Piscataway, NJ, USA, 2001; Volume 1, pp. 26–30. [CrossRef]
20. Shyam, G.K.; Manvi, S.S.; Bharti, P. Smart waste management using Internet-of-Things (IoT). In Proceedings of the 2017 2nd International Conference on Computing and Communications Technologies (ICCCT), Chennai, India, 23–24 February 2017; Institute of Electrical and Electronics Engineers, Inc.: Piscataway, NJ, USA, 2017; pp. 199–203. [CrossRef]
21. Fernandez, G.P.; Acosta, A.M. Migración de Aluminio a los Alimentos Provenientes de Envases y Utensilios de Cocina Nacionales e Importados, Comercializados en Nuestro País. Seminario de Difusión de Resultados del Proyecto de Investigacion. Published Online 2020. Available online: http://repositorio.conacyt.gov.py/handle/20.500.14066/3994 (accessed on 18 August 2023).
22. Cardona, L.; Jiménez, J.; Vanegas, N. Caracterización de materiales empleados en la fabricación de artefactos explosivos improvisados. *Rev. Colomb. Mater.* **2014**, 13–19. [CrossRef]

23. Shikha, U.S.; James, R.K.; Pradeep, A.; Baby, S.; Jacob, J. Threshold Voltage Modeling of Negative Capacitance Double Gate TFET. In Proceedings of the 2022 35th International Conference on VLSI Design and 2022 21st International Conference on Embedded Systems (VLSID), Bangalore, India, 26 February–2 March 2022; Institute of Electrical and Electronics Engineers, Inc.: Piscataway, NJ, USA, 2022; pp. 287–291. [CrossRef]
24. González Manteiga, M.T.; Pérez de Vargas, A. *Estadística Aplicada una Visión Instrumental*; Ediciones Díaz de Santos: Madrid, Spain, 2010; Available online: https://books.google.com.co/books?id=8tocMTUkICkC&printsec=frontcover (accessed on 11 December 2017).
25. Domínguez, J.; Castaño, E. *Diseño de Experimentos—Estrategias y Análisis en Ciencias e Ingeniería*; Alfaomega: Mexico City, Mexico, 2016.
26. Walpole, R.; Myers, R.; Myers, S. *Probabilidad y Estadística para Ingeniería y Ciencias*, 9th ed.; Pearson: London, UK, 2012.
27. Rico Martínez, M.; Vesga Ferreira, J.C.; Carroll Vargas, J.; Rodríguez, M.C.; Toro, A.A.; Cuevas Carrero, W.A. Shapley's Value as a Resource Optimization Strategy for Digital Radio Transmission over IBOC FM. *Appl. Sci.* **2024**, *14*, 2704. [CrossRef]

MDPI AG
Grosspeteranlage 5
4052 Basel
Switzerland
Tel.: +41 61 683 77 34

Applied Sciences Editorial Office
E-mail: applsci@mdpi.com
www.mdpi.com/journal/applsci

www.ingramcontent.com/pod-product-compliance
Lightning Source LLC
Chambersburg PA
CBHW040751250926
51850CB00007BB/1465
* 9 7 8 3 7 2 5 8 5 1 0 5 8 *